Lecture Notes in Artificial Intelligence   8775

Subseries of Lecture Notes in Computer Science

Jorge Baptista   Nuno Mamede   Sara Candeias
Ivandré Paraboni   Thiago A.S. Pardo
Maria das Graças Volpe Nunes (Eds.)

Computational Processing of the Portuguese Language

11th International Conference, PROPOR 2014
São Carlos/SP, Brazil, October 6-8, 2014
Proceedings

 Springer

Volume Editors

Jorge Baptista
Universidade do Algarve – FCHS, Faro, Portugal
E-mail: jbaptis@ualg.pt

Nuno Mamede
INESC-ID Lisboa, Lisbon, Portugal
E-mail: nuno.mamede@inesc-id.pt

Sara Candeias
IT-University of Coimbra, Coimbra, Portugal
E-mail: candeias.sara@gmail.com

Ivandré Paraboni
USP-EACH, São Paulo/SP, Brazil
E-mail: ivandre@usp.br

Thiago A.S. Pardo
Maria das Graças Volpe Nunes
USP-ICMC, São Carlos/SP, Brazil
E-mail: {taspardo,gracan}@icmc.usp.br

ISSN 0302-9743   e-ISSN 1611-3349
ISBN 978-3-319-09760-2   e-ISBN 978-3-319-09761-9
DOI 10.1007/978-3-319-09761-9
Springer Cham Heidelberg New York Dordrecht London

Library of Congress Control Number: 2014944798

LNCS Sublibrary: SL 7 – Artificial Intelligence

Typesetting: Camera-ready by author, data conversion by Scientific Publishing Services, Chennai, India

Printed on acid-free paper

Springer is part of Springer Science+Business Media (www.springer.com)

Preface

The International Conference on Computational Processing of Portuguese (PRO-POR), in its 2014 edition, continued being the most important natural language processing scientific event dedicated to the Portuguese language and its theoretical and technological advances, simultaneously dealing both with its spoken and written dimensions. This event is hosted every two years, alternating between Brazil and Portugal. Previous events were held in Lisbon/Portugal (1993), Curitiba/Brazil (1996), Porto Alegre/Brazil (1998), Évora/Portugal (1999), Atibaia/Brazil (2000), Faro/Portugal (2003), Itatiaia/Brazil (2006), Aveiro/Portugal (2008), Porto Alegre/Brazil (2010), and Coimbra/Portugal (2012). The conference brings together researchers and practitioners in the field, sharing their expertise, divulging their results, promoting methodologies and exchanging resources, forming a very active and vibrant community, internationally recognized for their excellence of standards and acutely aware of their role in the growing impact of Portuguese in international fora. This 11th edition of PROPOR took place in São Carlos/SP, in the south-east of Brazil, under the aegis of the Institute of Mathematical and Computer Sciences (ICMC) of the University of São Paulo (USP) and the Interinstitutional Center for Computational Linguistics (NILC). The event also featured the third edition of the MSc/MA and PhD Dissertation Contest, which selects the best new academic research in Portuguese NLP, as well as a workshop for demonstration of software and resources for Portuguese processing.

Three keynote speakers honored the event with their lectures: Dr. Advaith Siddharthan (University of Aberdeen), Dr. Andreas Stolcke (Microsoft Research), and Dr. Michael Riley (Google Inc.).

A total of 63 submissions were received for the main event, totaling 91 authors from many institutions worldwide, such as Algeria, Brazil, France, Norway, Portugal, The Netherlands, and USA.

This volume brings together a selection of the 35 best papers accepted at this meeting: 14 full papers and 19 short papers. The acceptance rate was 22%. To these, the two best papers from the Phd and MSc/MA dissertations contest were added.

In this volume, the papers are organized thematically and include the most recent developments in speech language processing and applications, linguistic description, syntax and parsing, ontologies, semantics and lexicography, corpora and language resources, natural language processing, tools and applications.

Our sincere thanks to every person and institution involved in the complex organization of this event, especially the members of the scientific committee of the main event, the dissertations contest and the associated workshops, the invited speakers, and the general organization staff.

We are also grateful to the agencies and organizations that supported and promoted the event, namely, the Brazilian Computer Society (SBC) and its Special Interest Group on NLP (CEPLN), the São Paulo Research Foundation (FAPESP), the Coordination for the Improvement of Higher Level Personnel (CAPES), the International Speech Communication Association (ISCA), and Samsung Eletrônica da Amazônia, Ltda.

October 2014

Jorge Baptista
Nuno Mamede
Sara Candeias
Ivandré Paraboni
Thiago A.S. Pardo
Maria das Graças Volpe Nunes

Organization

General Chairs

Thiago Alexandre
   Salgueiro Pardo   USP-ICMC, Brazil
Maria das Graças Volpe Nunes USP-ICMC, Brazil

Technical Program Chairs

Sara Candeias   IT-University of Coimbra, Portugal
Ivandré Paraboni   USP-EACH, Brazil

Editorial Chairs

Jorge Baptista   University of Algarve, Portugal
Nuno Mamede   IST/INESC-ID, Portugal

Workshop Chair

Vera Lúcia Strube de Lima   PUCRS, Brazil

Demos Chairs

Arnaldo Candido Jr.   UFSCar, Brazil
Fábio Natanael Kepler   UNIPAMPA, Brazil

PhD and MSc/MA Dissertation Contest Chairs

Helena de Medeiros Caseli   UFSCar-DC, Brazil
David Martins de Matos   IST/INESC-ID, Portugal

Local Organizing Committee

Sandra Aluísio   USP-ICMC, Brazil
Lúcia Helena Machado Rino   UFSCar-DC, Brazil
Magali Sanches Duran   USP-ICMC, Brazil
Pedro Paulo Balage Filho   USP-ICMC, Brazil

Program Committee

Albert Gatt	University of Malta, Malta
Alberto Abad	INESC-ID, Portugal
Alberto Simões	Universidade do Minho, Portugal
Alexandre Agustini	PUCRS, Brazil
Aline Villavicencio	UFRGS, Brazil
Amália Andrade	Univ. de Lisboa, Portugal
Amália Mendes	Universidade de Lisboa, Portugal
Ana Luís	Universidade de Coimbra, Portugal
Anabela Barreiro	INESC-ID, Portugal
Andreia Bonfante	UFMT, Brazil
Andreia Rauber	Appen, USA
António Branco	Universidade de Lisboa, Portugal
António Joaquim Serralheiro	AM/INESC-ID, Portugal
António Teixeira	Universidade de Aveiro, Portugal
Ariadne Carvalho	UNICAMP, Brazil
Ariani Di Felippo	UFSCAR, Brazil
Bento da Silva	UNESP, Brazil
Berthold Crysmann	CNRS, France
Brett Drury	USP-SC, Brazil
Carla Alexandra Calado Lopes	IT/IPL, Portugal
Carlos Prolo	UFRN, Brazil
Caroline Gasperin	TouchType, UK
Daniela Braga	VoiceBox, USA
Diana Santos	University of Oslo, Norway
Doroteo Torre Toledano	UAM, Spain
Eduardo Lleida	UZ, Spain
Eric Laporte	Université Paris Est, France
Fábio Natanael Kepler	UNIPAMPA, Brazil
Fernando Batista	INESC-ID/ISCTE-IUL, Portugal
Fernando Perdigão	IT/Universidade de Coimbra, Portugal
Fernando Resende	UFRJ, Brazil
Gaël Harry Dias	Université de Caen, Basse-Normandie, France
Gladis Almeida	UFSCAR, Brazil
Helena de Medeiros Caseli	UFSCAR, Brazil
Helena Moniz	L2F/INESC-ID, Portugal
Hugo Meinedo	INESC-ID, Portugal
Irina Temnikova	QCRI, Qatar
Isabel Falé	Universidade Aberta, Portugal
Isabel Trancoso	IST/INESC-ID, Portugal
João Balsa	Universidade de Lisboa, Portugal
João Luís Rosa	USP-ICMC, Brazil
João Paulo Neto	IST/INESC-ID, Portugal

Joaquim Llisterri	UAB, Spain
Jorge Baptista	Universidade do Algarve, Portugal
José João Almeida	Universidade do Minho, Portugal
Kees van Deemter	University of Aberdeen, UK
Laura Alonso Alemany	Universidad Nacional de Córdoba, Argentina
Leandro Oliveira	Embrapa, Brazil
Luciana Benotti	Universidad Nacional de Córdoba, Argentina
Luis A. Pineda	UNAM, Mexico
Luiz Pizzato	University of Sydney, Australia
Magali Sanches Duran	USP-SC, Brazil
Marcelo Finger	USP, Brazil
Maria das Graças Volpe Nunes	USP-SC, Brazil
Maria José Finatto	UFRGS, Brazil
Mário J. Silva	IST/INESC-ID, Portugal
Michel Gagnon	Ecole Polytechnique, Canada
Norton Trevisan Roman	USP-EACH, Brazil
Nuno Cavalheiro Marques	UNL, Portugal
Nuno Mamede	IST/INESC-ID, Portugal
Palmira Marrafa	Universidade de Lisboa, Portugal
Paulo Gomes	Universidade de Coimbra, Portugal
Paulo Quaresma	Universidade de Évora, Portugal
Plínio Barbosa	Unicamp, Brazil
Renata Vieira	PUCRS,Brazil
Ricardo Ribeiro	INESC-ID/ISCTE-IUL, Portugal
Ronaldo Martins	Univas, Brazil
Rubén San-Segundo	UPM, Spain
Ruy Luiz Milidiú	PUC-Rio, Brazil
Sandra Aluísio	USP-SC, Brazil
Sara Candeias	IT/Universidade de Coimbra, Portugal
Solange Rezende	USP-SC, Brazil
Steven Bird	University of Melbourne, Australia
Ted Pederson	University of Minnesota, USA
Thiago A.S. Pardo	USP-SC, Brazil
Thomas Pellegrini	Université de Toulouse III-Paul Sabatier, France
Valéria Feltrim	UEM, Brazil
Vera Lúcia Strube de Lima	PUCRS, Brazil
Violeta Quental	PUC-Rio, Brazil
Vitor Rocio	Universidade Aberta, Portugal
Wilker Aziz	University of Sheffield, UK

Additional Reviewers

David Batista	INESC-ID, Portugal
Denise Hogetop	PUCRS, Brazil
Derek Wong	University of Macau, China
Gracinda Carvalho	Universidade Aberta, Portugal
João Filgueiras	INESC-ID, Portugal
Marcelo Criscuolo	USP, Brazil
Mário Rodrigues	Universidade de Aveiro, Portugal
Raquel Amaro	Universidade de Lisboa, Portugal
Sara Mendes	Universidade de Lisboa, Portugal

Steering Committee

Jorge Baptista	Universidade do Algarve, Portugal (Chair)
Cláudia Freitas	PUC-Rio/Linguateca, Brazil
Fernando Perdigão	IT/Universidade de Coimbra, Portugal
Renata Vieira	PUCRS, Brazil
Thiago Alexandre Salgueiro Pardo	USP-ICMC, Brazil

Table of Contents

Speech Language Processing and Applications

Linguistic Description, Syntax and Parsing

Ontologies, Semantics and Lexicography

Corpora and Language Resources

Natural Language Processing, Tools and Applications

Automatically Recognising European Portuguese Children's Speech

Pronunciation Patterns Revealed by an Analysis of ASR Errors

Annika Hämäläinen[1,2], Hyongsil Cho[1,2], Sara Candeias[1,3], Thomas Pellegrini[4], Alberto Abad[5], Michael Tjalve[7], Isabel Trancoso[5,6], and Miguel Sales Dias[1,2]

[1] Microsoft Language Development Center, Lisbon, Portugal
[2] ISCTE - University Institute of Lisbon, Lisbon, Portugal
[3] Instituto de Telecomunicações-pole of Coimbra, Coimbra, Portugal
[4] IRIT - Université Toulouse III - Paul Sabatier, Toulouse, France
[5] INESC-ID Lisboa, Lisbon, Portugal [6] Instituto Superio Técnico, Lisbon, Portugal
[7] Microsoft & University of Washington, Seattle, WA, USA
{t-anhama,t-hych,michael.tjalve,miguel.dias}@microsoft.com,
saracandeias@co.it.pt, pellegri@irit.fr,
{alberto.abad,isabel.trancoso}l2f.inesc-id.pt

Abstract. This paper reports findings from an analysis of errors made by an automatic speech recogniser trained and tested with 3-10-year-old European Portuguese children's speech. We expected and were able to identify frequent pronunciation error patterns in the children's speech. Furthermore, we were able to correlate some of these pronunciation error patterns and automatic speech recognition errors. The findings reported in this paper are of phonetic interest but will also be useful for improving the performance of automatic speech recognisers aimed at children representing the target population of the study.

Keywords: Automatic speech recognition, children's speech, error analysis, European Portuguese, fricatives, pronunciation, vowel formants.

1 Introduction

Speech interfaces have tremendous potential in the education of children. Speech provides a natural modality for child-computer interaction and can, at its best, contribute to a fun, motivating and engaging way of learning [1]. However, it is well known that automatically recognising children's speech is a very challenging task. Recognisers trained on adult speech tend to perform substantially worse when used by children [1-6]. Moreover, word error rates (WERs) on children's speech are usually much higher than those on adult speech, even when using a recogniser trained on children's speech, and they show a gradual decrease as the children get older [1-7].

The difficulty of automatically recognising children's speech can be attributed to it being acoustically and linguistically very different from adult speech [1, 2]. For

J. Baptista et al. (Eds.): PROPOR 2014, LNAI 8775, pp. 1–11, 2014.

instance, due to their smaller vocal tracts, the fundamental and formant frequencies of children's speech are higher [1, 2, 7-9]. What is particularly characteristic of children's speech is its higher variability as compared with adult speech, both within and across speakers [1, 2]. This variability is caused by rapid developmental changes in their anatomy, speech production etc., and manifests itself, for example, in speech rate, in the degree of spontaneity, in the frequency of disfluencies, in the values of fundamental and formant frequencies, as well as in pronunciation quality [1, 2, 7-11]. The highly variable values of acoustic parameters converge to adult levels at around 13-15 years of age [9]. Research on age-related pronunciation error patterns, so-called phonological processes or deviations, have also been carried out widely (e.g. [12-14]). Studying and understanding the acoustic and linguistic patterns of children's speech is important for designing and implementing well-functioning speech interfaces for children.

This study focuses on European Portuguese (EP) children's speech in the context of automatic speech recognition (ASR). From the point of view of phonetics, EP has characteristics that make the study of children's speech very interesting. Examples of such characteristics include its high frequency of vowel reduction and consonantal clusters, both within words and across word boundaries [15]. These two characteristics make EP difficult for young speakers to pronounce because their articulatory muscles are not sufficiently developed yet for skilfully articulating all the speech sounds and clusters of speech sounds of the language. In fact, when children attempt to imitate adult speech, they use certain processes to simplify the production of speech sounds. Such simplification may have a negative effect on ASR performance [2].

Previous work on EP children's speech includes several linguistic research projects focused on children's language, especially on language acquisition [16-21]. Studies have also been carried out to identify common age-related phonological processes in EP children's speech [15, 22-23]. To the best of our knowledge, no other studies on the characteristics of EP children's speech have been published in the context of ASR.

In this paper, we report findings from a detailed analysis of errors made by an automatic speech recogniser trained and tested with 3-10-year-old EP children's speech. The goal of the study was to identify pronunciation patterns in children's speech that might be important from the point of view of ASR performance. The results of the study will allow us to understand the mechanisms of EP children's pronunciation and to find ways of improving the accuracy of ASR systems aimed at them and, hence, to improve their experience with speech-enabled applications.

2 Methodology

To reach our goal, we analysed EP children's speech with specific reference to a speech recogniser built for a multimodal educational game aimed at 3-10-year-old Portuguese children [24]. The recogniser was trained and tested with speech extracted from a corpus of EP children's speech, which was specifically collected for this purpose. When carrying out the analysis, we focused on utterances that had not been recognised correctly, as well as on utterances that had been recognised correctly but with a low confidence score. This chapter describes the speech material, the automatic speech recogniser, and the methodology used in our study. The results of our analysis are reported in Section 3.

2.1 Speech Material

We used speech extracted from the CNG Corpus of European Portuguese Children's Speech [24]. The corpus contains four types of utterances recorded from children aged 3-10: phonetically rich sentences, musical notes (e.g. *dó*), isolated cardinals (e.g. *44*), and sequences of cardinals (e.g. *28, 29, 30, 31*). The children were divided into two groups when developing the corpus: 3-6-year-olds and 7-10-year-olds. The prompts for both the cardinals and the cardinal sequences were designed to be easier in the case of the 3-6-year-olds, who were also asked to produce fewer prompts. Depending on their age and reading skills, the children either read the prompts, or repeated them after a recording supervisor. The corpus comes with manually verified transcriptions, as well as annotations for filled pauses, noises, and incomplete, mispronounced and unintelligible words. Table 1 presents the main statistics of the training and test data used in this study.

Table 1. The main statistics of the speech material

	Training	Test
#Speakers	432	52
#Word types	605	521
Ages 3-6	*557*	*319*
Ages 7-10	*585*	*494*
#Word tokens	102,537	12,029
Ages 3-6	*9553*	*1148*
Ages 7-10	*92,984*	*10,881*
hh:mm:ss	17:42:22	02:05:34
Ages 3-6	*02:30:24*	*00:18:31*
Ages 7-10	*15:11:58*	*01:47:03*

2.2 Automatic Speech Recognition

In [24], several different Hidden Markov Model (HMM) -based speech recognisers for EP children's speech were trained and tested. Table 1 summarises the datasets used for training and testing the recognisers. The best-performing recogniser, which we are also using in this study, was a cross-word triphone recogniser trained using a standard acoustic model training procedure with decision tree state tying (see e.g. [25]). Thirty-eight phone labels were used for training the triphones, which have 14 Gaussian mixtures per state. The recogniser also comprises a silence model, a hesitation model and a noise model; the last two were trained utilising the annotations for filled pauses and noises that are available in the corpus. The recogniser was specifically trained for a multimodal educational game, which was developed in the

Contents for Next Generation Networks (CNG) project and expects isolated cardinals, sequences of cardinals and musical notes as speech input [24]. Therefore, [24] used constrained grammars for language modelling purposes: a list grammar for the musical notes, and structure grammars for the isolated cardinals and the cardinal sequences. The grammar for the isolated cardinals allowed cardinals from 0 to 999, whereas the grammar for the cardinal sequences allowed sequences of 2-4 cardinals ranging from 0 to 999; the grammars corresponded both to the recorded data and to the expected speech input. During the experimentation phase, [24] recognised the phonetically rich sentences using a list grammar consisting of the phonetically rich sentences recorded for the corpus; the CNG game itself does not use this type of speech input.

Table 2. WERs (%) with a 95% confidence interval for all, for 3-6-year-old, and for 7-10-year-old speakers in the evaluation test set

	Full Test Set	Ages 3-6	Ages 7-10
Children's ASR	10.0 ± 0.5	27.1 ± 2.6	8.2 ± 0.5

Table 3. The WERs (%) of the children's speech recogniser per utterance type

	Full Test Set	Ages 3-6	Ages 7-10
Phonetically rich	10.4	25.6	6.6
Musical notes	4.2	13.3	2.2
Isolated cardinals	6.3	27.4	3.9
Sequences of cardinals	10.6	33.3	9.7
Overall (excl. phon. rich)	9.8	29.3	8.7

Table 4. The number of word substitution, insertion and deletion errors made by the children's speech recogniser, excluding the phonetically rich sentences

	Full Test Set	Ages 3-6	Ages 7-10
Substitutions	345	60	285
Insertions	198	15	183
Deletions	303	60	243

Table 2 summarises the speech recognition results obtained with the children's speech recogniser. Similar to other studies [3-5, 7], the WERs were considerably higher in the case of the younger children.

Table 3 lists the WERs of the children's speech recogniser for each of the recorded utterance types. It also includes the overall WERs without phonetically rich sentences, which represent a prompt type that is not applicable to the CNG game. Table 4 presents the corresponding number of substitution, insertion and deletion errors made by the children's speech recogniser; the higher number of errors in the case of the 7-10-year-olds reflects the larger amount of test data in their case. The results in Table 3 make it clear that the recognition performance of 3-6-year-olds leaves much to be desired. While the recognition performance of the different types of prompts also leaves room for improvement in the case of 7-10-year-olds, it may already be acceptable for the CNG game – in particular in the case of musical notes and isolated cardinals.

2.3 Auditory Analysis

We analysed the word substitution, insertion and deletion errors made by the children's speech recogniser on the set of test utterances excluding the phonetically rich sentences (see Section 2.2 and Table 4). In total, we analysed 87 errors made in the case of the 3-6-year-olds and 39 errors made in the case of the 7-10-year-olds. In some cases, the recogniser did not output any words for the whole utterance. A preliminary analysis of the utterances with recognition errors suggested that the word substitution errors were the most interesting errors for a thorough auditory phonetic analysis, so we focussed on those types of errors in particular. To get a better overall picture of the pronunciation patterns that might be important from the point of view of ASR performance, we also analysed utterances that had been recognised correctly but with a low confidence score (51 utterances from the 3-6-year-olds and 51 utterances from the 7-10-year-olds).

Two qualified phoneticians, one an expert in Portuguese phonetics and another an expert in general auditory phonetics, carefully listened to all the test utterances that had been misrecognised by the children's speech recogniser. They transcribed the children's phonetic realisations of the misrecognised words using SAMPA (Speech Assessment Methods Phonetic Alphabet; [27]), compared their transcriptions with the standard transcriptions of the words in question, and categorised the differences between the two. The results of this analysis are reported in Section 3.

2.4 Acoustic Analysis of Vowel Formants

The auditory analysis carried out by the phoneticians suggested that vowels are usually pronounced correctly by the children in the corpus and do not play a role when it comes to ASR performance. However, for a more complete analysis of EP children's speech, we computed the average formant values for the 3-6-year-old and the 7-10-year-old children. As there was no obvious correlation between the ASR errors and the realisation of the vowels, we did not limit this analysis to the utterances with ASR errors but extended it to all the vowels in the phonetically rich sentences of the corpus (1848 and 7077 phonetically rich sentences recorded from 3-6-year-olds and 7-10-year-olds, respectively).

To be able to compute the average formant values for vowels, we obtained phoneme-level segmentations by carrying out a forced alignment of the phonetically rich sentences using an in-house (adult) speech recogniser [28]. We used context-independent acoustic models for the forced alignment, as they are considered more suitable for linguistically motivated research than context-dependent models (e.g. [29]). We extracted the formant values, filtered out aberrant values, and drew the vowel charts using the Praat software [30]. To define the threshold values for filtering, we used the average formant values for EP adult females [31] as a reference (cf. Section 2.2). Formant values that were 400 Hz below or above the reference values were considered as artefacts and were discarded. After filtering, we were left with a set of 5,100 and a set of 24,100 vowels for computing the average F1/F2 values for the 3-6-year-olds and the 7-10-year-olds, respectively. Figure 1 illustrates the F1/F2 values for the nine oral vowels of EP, showing the expected shift in formant frequencies. The F1/F2 chart is discussed in more detail in Section 3.2.

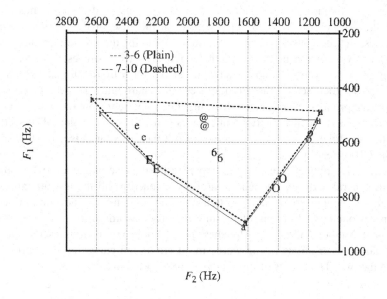

Fig. 1. F1/F2 chart for 3-6-year-olds and 7-10-year-olds

3   Pronunciation Patterns of European Portuguese Children's Speech

This section describes the findings from the auditory analysis (see Section 2.3) and the analysis of vowel formants (see Section 2.4). Before describing any pronunciations in this section, we must clarify an aspect concerning the phonetic/phonological binomial (see [32, 33]): we have here adopted a phonetic representation of sound patterns because it is closer to the physical reality of language than a phonological representation.

3.1   Consonants

ASR errors were often related to the reduction or truncation of consonant clusters, especially in the case of liquids. Previous studies [22, 23] have shown a high occurrence of consonant cluster reductions in European Portuguese children's speech.

We observed the same phenomenon in our data. For example, the word *três* ('three') was often pronounced as [t"eS] instead of the standard pronunciation [tr"eS]. This mispronunciation accounted for 10% of the misrecognitions and 10% of the correct recognition results with a low confidence score that we analysed. Considering the fact that children acquire the ability to accurately produce liquid consonants, such as [l] and [r], at the latter stage of their language acquisition process (at around 4 or 5 years old of age), this finding is not surprising.

The word *um* (["u~]; 'one') was sometimes incorrectly recognised as the word *onze* (["o~z@]; 'eleven'). We hypothesise that these ASR errors were related to background noise in the recordings or to the audible breathing of the speakers right after the production of the word *um*, which might have led the recogniser to output the word *onze* whose pronunciation includes the alveolar fricative [z].

As for fricative consonants, the substitution of the phones [s] and [z] with their palatal equivalents [S] and [Z] was common in the case of the 3-6-year-olds. Examples of such substitutions include:

— *sete* ('seven'): [s"Et@] → [S"Et@]
— *cinco* ('five'): [s"i~ku] → [S"i~ku]
— *dezasseis* ('sixteen'): [d@z6s"6jS] → [dZ6S"6jS]
— *dezassete* ('seventeen'): [d@z6s"Et@] → [dZ6S"Et@]
— *dezoito* ('eighteen'): [d@z"Ojtu] → [dZ"Ojtu]

There was some evidence suggesting that these phone substitutions might be correlated with ASR errors but the analysis did not provide conclusive results yet. Therefore, we must look into these phone substitutions again if and when more data becomes available in the future.

When analysing the pronunciation of plosives, we observed the velar consonant /k/ often being substituted with an alveolar stop in words like *quinze* ('fifteen'; [k"i~z@] → [t"i~z@]). This pronunciation pattern has also been reported in the literature [22] as one of the most common pronunciation patterns in EP children's speech. Interestingly, this phone substitution, which crosses phonological categories, did not seem to have any major impact on ASR performance.

We also found a devoicing deviation for the alveolar fricative [z] in words like *zero* and *doze*:

— *zero* ('zero'): [z"Eru] → [s"Eru] or [z"Eru] → [S"Eru]
— *doze* ('twelve'): [d"oz@] → [d"os@]

To further analyse devoicing deviations in EP children's speech, we will carry out an acoustic analysis of VOT (Voice Onset Time) in future research, and analyse whether or not there is a correlation with ASR performance.

3.2 Vowels

We could not identify any word substitution errors caused by deviations in the pronunciation of vowels. As for word deletion errors, one specific word caught our attention: the word *e* ('and') was often deleted by the recogniser in the case of cardinals between 22 and 99. Although monosyllabic function words are known to be a common source of ASR errors, these errors also seemed to correlate with a pronunciation pattern that we could observe in the children's speech. In Portuguese, the orthographic form of these cardinals includes *e* between the tens and the units (e.g. *vinte e cinco* ('twenty-five')). However, there are two alternative ways of pronouncing these cardinals: one with the *e* (pronounced as an unstressed [i]) and another without. The speakers in the corpus often merged the pronunciation of *e* into the final vowel of the previous word. This phenomenon, which is typical of EP continuous speech also in the case of adult speakers, gives rise to a change in the syllable structure of the syntagm, which seemed to cause the children's speech recogniser to make a number of word deletion errors. Examples of this phenomenon include, for instance:

— *vinte e cinco* ('twenty-five'): [v"i~t@ i s"i~ku] → [v"i~t i s"i~ku]
— *cinquenta e quatro* ('fifty-four'):
   [si~k"we~t6 i k"watru] → [s"i~kwe~t i k"watru]

The vowel formants F1 and F2 (see Figure 1) showed age-related tendencies that did not seem to correlate with ASR errors. Although the vowel triangles of the 3-6-years-olds are very similar to those of the 7-10-year-olds, the triangle of the 3-6-year-olds has higher F1 values, mainly for close and mid-close vowels. This slight increase in F1 values could be expected as the "closer" articulation of the 3-6-year-olds is related to their vocal tracts being smaller than those of the 7-10-year-olds. The centralization of the front vowels [i], [e] and [E] is reinforced by the total absence of lip rounding, showing that children become more skilled in their ability to control the articulators with age. This is a view shared by many experts in child language acquisition [12, 14].

3.3   Other Characteristics of EP Children's Speech

We also observed other linguistic events, such as truncated words and repetitions (e.g. [k"wa k"watru] for *qua- quatro* ('fo- four')), especially in the case of the 3-6-years-olds. We expected to observe these events, well-known as hesitations or disfluencies, as they are a characteristic of read speech [34]. However, similarly to [10], they did not have an impact on ASR performance.

Compared with adult speech corpora, some children in this study uttered words with a reduced duration and/or a quiet voice. We believe that there is a psychological explanation for this: especially the younger children often reacted to the recording situation with shyness [24]. The words with a short duration and/or a low volume - in particular monosyllabic words with a simple syllable structure, accounted for a large part of the word deletion errors made by the recogniser. Examples of words that were frequently deleted include, for instance, *e* ('and'; ["i]), *um* ('one'; ["u~]), and *sim* ('yes'; [s"i~]).

4   Conclusions and Discussion

This paper reported our findings from a detailed analysis of errors made by an automatic speech recogniser trained and tested with 3-10-year-old European Portuguese children's speech. The goal of the study was to identify pronunciation patterns in children's speech that might be important from the point of view of ASR performance. The analysis confirmed the general tendencies in EP children's pronunciation that have been described by others but it also provided us with valuable information on the pronunciation patterns that actually have an impact on ASR performance. Using the findings from the analysis, we intend to derive pronunciation rules for adding relevant pronunciation variants into a pronunciation lexicon used by the children's speech recogniser. Such an approach has previously led to significant decreases in WER when automatically recognising preschool children's speech [35].

Due to the nature of the corpus and the methodology used in this study, the analysis reported in this paper has its limitations. Because the types of utterances recorded are not fully representative of everyday language use, the findings of the study are difficult to generalise. Furthermore, the data in the children's speech corpus is read or repeated speech and, as such, not fully representative of the speech input expected in the multimodal educational game that the children's speech recogniser was built for. Therefore, future studies will have to focus on collecting speech data with a wider variety of utterance types to ensure the diversity of the data from the phonetic and phonological point of view. In addition to that, the setting of future recordings will need to be reviewed to make sure that the recorded data is more representative of the type of speech that is of interest to us (spontaneous speech instead of read or repeated speech). The best option would be to collect more speech data by recording children's verbal interaction with the multimodal educational game itself.

Acknowledgements. Microsoft Language Development Center carried this work out in the scope of the following projects: (1) QREN 5329 Fala Global, co-funded by Microsoft and the European Structural Funds for Portugal (FEDER) through POR Lisboa (Regional Operational Programme of Lisbon), as part of the National Strategic Reference Framework (QREN), the national program of incentives for Portuguese businesses and industry; (2) QREN 7943 CNG – Contents for Next Generation Networks, co-funded by Microsoft and FEDER through COMPETE (Operational Program for Competitiveness Factors), as part of QREN. Sara Candeias is supported by the FCT grant SFRH/BPD/36584/2007 and is involved in the LetsRead: Automatic Assessment of Reading Ability of Children project of Instituto de Telecomunicações, co-funded by FCT. This work has also been partially been supported by FCT grant PEst-OE/EEI/LA0021/2013.

References

1. Gerosa, M., Giuliani, D., Narayanan, S., Potamianos, A.: A Review of ASR Technologies for Children's Speech. In: Workshop on Child, Computer and Interaction, Cambridge, MA (2009)

2. Russell, M., D'Arcy, S.: Challenges for Computer Recognition of Children's Speech. In: Workshop on Speech and Language Technology in Education, Farmington, PA (2007)
3. Potamianos, A., Narayanan, S.: Robust Recognition of Children's Speech. IEEE Speech Audio Process 11(6), 603–615 (2003)
4. Wilpon, J.G., Jacobsen, C.N.: A Study of Speech Recognition for Children and Elderly. In: IEEE International Conference on Acoustics, Speech, and Signal Processing, Atlanta, GA, pp. 349–352 (1996)
5. Elenius, D., Blomberg, M.: Adaptation and Normalization Experiments in Speech Recognition for 4 to 8 Year Old Children. In: Interspeech, Lisbon (2005)
6. Gerosa, M., Giuliani, D., Brugnara, F.: Speaker Adaptive Acoustic Modeling with Mixture of Adult and Children's Speech. In: Interspeech, Lisbon (2005)
7. Gerosa, M., Giuliani, D., Brugnara, F.: Acoustic Variability and Automatic Recognition of Children's Speech. Speech Commun. 49(10-11), 847–860 (2007)
8. Huber, J.E., Stathopoulos, E.T., Curione, G.M., Ash, T.A., Johnson, K.: Formants of Children, Women and Men: The Effects of Vocal Intensity Variation. J. Acoust. Soc. Am. 106(3), 1532–1542 (1999)
9. Lee, S., Potamianos, A., Narayanan, S.: Acoustics of Children's Speech: Developmental Changes of Temporal and Spectral Parameters. J. Acoust. Soc. Am. 10, 1455–1468 (1999)
10. Narayanan, S., Potamianos, A.: Creating Conversational Interfaces for Children. IEEE Speech Audio Process. 10(2), 65–78 (2002)
11. Eguchi, S., Hirsh, I.J.: Development of Speech Sounds in Children. Acta Otolaryngol. Suppl. 257, 1–51 (1969)
12. Bowen, C.: Children's Speech Sound Disorders. Wiley-Blackwell, Oxford (2009)
13. Grunwell, P.: Clinical Phonology, 2nd edn. Wiliams & Wilkins, Baltimore (1987)
14. Miccio, A.W., Scarpino, S.E.: Phonological Analysis, Phonological Processes. In: Ball, M.J., Perkins, M.R., Muller, N., Howard, S. (eds.) The Handbook of Clinical Linguistics. Wiley-Blackwell, Malden (2008)
15. Candeias, S., Perdigão, F.: Syllable Structure in Dysfunctional Portuguese Children Speech. Clinical Linguistics & Phonetics 24(11), 883–889 (2010)
16. Freitas, M.J.: Acquisition in European Portuguese: Resources and Linguistic Results. Project funded by FCT: PTDC/LIN/68024/2006, Centro de Linguística da Universidade de Lisboa (CLUL) (2006)
17. Vigário, M.: Development of Prosodic Structure and Intonation (DEPE). Project funded by FCT: PTDC/CLELIN/108722/2008, Centro de Linguística da Universidade de Lisboa (CLUL) (2008)
18. Costa, J.: Syntactic Dependencies from 3 to 10. Project funded by FCT: PTDC/CLELIN/099802/2008, Centro de Linguística da Universidade Nova de Lisboa (CLUNL) (2008)
19. Freitas, M.J., Gonçalves, A., Duarte, I.: Avaliação da Consciência Linguística: Aspectos fonológicos e sintácticos do Português. Ed. Colibri, Lisbon (2011)
20. Faria, M.I.H.: Reading Comprehension. Word, Sentence and Text processing. Project funded by FCT: PTDC/LIN/67854/2006, Centro de Linguística da Universidade (2006)
21. Frota, S., Correia, S., Severino, C., Cruz, M., Vigário, M., Cortês, S.: PLEX5 A Production Lexicon of Child Speech for European Portuguese / Um léxico infantil para o Português Europeu. Laboratório de Fonética CLUL/FLUL, Lisbon (2012)
22. Guerreiro, H., Frota, S.: Os processos fonológicos na fala da criança de cinco anos: tipologia e frequência, vol. 3. Instituto de Ciências da Saúde, UCP (2010)

23. Almeida, L., Costa, T., Freitas, M.J.: Estas portas e janelas: O caso das sibilantes na aquisição do português europeu. In: Conferência XXV Encontro Nacional da Associação Portuguesa de Linguística, Porto (2010)
24. Hämäläinen, A., Miguel Pinto, F., Rodrigues, S., Júdice, A., Morgado Silva, S., Calado, A., Sales Dias, M.: A Multimodal Educational Game for 3-10-year-old Children: Collecting and Automatically Recognising European Portuguese Children's Speech. In: Workshop on Speech and Language Technology in Education, Grenoble (2013)
25. Young, S., Evermann, G., Hain, T., Kershaw, D., Moore, G., Odell, J., Ollason, D., Povey, D., Valtchev, V., Woodland, P.: The HTK Book (for HTK Version 3.2.1). Cambridge University, Cambridge (2002)
26. Microsoft Speech Platform Runtime (Version 11),
    http://www.microsoft.com/en-us/download/
    details.aspx?id=27225 (accessed March 25, 2013)
27. Wells, J.C.: Portuguese (1997),
    http://www.phon.ucl.ac.uk/home/sampa/portug.htm
28. Meinedo, H., Abad, A., Pellegrini, T., Neto, J., Trancoso, I.: The L2F Broadcast News Speech Recognition System. In: FALA, Vigo, pp. 93–96 (2010)
29. Vieru, B., Boula de Mareüil, P., Adda-Decker, M.: Characterisation and Identification of Non-Native French Accents. Speech Commun. 53(3), 292–310 (2011)
30. Boersma, P.: Praat, a System for Doing Phonetics by Computer. Glot International 5(9/10), 341–345 (2001)
31. Pellegrini, T., Hämäläinen, A., Boula de Mareüil, P., Tjalve, M., Trancoso, I., Candeias, S., Sales Dias, M., Braga, D.: A Corpus-Based Study of Elderly and Young Speakers of European Portuguese: Acoustic Correlates and Their Impact on Speech Recognition Performance. Interspeech, Lyon (2013)
32. Mateus, M.H., d'Andrade, E.: The Phonology of Portuguese. Oxford University Press, Oxford (2000)
33. Barbosa, J.M.: Introdução ao Estudo da Fonologia e Morfologia do Português. Almedina, Coimbra (1994)
34. Veiga, A., Celorico, D., Proença, J., Candeias, S., Perdigão, F.: Prosodic and Phonetic Features for Speaking Styles Classification and Detection. In: Toledano, D.T., Ortega, A., Teixeira, A., Gonzalez-Rodriguez, J., Hernandez-Gomez, L., San-Segundo, R., Ramos, D. (eds.) IberSPEECH 2012. CCIS, vol. 328, pp. 89–98. Springer, Heidelberg (2012)
35. Cincarek, T., Shindo, I., Toda, T., Saruwatari, H., Shikano, K.: Development of Preschool Children Subsystem for ASR and Q&A in a Real-Environment Speech-Oriented Guidance Task. In: Interspeech, Antwerp (2007)

Improving Speech Recognition through Automatic Selection of Age Group – Specific Acoustic Models

Annika Hämäläinen[1], Hugo Meinedo[2], Michael Tjalve[4], Thomas Pellegrini[5], Isabel Trancoso[2,3], and Miguel Sales Dias[1]

[1] Microsoft Language Development Center & ISCTE - University Institute of Lisbon, Lisbon, Portugal
[2] INESC-ID Lisboa, Lisbon, Portugal
[3] Instituto Superio Técnico, Lisbon, Portugal
[4] Microsoft & University of Washington, Seattle, WA, USA
[5] IRIT - Université Toulouse III - Paul Sabatier, Toulouse, France
{t-anhama,michael.tjalve,miguel.dias}@microsoft.com,
{hugo.meinedo,isabel.trancoso}@inesc-id.pt, pellegri@irit.fr

Abstract. The acoustic models used by automatic speech recognisers are usually trained with speech collected from young to middle-aged adults. As the characteristics of speech change with age, such acoustic models tend to perform poorly on children's and elderly people's speech. In this study, we investigate whether the automatic age group classification of speakers, together with age group –specific acoustic models, could improve automatic speech recognition performance. We train an age group classifier with an accuracy of about 95% and show that using the results of the classifier to select age group –specific acoustic models for children and the elderly leads to considerable gains in automatic speech recognition performance, as compared with using acoustic models trained with young to middle-aged adults' speech for recognising their speech, as well.

Keywords: Age group classification, acoustic modelling, automatic speech recognition, children, elderly, paralinguistic information.

1 Introduction

Currently available speech recognisers do not usually work well with children's or elderly people's speech. This is because several parameters of the speech signal (e.g. fundamental frequency, speech rate) change with age [1-4] and because the acoustic models (AMs) used for automatic speech recognition (ASR) have typically been trained with speech collected from young to middle-aged adults only, to serve mainstream business and research requirements. Furthermore, both children and elderly people are more likely to interact with computers using everyday language and their own commands, even when a specific syntax is required [5-9]. As compared with young to middle-aged adults, significantly higher word error rates (WERs) have been reported both for children [10-12] and for elderly speakers [10, 13, 14]. Improvements

J. Baptista et al. (Eds.): PROPOR 2014, LNAI 8775, pp. 12–23, 2014.

in ASR performance have been reported when using AMs adapted to children [10-12] and to the elderly [10, 13, 14], respectively, and more and more children's (e.g. [15-17]) and elderly speech corpora suitable for training AMs (e.g. [17-19]) are gradually becoming available. However, in many speech-enabled applications, the age of the user is unknown in advance. So, if multiple sets of AMs tailored to different age groups are available, the optimal set must either be selected manually by the user, or an automatic method must be devised for selecting it.

A real-life example of a speech-enabled application that is used by people of widely varying ages is the Windows Phone app World Search, which allows users to perform web searches via the Bing search engine using their voice. The European Portuguese version of the app (currently the only version available), uses three sets of AMs optimised for three age groups: children, young to middle-aged adults and elderly people. The models optimised for young to middle-aged adults are used by default. However, through a setting in the application, users have the option of manually selecting the set of AMs that they think is the most appropriate for them. Using the default models in the case of children and the elderly is expected to deteriorate the ASR performance dramatically (cf. [10-14]). However, having to make a manual selection is rather cumbersome from the usability point of view. An accurate age group classifier would, on the other hand, allow the optimal set of AMs to be selected automatically. Similarly, it could be used to automatically select a language model and a lexicon that represents the typical human-computer interaction (HCI) of users belonging to a given age group. In spoken dialogue systems, an age group classifier might be useful for selecting dialogue strategies or different ways of interacting with the user. For example, the persona and verbosity of the responses could be adapted to better match the typical preferences of the age group of the active user. A more fun and engaging way of addressing the user could be used if (s)he were recognised as a child, whereas a more polite way, which the elderly might prefer in HCI (cf. [8]), could be applied in the case of the elderly.

The goal of this paper is to investigate whether the automatic age group classification of speakers, together with age group –specific AMs, could improve ASR performance. Although much research has been done on automatic age estimation (e.g. [20-22]), we are not aware of other studies that would have used the results of automatic age estimation to select age group –specific AMs for improving ASR performance. After describing the speech material in Section 2, we present our age group classifier and the results of our age group classification experiments in Section 3. Section 4 presents the automatic speech recogniser and the age group –specific AMs used in this study, together with ASR results obtained using the default models, and age group –specific models selected using the speakers' real age and automatically detected age. We present our conclusions in Section 5.

2   Speech Material

The speech material used in this study originates from four different corpora of European Portuguese: the CNG Corpus of European Portuguese Children's Speech [16] (hereafter "CNG"); the EASR Corpus of European Portuguese Elderly Speech [18] (hereafter "EASR"); the BD-PUBLICO corpus [23], which contains young to middle-aged adults' speech; and a small corpus of European Portuguese young to middle-aged adults' speech collected in Lisbon in the summer of 2013 (hereafter "YMA").

The speakers in the CNG Corpus are 3-10 years of age, whereas the speakers in the EASR Corpus are aged 60 or over. Both corpora contain prompted speech: read or repeated speech in the case of the CNG Corpus, and read speech in the case of the EASR Corpus. They come with manually verified transcriptions, as well as annotations for filled pauses, noises and "damaged" words (e.g. mispronunciations, false starts). These two corpora were used for training and testing the age group classifier, for training AMs optimised for children's and elderly speech, as well as for the ASR experiments. While the training data extracted from both corpora contains phonetically rich sentences, different types of number expressions etc., we only used a subset of utterance types for testing purposes. Our development test set, which was used in the automatic age group classification experiments, only contained the longest utterance types because very short utterances are difficult to accurately estimate age (and other speaker characteristics) from. In the case of the CNG Corpus, the development test set included sequences of cardinal numbers and phonetically rich sentences. In the case of the EASR Corpus, they comprised phonetically rich sentences only. We only used phonetically rich sentences in the evaluation test set, both in the age group classification and in the ASR experiments. This is because we wanted to maximise the comparability of test data extracted from different corpora.

The BD-PUBLICO Corpus contains newspaper sentences read out by 18-48-year-old speakers, i.e., young to middle-aged adults. The transcriptions in this corpus have not been verified manually or annotated for noises etc. We used data from this corpus in the age classification experiments, both in the training and development test sets. We tested the age group classifier and carried out the ASR experiments using the YMA Corpus, which contains phonetically rich sentences read out by speakers aged 25-59. Each speaker in the YMA Corpus uttered 80 phonetically rich sentences, 20 of which originate from the same pool of phonetically rich sentences that were used for recording the CNG Corpus and 60 of which originate from the same pool of phonetically rich sentences that appear in the EASR Corpus. This makes the evaluation test sets used in the ASR experiments comparable across all three age groups. The transcriptions in the YMA Corpus are not verified manually nor annotated. However, the recordings were monitored closely and speakers were asked to reread sentences that they did not read correctly or that included filled pauses or noises. In the age group classification experiments, we used an additional set of speakers (hereafter "YMA-a") recorded during the YMA data collection. These speakers were left out from the final YMA Corpus because they did not record the full set of 80 utterances. However, they were useful for increasing the number of speakers aged up to 54 in the training and development test sets.

The training sets, development test sets and evaluation test sets are summarised in Tables 1, 2 and 3, respectively. Before the age group classification experiments, the data were automatically pre-processed to boost the energy levels and to remove unwanted silences. The feature extraction of the training data from the CNG and EASR corpora was carried out at a frame rate of 10 ms using a 25-ms Hamming window and a pre-emphasis factor of 0.98. 12 Mel Frequency Cepstral Coefficients (MFCCs) and log-energy with corresponding first, second and third order time derivatives were calculated, and the total number of features was reduced to 36 using Heteroscedastic Linear Discriminant Analysis (HLDA). These features were used for building AMs optimised for children's and elderly speech (see Section 4.1).

Table 1. The main statistics of the data used to train the age group classifier for children (CNG), young to middle-aged adults (BD-PUBLICO and YMA-a) and the elderly (EASR), and to optimise acoustic models for children (CNG) and the elderly (EASR)

	CNG	BD-PUBLICO	YMA-a	EASR
#Speakers	432	109	13	778
#Male+#Female	190 + 242	55 + 54	6 + 7	203 + 575
#Word types	605	16,517	1663	4905
#Word tokens	102,537	195,169	5688	482,208
#Utterances	18,569	9320	795	44,033

Table 2. The main statistics of the development test sets used in the age group classification experiments

	CNG	BD-PUBLICO	YMA-a	EASR
#Speakers	26	10	2	48
#Male+#Female	12 + 14	5 + 5	1 + 1	16 + 32
#Word types	480	2783	644	3492
#Word tokens	6221	16,758	1550	31,565
#Utterances	866	584	160	2836

Table 3. The main statistics of the evaluation test sets used in the age group classification and in the ASR experiments

	CNG	YMA	EASR
#Speakers	51	68	96
#Male+#Female	22 + 29	36 + 32	29 + 67
#Word types	747	4485	5728
#Word tokens	3439	46,987	49,580
#Utterances	1735	5440	5351

3 Age Group Classifier

One of the goals of our study was to develop an age group classifier for automatically determining the age group of speakers belonging to one of the following three age groups: children, young to middle-aged adults, or elderly people. To achieve this goal, we developed an age group classification approach that uses two modules. First, it extracts relevant acoustic features from the speech signal, effectively transforming and reducing the dimensionality space of the input data. Second, it tries to determine which output class (i.e. age group) the speech input belongs to. The following subsections present the feature extraction frontends and the age group classification experiments, and discuss the results obtained.

Table 4. The acoustic feature set used in the age group classifiers: 65 Low Level Descriptors (LLDs)

4 energy related LLD	Group
Sum of auditory spectrum (loudness)	prosodic
Sum of RASTA-filtered auditory spectrum	prosodic
RMS Energy, Zero-Crossing Rate	prosodic
55 spectral LLDs	Group
MFCC 1-14	cepstral
RASTA-filtered auditory spectrum	spectral
Spectral energy 250–650 Hz, 1 k–8 kHz	spectral
Spectral Roll-Off Pt. 0.25, 0.5, 0.75, 0.9	spectral
Spectral Flux, Centroid, Entropy, Slope	spectral
Psychoacoustic Sharpness, Harmonicity	spectral
Spectral Variance, Skewness, Kurtosis	spectral
6 voicing related LLDs	Group
F0 (SHS & Viterbi smoothing)	prosodic
Voicing Probability	voice quality
log HNR, Jitter (local & δ), Shimmer (local)	voice quality

Table 5. The acoustic feature set used in the age group classifiers: Statistic functionals applied to the LLDs

Functionals applied to LLD / Δ LLD	Group
quartiles 1–3, 3 inter-quartile ranges	percentiles
1 % percentile (≈ min), 99 % pctl. (≈ max)	percentiles
percentile range 1 %–99 %	percentiles
position of min / max, range (max – min)	temporal
arithmetic mean, root quadratic mean	moments
contour centroid, flatness	temporal
standard deviation, skewness, kurtosis	spectral
rel. dur. LLD is above 25/50/75/90% range	temporal
relative duration LLD is rising	temporal
rel. duration LLD has positive curvature	temporal
gain of linear prediction (LP), LP Coeff. 1–5	modulation
Functionals applied to LLD only	Group
mean value of peaks	peaks
mean value of peaks – arithmetic mean	peaks
mean / std.dev. of inter peak distances	peaks
amplitude mean of peaks, of minima	peaks
amplitude range of peaks	peaks
mean / std. dev. of rising / falling slopes	peaks
linear regression slope, offset, quadratic error	regression
quadratic regression a, b, offset, quadratic error	regression

3.1   Feature Extraction

We extracted features from the speech signal using TUM's open-source openSMILE toolkit [24]. This feature extraction toolkit is capable of producing a wide range of acoustic speech features and has been used in many paralinguistic information and speaker trait detection tasks [25, 26]. The feature set used in our age group classifier contains 6015 static features obtained by applying statistic functionals to the utterance contours of 65 Low-Level Descriptors (LLDs) and their deltas estimated from the speech signal every 10 ms. Table 4 summarizes the LLDs included as frame-level features. The set of statistic functionals applied to the LLD contours at the utterance level includes percentiles, modulations, moments, peaks and regressions, and is presented in Table 5. The LLDs and functionals are described in detail in [27].

In an attempt to preserve the features that are the most relevant to the task at hand and to reduce the complexity of the classification stage, we applied a correlation-based feature subset selection evaluator with a best-first search method [28]. This is a supervised dimensionality reduction technique that evaluates the worth of a subset of features by considering their individual predictive ability along with the degree of redundancy between features. It generally chooses subsets that have low intercorrelation and are highly correlated with the expected classification. We selected the feature subset using the training set and were left with 221 of the original 6015 static features. As the selection procedure resulted in a substantial reduction in the total number of features, we tested classifiers with both the original and the reduced set of features.

3.2   Age Group Classification Experiments

We implemented age group classifiers using linear kernel Support Vector Machines (SVMs) [29, 30] trained with the Sequential Minimal Optimisation (SMO) algorithm [31]. This combination is known to be robust against overfitting when used in tasks with high-dimensional feature spaces and unbalanced class distributions. We normalised feature values to be in the range [0, 1] prior to classifier training, estimating the normalisation parameters on the training set and then applying them to the training, development and evaluation test sets. We investigated SVM optimisation by training models with different values for the complexity parameter "C" of the SMO algorithm and by choosing the one that obtained the highest performance on the development test set. As we had two feature sets, one containing all the features and the other containing the automatically selected subset of features, we trained SVMs using both sets and chose the complexity parameter independently. Fig. 1 represents the classification results on the development and evaluation test sets with different complexity values for models trained with all the features and for models with the automatically selected subset of feature.

3.3   Results and Discussion

To evaluate the age group classifiers, we used the Unweighted Average Recall (UAR) metric. Compared with the standard accuracy metric, this metric allows a more meaningful assessment when evaluating datasets with unbalanced class distributions. For our three-class problem at hand – children (C), young to middle-aged adults (A) and elderly people (E), the UAR metric is calculated as (Recall(C)+Recall(A)+Recall(E))/3. That is, the number of instances per class is intentionally ignored.

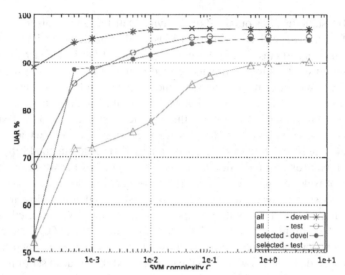

Fig. 1. Age group classification results (UAR %) on the development and evaluation test sets for the SVM models trained with all the features and with the automatically selected subset of features

After training 20 classifier models with different "C" values for the SMO complexity, we chose the model with the highest UAR on the development test set as our age group classifier. Table 6 shows that this model, trained with all the features, obtains a very high performance on the evaluation test set. Table 7 presents the corresponding age group confusion matrix for the evaluation test set.

Table 6. Age group classification results (UAR%) obtained on the development and evaluation test sets

SVM Classifier	C	Devel	Eval
all features	0.05	97.13	95.24
selected features	0.5	95.00	89.33

Table 7. Age group confusion matrix for the evaluation test set. Values in the table indicate the percentage of automatically classified utterances.

	CNG	YMA	EASR
CNG	**98.6**	0.8	0.6
YMA	0.8	**92.0**	7.2
EASR	1.8	3.2	**95.0**

Like most human physiological processes, aging and the consequent vocal aging are gradual processes. The aging of speech is not only affected by chronological age but people's lifestyles (e.g. smoking, alcohol consumption, psychological stress, abuse and overuse of vocal cords) [32], as well. While it might be impossible to determine the exact age from which an individual's speech could be considered elderly, studies usually regard 60-70 years of age as the minimum age range for elderly speech [10]. For our experiments, we decided to consider 60 years of age as the boundary between young to middle-aged adults and the elderly. Our choice is in line with the choice that was made for selecting speakers for the EASR Corpus (see [18]). However, the "fuzziness" of the age boundary is reflected in the number of errone-ously classified utterances from young to middle-aged and elderly speakers. At the other end of the age range, the number of incorrectly classified utterances from children is very small. The performance is undoubtedly boosted by the complete lack of test data from speakers aged 11- 24.

4   ASR Experiments

This section describes the ASR experiments carried out to test the potential benefits of automatic age group classification in ASR. We used three different sets of Hidden Markov Model (HMM) models for the experiments: "standard" AMs trained using young to middle-aged adults' speech (hereafter "SAM"), as well as two separate sets of AMs specifically optimised for children's and elderly people's speech (hereafter "CAM" and "EAM"), respectively. These three sets of AMs are discussed in Section 4.1.

The pronunciation lexicon and language model remained the same across the experiments. The lexicon contained an average of one pronunciation per word, represented using a set of 38 phone labels. For language modelling purposes, we authored a grammar that allowed the words in the phonetically rich sentences of our corpora (see Section 2) to appear 1 to 33 times; the phonetically rich sentences in our data have a minimum of one and a maximum of 33 words. The grammar can be considered as a very simplified language model and, although it yields unrealistically high WERs, the ASR results are comparable across the different evaluation test sets used in our experiments.

4.1   Acoustic Modelling

"Standard" Acoustic Models (SAM). The "standard" AMs originate from the European Portuguese language pack that comes with Microsoft Speech Platform Runtime (Version 11) [33]; a language pack incorporates the language-specific components necessary for ASR: AMs, a pronunciation lexicon, and grammars or a language model. The AMs comprise a mix of gender-dependent whole-word models and cross-word triphones trained using several hundred hours of read and spontaneous speech collected from young to middle-aged adult speakers. In other words, children and the elderly fall outside of this target demographic. The models also include a silence model, a hesitation model for modelling filled pauses, and a noise model for modelling human and non-human noises. More detailed information about the "standard" AMs is not publicly available, it being commercially sensitive information.

Acoustic Models Optimised for Children's Speech (CAM). The "standard" AMs were optimised for children's speech by retraining the female AMs with the training set extracted from the CNG Corpus (see Table 1), regardless of the children's gender. The motivation for only retraining the female AMs with children's speech stems from the fact that the acoustic characteristics of children's speech are more similar to adult female speech than to adult male speech [1, 2]. The hesitation and noise models of the baseline recogniser were retrained utilising the annotations for filled pauses and noises that are available in the corpus (see Section 2). The children's speech models are discussed in more detail in [12].

Acoustic Models Optimised for Elderly Speech (EAM). The "standard" AMs were optimised for elderly speech by retraining the male and female AMs with the male and female data in the training set extracted from the EASR Corpus (see Table 1), respectively. The hesitation and noise models of the baseline recogniser were again retrained utilising the annotations available in the corpus (see Section 2).

4.2 Experimental Set-Up

We carried out ASR experiments on the children's, young to middle-aged adults', and elderly people's speech in our test sets using three different set-ups: 1) Speech recognised using the "standard" AMs (SAM) regardless of the age group of the speaker, 2) Speech recognised using the AMs corresponding to the known age group of the speaker (CAM, SAM or EAM), and 3) Speech recognised using the AMs corresponding to the automatically determined age group of the speaker (AM-Auto). The results of set-up 1) represent a situation in which we have no way of knowing the speaker's age and must use the "standard" AMs. These results represent our baseline; should no alternative, age group –specific AMs and ways of selecting the correct set of age group –specific AMs be available, we would have to use the "standard" AMs. The results of set-up 2) represent the best achievable results ("oracle") because we are using AMs that have been selected using the known age groups of the speakers. The results of set-up 3) represent the results achieved using the automatic age group classifier described in Section 3.

4.3 Results and Discussion

Table 8 illustrates the ASR results for the experimental set-ups described in Section 4.3. Because of the simplified language model that we used (see Section 4), the overall WERs are high. However, the results show that recognising children's and elderly people's speech using AMs optimised for their own age group can lead to significant improvements in ASR performance over the baseline performance. The results also show that, although the ASR results achieved using the automatic age group classifier are worse than the best achievable ASR results ("*oracle*") in the case of children's and elderly speech, as could be expected, the delta is very small. The ASR results achieved using the automatic age group classifier on young to middle-aged adults' speech are slightly better than the ASR results achieved using the baseline recogniser. This means that, for some of those speakers' voices, the age group –specific models were acoustically a better match than the default models. We intend to analyse if this is, for instance, related to them being close to the age group boundary that we selected.

Table 8. Automatic speech recognition results (WERs).

Eval Set	SAM	CAM	EAM	AM-Auto
CNG	78.3%	46.1%		47.8%
YMA	56.4%			56.3%
EASR	55.9%		48.2%	48.9%

5   Conclusions and Future Work

This paper presented an age group classification system that automatically determines the age group of a speaker from an input speech signal. There were three possible age groups: children, young to middle-aged adults and the elderly. What sets our study apart from other studies on age classification is that we used our age group classifier together with an automatic speech recogniser. More specifically, we carried out ASR experiments in which the automatically determined age group of speakers was used to select age group –specific acoustic models, i.e., acoustic models optimised for children's, young to middle-aged adults' and elderly people's speech. The ASR results showed that using the results of the age group classifier to select age group –specific acoustic models for children and the elderly leads to considerable gains in automatic speech recognition performance, as compared with using "standard" acoustic models trained with young to middle-aged adults' speech for recognising their speech, as well. This finding can be used to improve the speech recognition performance of speech-enabled applications that are used by people of widely varying ages. What makes the approach particularly interesting is that it is a user-friendly alternative for speaker adaptation, which requires the user to spend time training the system.

Both children and elderly people are more likely to interact with computers using everyday language and their own commands, even when a specific syntax is required [5-8]. In future research, we will attempt building age group –specific language models and lexica, and select the optimal language model and lexicon using the results of our age group classifier. We hypothesise that such an approach could lead to further gains in ASR performance.

One of the limitations of this study is that we did not have any acoustic model training data or test data from 11-24-year-old speakers. This probably led to unrealistically good age group classification performance in the case of children and young to middle-aged adults. Therefore, in future research, we also intend to record test data from 11-24-year-old European Portuguese speakers, optimise acoustic models for representatives of that age group, and rerun age group classification and ASR experiments. We expect this to be a challenging age group to work with, as the values of children's acoustic parameters converge to adult levels at around 13-15 years of age [1].

Acknowledgements. This work was partially supported by: (1) the QREN 5329 Fala Global project, which is co-funded by Microsoft and the European Structural Funds for Portugal (FEDER) through POR Lisboa (Regional Operational Programme of Lisbon), as part of the National Strategic Reference Framework (QREN), the national program of incentives for Portuguese businesses and industry; (2) the EU-IST FP7 project SpeDial under contract 611396 and (3) Fundação para a Ciência e a Tecnologia, through project PEst-OE/EEI/LA0008/2013.

References

[1] Lee, S., Potamianos, A., Narayanan, S.: Acoustics of Children's Speech: Developmental Changes of Temporal and Spectral Parameters. J. Acoust. Soc. Am. 10, 1455–1468 (1999)
[2] Huber, J.E., Stathopoulos, E.T., Curione, G.M., Ash, T.A., Johnson, K.: Formants of Children, Women and Men: The Effects of Vocal Intensity Variation. J. Acoust. Soc. Am. 106(3), 1532–1542 (1999)

[3] Xue, S., Hao, G.: Changes in the Human Vocal Tract Due to Aging and the Acoustic Correlates of Speech Production: A Pilot Study. Journal of Speech, Language, and Hearing Research 46, 689–701 (2003)

[4] Pellegrini, T., Hämäläinen, A., Boula de Mareüil, P., Tjalve, M., Trancoso, I., Candeias, S., Sales Dias, M., Braga, D.: A Corpus-Based Study of Elderly and Young Speakers of European Portuguese: Acoustic Correlates and Their Impact on Speech Recognition Performance. In: Interspeech, Lyon (2013)

[5] Narayanan, S., Potamianos, A.: Creating Conversational Interfaces for Children. IEEE Speech Audio Process. 10(2), 65–78 (2002)

[6] Strommen, E.F., Frome, F.S.: Talking Back to Big Bird: Preschool Users and a Simple Speech Recognition System. Educ. Technol. Res. Dev. 41(1), 5–16 (1993)

[7] Anderson, S., Liberman, N., Bernstein, E., Foster, S., Cate, E., Levin, B., Hudson, R.: Recognition of Elderly Speech and Voice-Driven Document Retrieval. In: IEEE International Conference on Acoustics, Speech, and Signal Processing, Phoenix, AZ, pp. 145–148 (1999)

[8] Takahashi, S., Morimoto, T., Maeda, S., Tsuruta, N.: Dialogue Experiment for Elderly People in Home Health Care System. In: Matoušek, V., Mautner, P. (eds.) TSD 2003. LNCS (LNAI), vol. 2807, pp. 418–423. Springer, Heidelberg (2003)

[9] Teixeira, V., Pires, C., Pinto, F., Freitas, J., Dias, M.S., Mendes Rodrigues, E.: Towards Elderly Social Integration using a Multimodal Human-computer Interface. In: Proc. of the 2nd International Living Usability Lab Workshop on AAL Latest Solutions, Trends and Applications, AAL 2012, Milan (2012)

[10] Wilpon, J.G., Jacobsen, C.N.: A Study of Speech Recognition for Children and Elderly. In: IEEE International Conference on Acoustics, Speech, and Signal Processing, Atlanta, GA, pp. 349–352 (1996)

[11] Potamianos, A., Narayanan, S.: Robust Recognition of Children's Speech. IEEE Speech Audio Process 11(6), 603–615 (2003)

[12] Hämäläinen, A., Miguel Pinto, F., Rodrigues, S., Júdice, A., Morgado Silva, S., Calado, A., Sales Dias, M.: A Multimodal Educational Game for 3-10-year-old Children: Collecting and Automatically Recognising European Portuguese Children's Speech. In: Workshop on Speech and Language Technology in Education, Grenoble (2013)

[13] Pellegrini, T., Trancoso, I., Hämäläinen, A., Calado, A., Sales Dias, M., Braga, D.: Impact of Age in ASR for the Elderly: Preliminary Experiments in European Portuguese. In: IberSPEECH, Madrid (2012)

[14] Vipperla, R., Renals, S., Frankel, J.: Longitudinal Study of ASR Performance on Ageing Voices. In: Interspeech, Brisbane, pp. 2550–2553 (2008)

[15] Batliner, A., Blomberg, M., D'Arcy, S., Elenius, D., Giuliani, D., Gerosa, M., Hacker, C., Russell, M., Steidl, S., Wong, M.: The PF_STAR Children's Speech Corpus. In: Interspeech, Lisbon (2005)

[16] Hämäläinen, A., Rodrigues, S., Júdice, A., Silva, S.M., Calado, A., Pinto, F.M., Dias, M.S.: The CNG Corpus of European Portuguese Children's Speech. In: Habernal, I. (ed.) TSD 2013. LNCS (LNAI), vol. 8082, pp. 544–551. Springer, Heidelberg (2013)

[17] Cucchiarini, C., Van Hamme, H., van Herwijnen, O., Smits, F.: JASMIN-CGN: Extension of the Spoken Dutch Corpus with Speech of Elderly People, Children and Non-natives in the Human-Machine Interaction Modality. In: Language Resources and Evaluation, Genoa (2006)

[18] Hämäläinen, A., Pinto, F., Sales Dias, M., Júdice, A., Freitas, J., Pires, C., Teixeira, V., Calado, A., Braga, D.: The First European Portuguese Elderly Speech Corpus. In: IberSPEECH, Madrid (2012)

[19] Hämäläinen, A., Avelar, J., Rodrigues, S., Sales Dias, M., Kolesiński, A., Fegyó, T., Nemeth, G., Csobánka, P., Lan Hing Ting, K., Hewson, D.: The EASR Corpora of European Portuguese, French, Hungarian and Polish Elderly Speech. In: Langauge Resources and Evaluation, Reykjavik (2014)

[20] Minematsu, N., Sekiguchi, M., Hirose, K.: Automatic Estimation of One's Age with His/Her Speech Basedupon Acoustic Modeling Techniques of Speakers. In: IEEE International Conference on Acoustics, Speech, and Signal Processing, Orlando, FL, pp. 137–140 (2002)

[21] Dobry, G., Hecht, R., Avigal, M., Zigel, Y.: Supervector Dimension Reduction for Efficient Speaker Age Estimation Based on the Acoustic Speech Signal. IEEE Transactions on Audio, Speech & Language Processing 19(7), 1975–1985 (2011)

[22] Bahari, M., McLaren, M., Van Hamme, H., Van Leeuwen, D.: Age Estimation from Telephone Speech Using i-Vectors. In: Interspeech, Portland, OR (2012)

[23] Neto, J., Martins, C., Meinedo, H., Almeida, L.: The Design of a Large Vocabulary Speech Corpus for Portuguese. In: European Conference on Speech Technology, Rhodes (1997)

[24] Eyben, F., Wollmer, M., Schuller, B.: openSMILE - The Munich Versatile and Fast Open-Source Audio Feature Extractor. In: ACM International Conference on Multimedia, Florence, pp. 1459–1462 (2010)

[25] Meinedo, H., Trancoso, I.: Age and Gender Detection in the I-DASH Project. ACM Trans. Speech Lang. Process. 7(4), 13 (2011)

[26] Schuller, B., Steidl, S., Batliner, A., Noeth, E., Vinciarelli, A., Burkhardt, F., van Son, R., Weninger, F., Eyben, F., Bocklet, T., Mohammadi, G., Weiss, B.: The Interspeech 2012 Speaker Trait Challenge. In: Interspeech 2012, Portland, OR (2012)

[27] Weninger, F., Eyben, F., Schuller, B.W., Mortillaro, M., Scherer, K.R.: On the Acoustics of Emotion in Audio: What Speech, Music and Sound Have in Common. Frontiers in Psychology, Emotion Science, Special Issue on Expression of Emotion in Music and Vocal Communication 4(Article ID 292), 1–12 (2013)

[28] Hall, M.: Correlation-Based Feature Subset Selection for Machine Learning. Hamilton, New Zealand (1998)

[29] Hall, M., Frank, E., Holmes, G., Pfahringer, B., Reutemann, P., Witten, I.: The WEKA Data Mining Software: An Update. SIGKDD Explorations 11 (2009)

[30] Platt, J.: Fast Training of Support Vector Machines Using Sequential Minimal Optimization. In: Schoelkopf, B., Burges, C., Smola, A. (eds.) Advances in Kernel Methods - Support Vector Learning (1998)

[31] Keerthi, S.S., Shevade, S.K., Bhattacharyya, C., Murthy, K.R.K.: Improvements to Platt's SMO Algorithm for SVM Classifier Design. Neural Computation 13(3), 637–649 (2001)

[32] Linville, S.E.: Vocal Aging. Singular, San Diego (2001)

[33] Microsoft Speech Platform Runtime (Version 11), http://www.microsoft.com/en-us/download/details.aspx?id=27225 (accessed March 25, 2013)

Characterizing Parkinson's Disease Speech by Acoustic and Phonetic Features

Jorge Proença[1,2], Arlindo Veiga[1,2], Sara Candeias[1,3], João Lemos[4], Cristina Januário[4], and Fernando Perdigão[1,2]

[1] Instituto de Telecomunicações, Coimbra, Portugal
[2] Electrical and Computer Eng. Department, University of Coimbra, Portugal
[3] Microsoft Language Development Centre, Lisbon, Portugal
{jproenca,aveiga,saracandeias,fp}@co.it.pt
[4] Coimbra's Hospital and University Centre, Coimbra, Portugal
merrin72@hotmail.com, cristinajanuario@gmail.com

Abstract. This study intends to identify acoustic and phonetic characteristics of the speech of Parkinson's Disease (PD) patients, usually manifesting hypokinetic dysarthria. A speech database has been collected from a control group and from a group of patients with similar PD severity, but with different degrees of hypokinetic dysarthria. First and second formant frequencies of vowels in continuous speech were analyzed. Several classifiers were built using phonetic features and a range of acoustic features based on cepstral coefficients with the objective of identifying hypokinetic dysarthria. Results show a centralization of vowel formant frequencies for PD speech, as expected. However, some of the features highlighted in literature for discriminating PD speech were not always found to be statistically significant. The automatic classification tasks to identify the most problematic speakers resulted in high precision and sensitivity by using two formant metrics simultaneously and in even higher performance by using acoustic dynamic parameters.

Keywords: Parkinson's speech, hypokinetic dysarthria, acoustic features, phonetic features.

1 Introduction

Parkinson is a degenerative disease of the central nervous system more common in the elderly, with most cases occurring after the age of 50 [1]. Some of the most obvious symptoms are related to movement impairment, including bradykinesia (or slowness of movement), resting tremor, rigidity and difficulty on walking [2]. The number of individuals that suffer from Parkinson's disease (PD) is growing worldwide [1], and typically 90% of patients with PD reveal disabilities in speech production [3]. The most common speech problems experienced fall under the term of hypokinetic dysarthria and involve hypophonia (or reduced volume), monotone, hoarse voice, difficulty with articulation of sounds and syllables (or reduced pitch range) [3,4]. In essence, people with PD cannot speak as loudly as others and find it

J. Baptista et al. (Eds.): PROPOR 2014, LNAI 8775, pp. 24–35, 2014.

harder to form words, reducing their ability to communicate. The patients also show difficulty to handle common computer peripherals (keyboard, mouse, touchscreen, etc.). In fact, they are often referred for physical rehabilitation due to the rigidity of the upper limbs. As a result of the reduced speech intelligibility and movement rigidity, people with PD often become isolated [5]; speech technologies can therefore offer a relevant contribution to improve their quality of life.

The present study aims to identify some acoustic and phonetic characteristics of the speech of PD patients for European Portuguese language. The ultimate goal is to adapt a speech recognizer for PD patients to be used for friendly technological interfaces.

Some studies have already examined the correlates of the PD voice disorders in fundamental frequency (F0) of the vowels pronounced, revealing that F0 variability decreases during syllable production and reading tasks [6,7,8,9,10,11]. In general, these studies noted a measurable decrease in F0 variability as the severity of PD increases. In [12] an increase of the mean speaking F0 is reported. Others have reported longer pause duration in the speech of PD patients [6], [8]. However, these conclusions are not entirely consensual. Some authors, such as [8], found no differences in the number of pauses and in the mean pause duration during a reading passage, whereas others, such as [10], have suggested that the percentage of pause duration in the speech of PD's disease is higher than that of healthy control subjects, and that pauses occur more frequently. Speaking rate characteristics have also been found to be somehow incoherent in patients with PD and it has been suggested that as the disease progresses, the speaking rate can either be strangely accelerated or slowed, [13].

In order to reveal possible correlations among vowel articulation and the stage of the disease, Skoda et al. [7] analyzed both first and second formants frequency values (F1 and F2) of the vowels [a], [i] and [u] (in IPA[1]), representing the triangle vertices of the vowel chart. Based on the values of the F1 and F2 of these three vowels, geometric calculations such as vowel space area (VSA) [7], [14] and vowel articulation index (VAI) [7], have been used as indicators of a dysarthric speech. Among the differences between PD and healthy speech related to the first two formants, differences in the patterns of variability for vowel production were marked in [15].

The extension of these findings for European Portuguese language awaits clarification so far. A previous work has already shown some differences in formant frequencies for PD speech and concluded that dynamic features may be highly important to consider [16]. Regarding the specific characteristics typically associated in the literature with PD speech, extending [16], this collaborative study first focuses on the vowel formant frequencies. Phonetic information relies mostly on F1 and F2 for the vowels [i], [ɛ], [a], [ɔ] and [u]. Finally, several acoustical measures based on Mel Frequency Cepstral Coefficients (MFCC), a standard parameterization of audio for speech and speaker recognition [17], were also considered in order to identify features that are well correlated with hypokinetic dysarthric speech. Using this information, several classifiers were built to compare the performance of different features in separating normal and affected/dysarthric speech.

---

[1]   International Phonetic Alphabet, http://www.langsci.ucl.ac.uk/ipa/

This paper is organized as follows. In the next section, the speech corpus of Parkinson disease patients collected for experimental findings is described. Section 3 describes the phonetic analysis to characterize speech in patients with PD in comparison to a healthy group control. In section 4 the additional acoustic features to be used in classification tasks are presented. Section 5 describes the procedure for the classification of hypokinetic dysarthria and presents the final results. The paper ends with the concluding remarks.

2   Parkinson Disease Speech Corpus

A corpus consisting of a series of 1002 utterances of read phonetically rich sentences and isolated words (mainly commands) was collected from 22 PD patients (12 females and 10 males) with degrees of the disease in stages 2 and 3 of the Hohen and Yahr (H-Y) scale [18]. They are aged between 50 and 80 years old. Recordings took place at the neurology department of Coimbra's Hospital and University Centre. In each recording session, the same common set-up was used, which consists of a laptop computer and three microphones. In order to better resemble the actual speech recordings to potential users, background noise conditions were not controlled particularly. Parkinson disease speech corpus represents 90 minutes of read speech recorded at a 48 kHz sampling rate. A healthy control group (3 females and 4 males between 25 and 51 years old) was also recruited for recording the same battery of speech production tasks under identical acoustic conditions. Segmentation at the phone-level was automatically done for each session through forced-alignment with a phone recognizer [19]. This process was further manually verified.

The H-Y stages of the patients are not very distinct, but it is possible to differentiate two PD speech subtypes on the basis on the fluency of the speech produced: normally articulated and rhythmical (named here as Low-PD), and slow and abnormally articulated speech (named here as High-PD). Hence, the patients were distinguished as belonging to one of these two levels of speech dysfunction. This classification was done through perceptual experiments by 6 individuals, all experts on speech processing, with a high level of agreement (86% mean). The distribution of age between the two groups shows a tendency of older individuals for most affected speech (High-PD: 72.00±7.60 years old), although there is a considerable overlap (Low-PD: 62.36±9.43 years old), meaning that the speech disorders encountered are not necessarily correlated to age or disease stage.

3   Phonetic Analysis

3.1   Vowel Formants

Given the characteristics of continuous speech, estimating formant frequency values is not a straightforward task. We observed several problematic vowel segments where formants are not clearly outlined, or where the vowel production is highly context dependent and perceived through rapid variations of formants. To minimize these problems we decided to make some restrictions on which segments would be proper for analysis, namely choosing long vowels in stressed position. Therefore, vowels [a], [ɛ], [i], [ɔ] and [u] in a stressed position were selected from the aligned transcriptions at phone level, including only those with duration above 50 ms. Furthermore, each segment was cut where the energy level was 20 dB below of the maximum energy, to specifically consider only the well-established part of the vowel.

The Praat tool [20] was used to automatically extract the first (F1) and second (F2) formant frequencies. In order to minimize the errors in the estimation of the formant values, a frequency ceiling for the formant analysis algorithm is recommended [21]. A similar method to [21] was then applied, given the foreknowledge that different vowels and speakers need different formant ceilings for the automatic calculation. An iterative calculation of formants was performed in 10 ms steps using ceilings in the 4000-5500Hz range (for males) or 4500-6500Hz (for females) in 50Hz steps, followed by a selection of the optimal ceiling. The optimal ceiling for a given vowel of a given speaker was the one that provided the minimum variance of F1 and F2 between the available samples of that vowel. This was calculated as the sum of the variances of 20log(F1) and 20log(F2), where F1 and F2 are the median formant frequency values of each segment. This method was applied for the vowels [a], [ɛ], [i] and [ɔ]; however, for [u], we decided to select not the median values of each segment, but the lower quartile of F2. Since [u] should have the lowest F2 of all other vowels, this method further eliminates context and variation problems. Segments with abnormally large formant variance were discarded to eliminate probable failures in correctly detecting F1 and F2.

Using the method described above, the optimal ceilings throughout the speakers varied too much for a given vowel. Through manual verification, we could conclude that some of the optimal ceilings obtained would indeed provide wrong estimations and were far from the expected ceilings for a given vowel. Therefore, it was deemed necessary to further restrict the ceiling ranges of the method, depending on vowel, ranges decided through empirical observations: (Female/Male) 5500-6000/4800-5200Hz for [i] and [ɛ], 4800-5200/4300-4700Hz for [a] and 4000-4500/4000-4200 for [ɔ] and [u]. The median optimal ceilings obtained were the following: (Female/Male) 5900/5200Hz for [i], 5550/4800Hz for [ɛ]; 5150/4300Hz for [a], 4350/4050Hz for [ɔ] and 4050/4050Hz for [u]. After calculations in 10 ms steps, we chose to only extract the median formant values for each vowel as an effective way to remove some outliers. The number of vowel tokens kept after restrictions was 4555 ([i]: 1233; [ɛ]: 666; [a]: 1941; [ɔ]: 376; [u]: 339).

F1 and F2 values for vowels [i], [ɛ], [a], [ɔ] and [u] show large variations but don't overlap much for the different vowels. Figure 1 shows an example of these values for males and females of the Low-PD group. The variation cloud for [u] is large, partially due to the logarithmical scale of the figure, but, even so, some of the F2 values are uncharacteristically high. Many of these cases were manually verified, and the formant estimations given were indeed approximately correct. This can be explained by the circumstance of continuous speech, where [u] was displaced to a centralized position, not given enough time or emphasis to reach a very low F2 value. All of the vowels may partially suffer from this centralization during continuous speech.

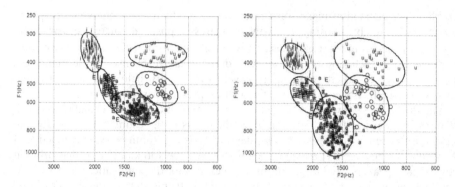

Fig. 1. First and second formant frequencies (F1 and F2) for vowels [i], [ɛ] (E in the picture), [a], [ɔ] (O in the picture) and [u] of Males of the Low-PD group (left) and Females of the Low-PD group (right), with corresponding 2σ concentration ellipses

Comparing the F1/F2 vowel triangles for High-PD and Low-PD speakers (Figure 2), it can be reported that as the speech abnormalities progresses the triangle of the vowels reduces, with lower F1 values, mainly for the open vowels [ɛ], [a] and [ɔ], and with a slight centralization of F2 values. This tendency could confirm the difficulty of PD patients on the movement of the tongue's body. The articulatory restriction as a result of the rigidity of the vocal muscle is considerably evident in the production of open central vowel [a] for both male and female speakers. For High-PD male speakers, the reduced [a] vowel is closer to the near-open central vowel [ɐ] area than to the [a] area. Male PD speakers also presented a larger centralization of [ɔ] and [u]. In general, the articulation of vowels of the low-PD group tends to be closer to the control group, indicating that mild forms of hypokinetic dysarthria may be hard to distinguish from normal speech through formant frequency analysis.

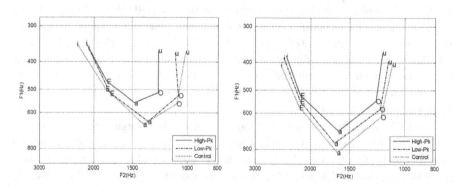

Fig. 2. First and second formant frequencies (F1 and F2) median values for [i], [ɛ] (E in the picture), [a], [ɔ] (O in the picture) and [u] of males (left) and females (right) in Control, Low-PD and High-PD groups (respectively, Low-Pk and High-Pk in the figure)

3.2   Formant Metrics

Vowel Space Area (VSA) and Vowel Articulatory Index (VAI) are two metrics that use vowel formant frequency values, which have been used for the study of speech production deficits and reductions in intelligibility [7], [14]. VSA is usually constructed by the Euclidean distances between the F1 and F2 coordinates of the corner vowels [i], [u], and [a] (triangular VSA) as described by equation 1 (lower VSA indicates that corner vowels are closer together), or the corner vowels [i], [u], [a], and [æ] (quadrilateral VSA) in the F1-F2 plane [22]. Recently, a new method was proposed to calculate VSA by exploring the entire vowel working space in continuous speech and not just the corner vowels [23].

$$VSA = \frac{\left| F_1 i \times (F_2 a - F_2 u) + F_1 a \times (F_2 u - F_2 i) + F_1 u \times (F_2 i - F_2 a) \right|}{2} \tag{1}$$

VAI was designed to be minimally sensitive to inter-speaker variability and maximally sensitive to vowel formant centralization [24], calculated through equation 2 (lower VAI also indicates that formant centralization is higher).

$$VAI = \frac{F_2 i + F_1 a}{F_1 i + F_1 u + F_2 u + F_2 a} \tag{2}$$

In this study, triangular VSA and VAI were calculated for each speaker. Table 1 presents the mean and standard deviation values for these two metrics. Although average values show an expected reduction of vowel space area and lower articulation for PD-patients, the only result of statistical significance is Male VAI for Control vs. High-PD with p=.038. Another usual metric is the Formant Centralization Ratio (FCR) [25] which is simply the reciprocal of VAI.

Table 1. Formant metrics for Control, Low-PD and High-PD groups: VSA and VAI means and standard deviations per group per gender

Gender	Control	Low-PD	High-PD
Male			
VSA (Hz^2×10^{-5})	1.84 ± 0.73	1.33 ± 0.31	1.02 ± 0.42
VAI	0.92 ± 0.07	0.85 ± 0.06	0.80 ± 0.07
Female			
VSA (Hz^2×10^{-5})	2.95 ± 0.31	2.32 ± 0.49	2.26 ± 0.62
VAI	0.95 ± 0.05	0.88 ± 0.02	0.89 ± 0.03

4   Acoustic Features

In order to identify acoustical characteristics which may be appropriate to classify speech from PD patients, two additional sets of features were obtained per speaker. These will be considered in the next section for classification tasks. One feature set is based on Gaussian Mixture Model (GMM) "supervectors" derived from Mel Frequency Cepstral Coefficients (MFCC), the standard acoustic feature used on speech recognition framework [26]. The other set is derived from the openSMILE toolkit [27] which corresponds to measurements (functionals) applied to MFCCs. We named this set as "Smile" features.

4.1   GMM Supervector Features

The GMM features were used using the same approach as in state-of-art speaker verification [26], but in this case to automatically classify the degree of the dysarthria. This approach uses a GMM as Universal Background Model (UBM-GMM) trained with MFCC features [26]. The model corresponds to a set of the means (μ_k), weights (λ_k) and covariance matrices (Σ_k) of a mixture of M Gaussian densities that best describes the data (Figure 3)   The UBM-GMM was adapted for each speaker and the mean vectors were concatenated to form a supervector (see Figure 3). A supervector per speaker was used to train and test classifiers. We computed 13 MFCC coefficients (statics coefficients) plus first and second order derivatives (dynamics coefficients) as base features (a vector of 39 parameters). Early tests confirmed that the dynamic features are the most relevant (we have found that dynamics coefficients were 20% more accurate on a simple Low-PD vs. High-PD classification), as was already reported in [28]. Therefore, for the classification task, we only considered the dynamic coefficients of the MFCC for subsequent tasks, a vector of 26 parameters. An UBM-GMM with 8 Gaussian components was trained and then adapted for each speaker using maximum a posteriori approach (MAP). Normalized means of the adapted model were used as the speaker features (208 parameters).

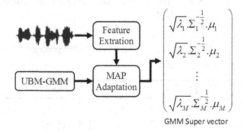

Fig. 3. Building a GMM Supervector. A GMM supervector is constructed by stacking the means of the adapted mixture components.

4.2   "Smile" Features

The "Smile" set, derived from the openSMILE toolkit [27], corresponds to 68 measures ("functionals") applied to a vector of 39 MFCC parameters (statics and dynamics MFCC coefficients). The "functionals" can be divided into six groups: Extremes, Moments, Percentiles, Times, Means and Peaks. Extremes contains 5 parameters (maximum, minimum, range, mean frames between maximum values and mean of frames between minimum values); Moments contains 4 parameters (variance, skewness, kurtosis and standard deviation); Percentiles contains 9 parameters (3 quartiles, 3 inter-quartile ranges, percentiles 1 and 99 and inter-percentile ranges), Times contains 10 parameters (4 up and down-level times, raise and fall times), Means contains 11 parameters (including arithmetic mean, arithmetic mean of absolute values, quadratic mean among others) and Peaks contains 29 parameters (including number of peaks, mean of frames between peaks, standard deviation of frames between peaks among others). A total of 2652 features were computed for each speaker.

5   Classification

5.1   The Classification Task

The final objective of this study was to characterize how well different features behaved in classification tasks to distinguish hypokinetic dysarthric speech of PD patients from normal speech. The already gathered information of formant metrics was considered separately, as were the GMM and Smile features. A two-class problem was considered, where a positive result would indicate the presence of hypokinetic dysarthria, and a negative result would indicate normal speech. However, it was decided to separate the data in two different ways:

- (A) Simply joining the Low-PD group with the control group, in contrast to the altered speech of the High-PD group;
- (B) Joining only some of the individuals of the Low-PD group with the control, those which the perceptual classification as normal speech was unanimous. This leaves those of Low-PD which were almost classified as High-PD to the altered speech group.

In essence, B offers an alternate dividing line for creating two groups out of the existing three, where we argue that some early indicators of altered speech may be present on certain PD subjects (from Low-PD) that were not classified as being speech disordered.

The features considered as parameters for individual classification tasks were the formant metrics, GMM parameters, and Smile features. The instance-based learning algorithm IB1 [29] and leave-one-out cross-validation paradigm where used to test classifiers. The IB1 is similar to the Nearest Neighbor algorithm and uses normalized Euclidean distance to compute similarity between instances. It is a simple classifier that does not require an elaborate tuning scheme, and we found it appropriate for a scenario such as ours with a low number of subjects. Every attribute was calculated per speaker, thus resulting in 29 instances to be classified. To deal with the unbalanced dataset (groups with different number of speakers) we use the Synthetic Minority Oversampling Technique [30] to oversample the minority class. The tests and dataset manipulation were carried out by WEKA data mining toolkit [31] for data separations A and B.

5.2   Features Selection

Since GMM and Smile use hundreds of parameters, it is foreseeable that selecting only the best parameters from these sets could provide improved results. Feature ranking is a method that allows us to discard less significant features from the dataset. We applied this method to the GMM and "Smile" features using WEKA to rank and to test the classifier with only the most significant features. Two attribute evaluation methodologies were used to select the features: one based on entropy and another based on correlation. The feature selection was performed in two steps. The first step used an information gain (IG) approach, discarding all features with residual IG. The second step used a correlation-based approach to select the best set of features [31]. The reduction of GMM attributes was from 208 to 48 to 7 for data separation A, and from 208 to 97 to 12 for data separation B. As for Smile, the reduction was even more drastic, from 2652 to 170 to 34 for A, and from 2652 to 135 to 22 for B.

Table 2 shows the classification results for the different features where, without the selection of the best attributes from GMM and Smile, the formant metrics provided the best results for the data separation B. These use as features VAI and VSA at the same time which seem to be acceptable indicators when used in this fashion with a large improvement over their individual analysis. They seemingly provide complementary information although based on some of the same formant values. However, with the feature selection, GMM and Smile classifications improved their results substantially, with Smile being the best overall for data separation A. As the GMM approach used only dynamic features and Smile being also mostly derivative of dynamic features, it is shown that the dynamics in speech of Parkinson's patients should also be of relevant consideration for its characterization.

Table 2. Classification performance of all features. Precision is the percentage of the correctly accepted as hypokinetic dysarthria over all the classified as such. Recall is the percentage of the correctly accepted over the entire original hypokinetic dysarthria group. F1-score is the harmonic mean of precision and recall.

Features	Data separation A			Data separation B		
	Precision	Recall	F1-score	Precision	Recall	F1-score
VSA	47,1%	44,4%	45,7%	62,5%	66,7%	64,5%
VAI	76,5%	72,2%	74,3%	66,7%	66,7%	66,7%
Both formant metrics	68,8%	61,1%	**64,7%**	86,7%	86,7%	**86,7%**
GMM	73,9%	94,4%	82,9%	64,7%	73,3%	68,8%
GMM 1st selection	75,0%	83,3%	78,9%	70,0%	93,3%	80,0%
GMM 2nd selection	85,7%	100,0%	**92,3%**	86,7%	86,7%	**86,7%**
Smile	64,3%	100,0%	78,3%	40,0%	40,0%	40,0%
Smile 1st selection	76,2%	88,9%	82,1%	100,0%	73,3%	84,6%
Smile 2nd selection	94,7%	100,0%	**97,3%**	92,9%	86,7%	**89,7%**
Feature Bundle	90,0%	100,0%	**94,7%**	87,5%	93,3%	**90,3%**

The attributes selected from hundreds of GMM and Smile parameters are elaborate and their individual analysis may be hard to comprehend. These chosen ones could apparently be related to our specific dataset and, ideally, further studies with a higher number of samples could confirm these findings.

An additional exercise is to agglomerate all the features as simultaneous input for a classification task. We bundled formant metrics and the features from the second step selection of GMM and Smile. The results, shown on Table 2 for "Feature Bundle", are similar to the selected Smile features, but show that the performance is very high, with even a slight improvement for data separation B.

As for the two distinct ways to separate the data, B included some of the early stage hypokinetic dysarthria subjects, not perceptually agreed to be highly affected speech, on the hypokinetic dysarthria group. This showed only improved results for the formant metrics, which could then be already slightly affected for these early-stage subjects. Data separation A, by only having the de-facto perceptually highly affected subjects as hypokinetic dysarthria, resulted in the best classification for the acoustic features.

6   Conclusions

In this paper we described our current research towards finding acoustic and phonetic characteristics in the hypokinetic dysarthric speech of Parkinson disease (PD) patients. A speech database with different levels of severity of hypokinetic dysarthria from Parkinson's disease patients has been collected for this purpose.

Formant frequencies F1 and F2 have shown differences especially in central vowels for more advanced speech disorder. Of the usual formant metrics to evaluate dysarthric speech, Vowel Space Area and Vowel Articulation Index, only the latter proved to have statistical significance for control males versus high-level PD males. From an acoustic analysis based on Mel Frequency Cepstral Coefficients we also found that some of the most significant features for PD speech discrimination are dynamic features. This is in accordance with the main impairment in PD patients, which reduces their articulating ability in a dynamic way. Furthermore, the results show that PD speech disorders can be detected using only a small number of acoustic features. Both phonetic and acoustic features have shown to have complementary information and should be taken into account when analyzing hypokinetic dysarthria.

As future work, we intend to obtain additional samples of PD speech and of an older control group in order to confirm these findings. We also plan to integrate these features for enhancing a in-house speech recognizer system adapted to PD patients.

Acknowledgements. This work is funded by FCT and QREN projects (PTDC/CLE-LIN/11 2411/2009; TICE.Healty13842) and partially supported by Instituto de Telecomunicações, co-funded by FCT strategic project PEst-OE/EEI/LA0008/2013. Sara Candeias and Jorge Proença are supported by the SFRH/BPD/36584/2007 and SFRH/BD/97204/2013 FCT grants, respectively. We would like to thank Coimbra's Hospital and University Centre and all the participants who provided their voice.

References

1. Fahn, S.: Parkinson's disease: 10 years of progress, 1997–2007. Movement Disorders 25 (suppl. 1), S2–S14 (2010)
2. Lees, A.J., Hardy, J., Revesz, T.: Parkinson's disease. Lancet 373, 2055–2066 (2009)
3. Ramig, L., Fox, C., Sapir, S.: Speech treatment for Parkinson's disease. Expert Review of Neurotherapeutics 8(2), 297–309 (2008)
4. Goberman, A., Coelho, C.: Acoustic analysis of Parkinsonian speech I: Speech characteristics and L-Dopa therapy. NeuroRehabilitation 17, 237–246 (2002)
5. Cote, L., Sprinzeles, L.L., Elliott, R., Kutscher, A.H.: Parkinson's disease and Quality of Life. Haworth Press, New York (2000)
6. Harela, B.T., Michael, S., Cannizzaroa, H.C., Reillya, N., Snyder, P.J.: Acoustic characteristics of Parkinsonian speech: a potential biomarker of early disease progression and treatment. Journal of Neurolinguistics 17(6), 439–453 (2004)
7. Skodda, S., Visser, W., Schlegel, U.: Vowel Articulation in Parkinson's Disease. Journal of Voice 25(4), 467–472 (2011)
8. Canter, G.J.: Speech characteristics of patients with Parkinson's disease: I. Intensity, pitch, and duration. Journal of Speech and Hearing Disorders 28, 221–229 (1963)
9. Canter, G.J.: Speech characteristics of patients with Parkinson's disease: II. Physiological support for speech. Journal of Speech and Hearing Disorders 30, 44–49 (1965)
10. Metter, J., Hanson, W.: Clinical and acoustical variability in hypokinetic dysarthria. Journal of Communication Disorders 19, 347–366 (1986)
11. Holmes, R.J., Oates, J.M., Phyland, D.J., Hughes, A.J.: Voice characteristics in the progression of Parkinson's disease. International Journal of Language and Communication Disorders 35, 407–418 (2000)
12. Goberman, A., Coelho, C., Robb, M.: Phonatory characteristics of Parkinsonian speech before and after morning medication: The ON and OFF states. Journal of Communication Disorders 35, 217–239 (2002)
13. Weismer, G.: Articulatory characteristics of Parkinsonian dysarthria: Segmental and phrase-level timing, spirantization, and glottal-supraglottal coordination. In: Iin, M., McNeil, J., Rosenbeck, Aronson, A. (eds.) The Dysarthrias: Physiology, Acoustics, Perception, Management, pp. 101–130. College-Hill Press, San Diego (1984)
14. Roy, N., Nissen, S.L., Dromey, C., Sapir, S.: Articulatory changes in muscle tension dysphonia: Evidence of vowel space expansion following manual circumlaryngeal therapy. Journal of Communication Disorders 42(2), 124–135 (2009)
15. Soares, M.F., de Paula: Vowel variability in speakers with parkinson's disease. In: Proc. ICPhS XVII, Hong Kong, pp. 1570–1573 (2011)
16. Proença, J., Veiga, A., Candeias, S., Perdigão, F.: Acoustic, Phonetic and Prosodic Features of Parkinson's disease Speech. In: STIL - IX Brazilian Symposium in Information and Human Language Technology, 2nd Brazilian Conference on Intelligent Systems (BRACIS 2013), Fortaleza/Ceará, Brazil, October 21-23 (2013)
17. Davis, S.B., Mermelstein, P.: Comparison of Parametric Representations for Monosyllabic Word Recognition in Continuously Spoken Sentences. IEEE Transactions on Acoustics, Speech, and Signal Processing 28(4), 357–366 (1980)
18. Hoehn, M., Yahr, M.: Parkinsonism: Onset, progression and mortality. Neurology 17(5), 427–442 (1967)

19. Lopes, J., Neves, C., Veiga, A., Maciel, A., Lopes, C., Perdigão, F., Sá, L.: Development of a Speech Recognizer with the Tecnovoz Database. In: Teixeira, A., de Lima, V.L.S., de Oliveira, L.C., Quaresma, P. (eds.) PROPOR 2008. LNCS (LNAI), vol. 5190, pp. 260–263. Springer, Heidelberg (2008)

20. Boersma, P., Weenink, D.: Praat: doing phonetics by computer. Computer program, Version 5.3.42, http://www.praat.org/ (retrieved on March 2, 2013)

21. Escudero, P., Boersma, P., Rauber, A.S., Bion, R.A.H.: A cross-dialect acoustic description of vowels: Brazilian and European Portuguese. Journal of the Acoustical Society of America 126(3), 1379–1393 (2009)

22. Kent, R., Kim, Y.: Toward an acoustic typology of motor speech disorders. Clinical Linguistics and Phonetics 17, 427–445 (2003)

23. Sandoval, S., Berisha, V., Utianski, R.L., Liss, J.M., Spanias, A.: Automatic assessment of vowel space area. Journal of the Acoustical Society of America 134(5), EL477–EL483 (2013)

24. Sapir, S., Ramig, L., Spielman, J., Fox, C.: Acoustic metrics of vowel articulation in Parkinson's disease: Vowel space area (VSA) vs. vowel articulation index (VAI). In: 7th International ISCA Workshop on Methods and Analysis of Vocal Emissions for Biomedical Applications (2011)

25. Sapir, S., Ramig, L., Spielman, J.: Formant Centralization Ratio: A Proposal for a New Acoustic Measure of Dysarthric Speech. Journal of Speech, Language, and Hearing Research 53, 114–125 (2010)

26. Campbell, W.M., Sturim, D.E., Reynolds, D.A., Solomonoff, A.: SVM Based Speaker Verification using a GMM Supervector Kernel and NAP Variability Compensation. In: Proc. ICASSP 2006, vol. 1, pp. 14–19 (2006)

27. Eyben, F., Wöllmer, M., Schuller, B.: openSMILE - The Munich Versatile and Fast Open-Source Audio Feature Extractor. In: Proc. of ACM Multimedia, pp. 1459–1462 (2010)

28. Bocklet, T., Noth, E., Stemmer, G., Ruzickova, H., Rusz, J.: Detection of persons with Parkinson's disease by acoustic, vocal, and prosodic analysis. In: Automatic Speech Recognition and Understanding (ASRU), pp. 478–483 (2011)

29. Aha, D., Kibler, D., Albert, M.: Instance-based Learning Algorithms. Machine Learning 6, 37–66 (1991)

30. Chawla, N., Bowyer, K., Hall, L., Kegelmeyer, W.: Synthetic Minority Over-sampling Technique. Journal of Artificial Intelligence Research 16, 321–357 (2002)

31. Hall, M., Frank, E., Holmes, G., Pfahringer, B., Reutemann, P., Witten, I.: The WEKA Data Mining Software: An Update. SIGKDD Explorations 11 (2009)

32. Wells, J.C.: SAMPA computer readable phonetic alphabet. In: Gibbon, D., Moore, R., Winski, R. (eds.) Handbook of Standards and Resources for Spoken Language Systems. Part IV, section B. Mouton de Gruyter, Berlin (1997)

Rule-Based Algorithms for Automatic Pronunciation of Portuguese Verbal Inflections

Vanessa Marquiafável[1], Christopher Shulby[2], Arlindo Veiga[3,4], Jorge Proença[3,4], Sara Candeias[3], and Fernando Perdigão[3,4]

[1] Universidade Estadual Paulista (IBILCE-UNESP), S. J. do Rio Preto, Brazil
[2] Universidade de São Paulo (ICMC-USP), S. Carlos, Brazil
[3] Instituto de Telecomunicações - Pole of Coimbra, Portugal
[4] University of Coimbra - DEEC, Coimbra, Portugal
{marquiafavel,chrisshulby}@gmail.com,
{aveiga,jproenca,saracandeias,fp}@co.it.pt

Abstract. The correct automatic pronunciation of words is a nontrivial problem, even for inflexions of Portuguese verbs, and has not been systematically solved yet, if verbal irregularity is taken into account. The purpose of this work is to enhance a grapheme-to-phoneme system with a verb pronunciation system for both varieties of Portuguese, Brazilian (BP) and European (EP), given only its infinitive form. The most common verbs for BP and EP (1000 and 2600 respectively) constituted our database to test the pronunciation system. A detailed and systematic analysis of regular and non-regular pronunciation forms of the inflected verbs was performed, and an index of irregularity for verb pronunciation is proposed. A rule-based algorithm to pronounce all inflexions according to verb paradigms is also described. The defined paradigms are, with a high level of certainty, representative of all the verbs for Portuguese.

Keywords: Brazilian Portuguese, European Portuguese, Verbs, Pronunciation, Rule-based algorithms.

1 Introduction

Sometimes, the decision as to which pronunciation of a word is the correct one is a nontrivial problem. Specifically, the cases in which we need to decide between two possible pronunciations, such as whether the vowel is closed [e/o] or open [ɛ/ɔ], have been widely responsible for many of the errors of automatic pronunciation systems. For instance, for Brazilian Portuguese (BP) - standard pronunciation of São Paulo - should the verbal form <*dorme*> 'he/she sleeps' be pronounced as [d'ɔrmɪ] or [d'ormɪ][1]? And the form <*requeri*> 'I required', should it be pronounced as [xek'ɛɾɪ] or [xek'eɾɪ]? For standard European Portuguese (EP), the verbal form <*eu coloria*>

---

[1] For phonetic annotation, we follow International Phonetic Alphabet (IPA) transcription symbols. As default, the examples within this paper are for Brazilian Portuguese, São Paulo standard pronunciation (see section 3 for a brief description of the subject).

J. Baptista et al. (Eds.): PROPOR 2014, LNAI 8775, pp. 36–47, 2014.

from <*colorir*> 'to color', exemplifies another case where the correct vowel pronunciation decision is unclear: is the correct pronunciation <*eu* [kul'urɐ]> or <*eu* [kɔl'urɐ]>? If those words are not accurately analyzed in terms of linguistic information, they may give rise to a wrong pronunciation even by native speakers of the language who are not familiar with the word to be pronounced.In a world dominated by rapid technological changes, commercial systems using speech applications are present in our daily life, allowing us to access information, create messages, and control devices, all with the power of our voice and we expect them to offer satisfactory performance. Industry and researchers in the field of speech technologies have been focused on minimizing the errors related to automatic pronunciation, combining linguistic-based knowledge and data-driven approaches. Previous work on grapheme to phone(me) conversion (g2p), including linguistic-based knowledge, can be pointed out for Portuguese, either for European [1, 2, 3, 4] or Brazilian [5, 6, 7] varieties. On the subject of heterophonic homographs, the issue of trying to cope with pronunciation problems related to words having the same spelling but different pronunciation was tackled by [8, 9] and [34].

The pronunciation of inflected Portuguese verb forms deals with several phonetic-phonological irregularities, which have been reported in the literature. Within the Generative theory, for EP in particular, [10] has explained thematic vowel changes in verbs, taking into account certain phonological processes such as vowel harmonization and the incidence of the stress. Within a structuralist approach, [11] presented radical vowel changes as dependent on phonological contexts as well. Thematic vowel changes like /'e/→/'i/ or /'e/→/'ɐj/ in a nasal context, were described as processes of assimilation, dissimilation and diphthongization by [10], [12] and occur in examples such as [bɐt'eɾ]→[bɐt'i] and [bɐt'iɐ] or [b'atɐ̃j] (different tense forms of 'to hit'). For BP, verbs' mid-vowels have been studied either in terms of phonological theory by [13, 14] (describing certain pre-stressed vowel behavior in verb pronunciation), or in a grammatical perspective, by [15]. In [15] part of the study was dedicated to describing vowel alternation in irregular verbs. In [16], those vowel alternations were described with detail and some pronunciation paradigms were proposed for both BP and EP. However, the findings reported in all of those studies derive from small corpora and the verb pronunciation analysis is still in need of a systematic study, useful to develop speech engines in the field of language technology. On the orthographic level, several inflected verb forms have been built for Portuguese and are freely available on the Web [17, 18, 19]. To the best of our knowledge, besides [20, 21], no other description of the pronunciation of inflected verb forms for Portuguese by using large corpora has been defined in the literature. In [20] irregularity paradigms accounting for the pronunciation of inflected verb forms for EP are present, using a rule-based approach in a way that can lead to the implementation of automatic pronunciation systems.

This study aims to identify pronunciation patterns for Portuguese inflected verb forms. The purpose is to enhance our g2p system with a verb pronunciation system for both varieties of Portuguese, BP and EP, given only its infinitive form. A detailed analysis of the regular and non-regular pronunciation forms of the inflected verbs was performed, contributing to propose a rule-based algorithm able to identify a paradigm for the pronunciation of all Portuguese verbs.

This work is organized as follows. Section 2 briefly describes the corpora. In section 3 the proposed methods for generating pronunciation of inflected verb forms is presented. In section 4 the pronunciation irregularity of inflected verb forms is examined and characterized, showing the differences found between BP and EP verb pronunciation. Finally, section 5 contains our conclusions.

2   Corpora

The source of common Portuguese verbs for this study was the text corpora CETEMPúblico for EP [22], and CETENFolha for BP, [23, 24]. These corpora are suitable for this purpose since they are large enough to obtain an accurate frequency representation of verb usage (at least in newspapers), and because all sentences are morphologically annotated in terms of PoS tags, from which it is easy to obtain verb forms for the same lemma. In the process of ranking the most frequent verbs all inflections were considered, except past participles, since they could be grammatically classified as verbs or adjectives.

As a result, two lists were obtained, one for EP, containing 2600 infinitive forms and another for BP, comprising 1000 infinitive forms. Both lists were manually verified in order to abridge the set of spurious verbs and then transcribed into the corresponding phonetic forms. For this purpose we used the g2p system called grɐfonə, [25], for EP, wheras for BP, UFPADic pronunciation dictionary (described in [26, 7]) was utilized. In the later situation, a further step was performed, mainly applying rules in order to convert the transcriptions into the phonetic standard variety for the state of São Paulo. Both lists have been carefully verified by experts in language processing from both varieties of Portuguese and serve as the input for the system of verb pronunciation.

3   Methodology

In this section we report all relevant technical procedures for generating the pronunciation of inflected verb forms. For EP, the process of identifying the necessary paradigms is already presented in [20]. For BP, we have adapted the process from EP to BP, by a rule-based approach. The adapted rules considered grammatical descriptions and linguistic studies on BP presented in [15, 16 27, 28, 29]. Some phonological and phonetic studies (presented in [28, 29, 30, 31]) for describing São Paulo standard pronunciation have been considered; specifically, the phonetic symbols of [ɐ], [ɪ] and [ʊ] were incorporated to represent the unstressed vowels /a/, /i/ and /u/ in word final positions, respectively, as in [s 'igɐ] 'go', [s 'ẽʧɪ] 'he/she feels' and [f'alʊ] 'I speak'. We simultaneously included a couple of adaptations at the phonetic level. We assumed the symbols [j] and [w] to represent glides in falling diphthongs (as in [ap 'ɔjʊ] 'I support', [bat'ẽjdʊ] 'beating' or [ẽdaɾ'ẽw̃] 'they shall walk') and we adopted the symbol [ẽ] to be used both with nasals and in nasalized phonological contexts, as in ['ẽmʊ] 'I love' and [k'ẽtʊ] 'I sing'. Every phonetic output was analyzed by a native Brazilian (São Paulo) phonetician, in order to ensure the correct transcription of the verb forms.

3.1   Verb Irregularity in Pronunciation

In this paper we intended to identify all situations of verb inflection irregularity for Portuguese. We define regular verb inflections (regular forms) as the concatenation of a verb radical with a regular verb suffix. The radical corresponds to the pronounced part of the infinitive form just before the thematic vowel [12]. The suffixes correspond to the last part of the forms and include the thematic vowel [20, 21]. We considered the usual 3 conjugations, in thematic vowels -a-, -e- and -i-, as well as -o-[2] as source of regular forms (see paradigms for those 4 conjugations in Tables 2 and 3, in its first four rows). In Portuguese. most verbs are regular, from which the regular suffixes can be predefined, both for EP as for BP. A verb form is regular if and only if it follows the pattern of radical plus regular suffix form; otherwise it is considered irregular (see the paradigms for irregular verbs (in pronunciation) listed in Tables 2 and 3, after row four).

3.2   Algorithm for Identifying Verb Irregularity

According to the definition of verbal irregularity, a simple algorithm can be defined in order to mark and count the irregularities in pronunciation. This is accomplished with the edit distance algorithm, sometimes called Levenshtein distance, [33], and consists of optimally aligning two forms of pronunciation: the actual form and the form pronounced as if the verb were regular. If the edit distance is not null, then one or two alterations are set: if the regular suffix is present, there is a change in the radical; otherwise there is a change in the suffix or in the radical and in the suffix. In this process, the different radicals needed to pronounce the verb correctly are gathered. A graphical display of the irregularities can be highlighted in order to further detect, validate or rectify pronunciations. For example, the correct form of the present tense of the verb <*ser*> 'to be' in the 3rd person, plural, is [s'omʊs] 'we are'. But if the verb were regular, the form would be [s]+['emʊs]=[s'emʊs]; thus, the phone alignment exhibits a substitution of [e] with [o], implying a change in the suffix. We have found this method of counting irregularities very important in order to establish new paradigms and find inaccuracies in transcriptions.

By counting all the occurrences of irregularities in all inflexions of a verb, an irregularity index can be defined. We have tested several methods of defining verb irregularity indices by counting, in different ways, the number of irregularities in radicals only, suffixes only and in both radicals and suffixes. We have found that using the number of total irregularities, regardless of place (radical, suffixes or both), jointly with the changes in suffixes and the number of different radicals, results in a simple and, apparently, accurate value of verb irregularity. The index is computed as:

---

[2] In this study, we considered a 4th conjugation with thematic vowel -o-, comprising the verb <*pôr*> 'to put', and all its derivate such as <*repor*> 'to replace', <*compor*> 'to compose', <*dispor*> 'to dispose', among others. A conjugation in -o- was already proposed in the literature by [16] and [20, 21].

$$Ind = IP + S + (R-1). \tag{1}$$

where IP is the number of irregular pronunciations in all inflexions of a verb; S is the number of inflexions with irregular suffixes; and R is the number of radicals needed to pronounce the verb. Consider for instance the verbs *<valer>* 'to worth' and *<escrever>* 'to write' in Table 3 (for BP), both with an index of 8. The verb *<valer>* needs only two radicals ([val] and [vaʎ]) to be pronounced, resulting in 7 changes but none in suffixes. On the other hand, the verb *<escrever>* needs 3 different radicals, [ɪskɾev], [ɪskɾɛv] and [ɪskɾit], the last one for the past participle which is irregular in both the radical and suffix. However, there are only a total of 5 irregularities. According to the given expression, both verbs are equally irregular because of the balance in the chosen variables, which in turn, seems to be reasonable.

Table 1. Templates of Tense, Mood and Persons (TMP) relevant to EP and BP verbal inflections. The Column **Ch.** indicates the number of changes.

TMP	Description	Tenses/Moods/Persons	Ch.
1	Stressed vowel in radical becomes opened or nasalized. E.g.: *<errar>*, <tocar>, <amar> (BP).	Pres. Indic.(1s,2s,3s,3p); Imp(2s) Pres. Subj.(1s,2s,3s,3p);	9
1a	Stressed vowel in radical becomes mid-open (EP). E.g.: *<beber>*, *<mover>*.	Pres. Indic.(1s); Pres. Subj. (1s,2s,3s,3s)	5
1b	Stressed vowel in radical becomes opened. E.g.: *<subir>* (BP), *<dormir>* (EP).	Pres. Indic. (2s,3s,3p); Imp(2s)	4
2a	Change from d→dʒ and t→tʃ in BP. E.g.: *<mudar>*, *<cantar>*.	Pres. Subj.(1s,2s,3s)	3
2b	Vowel+glide combination is dropped (BP). Change dʒ→d in <medir> (BP).	Pres. Indic.(3p)	1
3	Change in unstressed vowels or radical consonants. E.g.: *<ouvir>*, *<ferir>*.	Pres. Indic.(1s); Pres. Subj. (1s,2s,3s,1p,2p,3p)	7
4	Change to vowel with vowel+glide combination. E.g.: *<construir>*, *<trair>*.	Pres. Indic.(2s,3s); Imp(2s)	3
4a	Change from [zə] or [zɪ] to [ʃ] (EP) or [s] (BP). E.g.: *<induzir>*, *<jazer>*.	Pres. Indic.(3s); Imp(2s)	2
5	Radical consonant change (dʒ→d) and (tʃ→t) in BP. E.g.: *<decidir>*, *<partir>*.	Pres. Indic. (1s,3p); Pres. Subj.(1s,2s,3s,1p,2p,3p)	8
6	Radical consonant change (t→tʃ) and (d→dʒ) in BP. E.g.: *<debater>*, *<aprender>*.	Pres. Indic. (2s,3s); Imp.Past.Ind.(1s,2s,3s,1p,2p,3p); Past.Perf.(1s); Imp(2s); pcp	11
7	Change from [ə] or [e] to [i] in some verbs ending in *<ir>*. E.g.: *<agredir>* (EP) or *<prevenir>* (BP).	Pres. Indic.(1s,2s,3s,3p); Pres.Subj.(1s,2s,3s,1p,2p,3p); Imp(2)	11
8	Insertion of [id] or [ed] in verb classes *<rir>*, *<ler>* and the verb *<prover>*.	Pres. Indic. (2p); imp(2p)	2

Table 2. Verb paradigms for EP sorted by index of irregularity, **Ind**. Column **IP** means the number of irregularities in the verbal pronunciations; **Rad** the number of verb radicals; **S** the number of changes in the suffixes; **Template** refers to Table 1. Asterisk note (*) means that the verb has an irregular past participle.

Verb	Pron.	IP	Rad	S	Ind	Phone Change	Template
pensar	pẽs'aɾ	0	1	0	0	-	-
viver	viv'eɾ	0	1	0	0	-	-
unir	un'iɾ	0	1	0	0	-	-
pôr	p'oɾ	0	1	0	0	-	-
induzir	ĩduz'iɾ	2	2	2	5	zə → ʃ	4a
dormir	duɾm'iɾ	4	2	0	5	u → ɔ	1b
erguer	eɾg'eɾ	4	2	0	5	e → ɛ	1b
frigir	friʒ'iɾ	4	0	2	5	i→ɛ	1b
aquecer	ɐkɛs'eɾ	5	2	0	6	ɛ → e	1a
afluir	ɐflu'iɾ	4	2	3	8	u → uj	1b
construir	kõʃtru'iɾ	4	2	3	8	u → ɔj	1b
cobrir	kubɾ'iɾ	5	3	1	8	u → ɔ (*)	1b
ouvir	ov'iɾ	7	2	0	8	v → s	3
refletir	ʁəflet'iɾ	7	2	0	8	ɛ → i	3
sentir	sẽt'iɾ	7	2	0	8	ẽ → ĩ	3
debater	dəbɐt'eɾ	9	2	0	10	ɐ → a	1
desejar	dəzɐʒ'aɾ	9	2	0	10	ə → ɐj	1
desenhar	dəzɐɲ'aɾ	9	2	0	10	ə → ɐ	1
errar	eʁ'aɾ	9	2	0	10	e → ɛ	1
lavar	lɐv'aɾ	9	2	0	10	ɐ → a	1
somar	sum'aɾ	9	2	0	10	u → o	1
agir	ɐʒ'iɾ	9	2	0	10	ɐ → a	1
chegar	ʃəg'aɾ	9	2	0	10	ə → e	1
negar	nəg'aɾ	9	2	0	10	ə → ɛ	1
tocar	tuk'aɾ	9	2	0	10	u → ɔ	1
ansiar	ẽsi'aɾ	9	2	0	10	i → ɐj	1
beber	bəb'eɾ	9	3	0	11	ə → e / ɛ	1a / 1b
mover	muv'eɾ	9	3	0	11	u → o / ɔ	1a / 1b
agredir	ɐgɾəd'iɾ	11	2	0	12	ə → i	7
abrir	ɐbɾ'iɾ	10	3	1	13	ɐ → a (*)	1
ferir	fəɾ'iɾ	11	3	0	13	ə → i / ɛ	3 / 1b
requerer	ʁəkəɾ'eɾ	11	3	0	13	ə → ɐj / ɛ	3 / 1b
jazer	ʒɐz'eɾ	9	3	2	13	ɐ → a/zə→ʃ	1 / 4a
roer	ʁu'eɾ	9	3	3	14	u → o / ɔj	1a / 1b
escrever	əʃkɾəv'eɾ	10	4	1	14	ə → e / ɛ (*)	1a / 1b
medir	məd'iɾ	11	4	0	14	ə → ɛ / d → s	1 / 3
perder	pəɾd'eɾ	11	4	0	14	ə → ɛ / d → k	1 / 3
valer	vɐl'eɾ	11	4	0	14	ɐ → a / l → ʎ	1 / 3
trair	tɾɐ'iɾ	11	3	3	16	ɐ → aj / ɐ → ɐj	1 / 3
rir	ʁ'iɾ	13	3	5	20	ʁ→ʁ'i /'idə	7/8
ler	l'eɾ	13	4	5	21	e→ej/l→l'e/ledə	3/1b/8
prover	pɾuv'eɾ	13	5	5	22	ɐjʒ/e/əʒ/edə	7+8

Table 3. Verb paradigms for BP sorted by irregularity index, **Ind**. Column **IP** means the number of irregularities in the verb pronunciations; **Rad** the number of verb radicals; **S** the number of changes in the suffixes; **Template** refers to Table 1. Asterisk note (*) means that the verb has an irregular past participle.

Verb	Pron.	IP	Rad	S	Ind	Phone Change	Template
pensar	pẽjs'aɾ	0	1	0	0	-	-
viver	viv'eɾ	0	1	0	0	-	-
unir	un'iɾ	0	1	0	0	-	-
pôr	p'oɾ	0	1	0	0	-	-
abrir	abɾ'iɾ	1	2	1	3	(*)	-
mudar	mud'aɾ	3	2	0	4	d→dʒ	2a
cantar	kẽt'aɾ	3	2	0	4	t→tʃ	2a
induzir	ĩduz'iɾ	2	2	2	5	zɪ→ s	4a
jazer	ʒaz'eɾ	2	2	2	5	zɪ → s	4a
beber	beb'eɾ	4	2	0	5	e → ɛ	1b
mover	mov'eɾ	4	2	0	5	o→ɔ	1b
frigir	fɾiʒ'iɾ	4	2	0	5	i→ɛ	1b
subir	sub'iɾ	4	2	0	5	u→ɔ	1b
ouvir	owv'iɾ	7	2	0	8	v→s	3
valer	val'eɾ	7	2	0	8	l → ʎ	3
escrever	ɪskɾev'eɾ	5	3	1	8	e→ɛ (*)	1b
decidir	desidʒ'iɾ	8	2	0	9	dʒ→d	5
partir	paɾtʃ'iɾ	8	2	0	9	tʃ→ t	5
construir	kõstɾu'iɾ	4	3	3	9	u→ɔj / ɔ	4/2b
roer	xo'eɾ	4	3	3	9	o→ɔj / ɔ	1b
errar	ex'aɾ	9	2	0	10	e→ɛ	1
amar	am'aɾ	9	2	0	10	a→ ẽ	1
sentir	sẽjtʃ'iɾ	8	3	0	10	ẽj→ĩ / tʃ →t	3/5
tocar	tok'aɾ	9	2	0	10	o→ɔ	1
apoiar	apoj'aɾ	9	2	0	10	o→ɔ	1
aprender	apɾẽjd'eɾ	11	2	0	12	d→dʒ	6
debater	debat'eɾ	11	2	0	12	t→tʃ	6
prevenir	pɾeven'iɾ	11	2	0	12	e→i	7
polir	pol'iɾ	11	2	0	12	o→u	7
dormir	doɾm'iɾ	11	3	0	13	o→u / ɔ	3/1b
ansiar	ẽsi'aɾ	9	2	3	13	i→ej	1
ferir	feɾ'iɾ	11	3	0	13	e→ɛ / e→i	1b/3
trair	tɾa'iɾ	10	2	3	14	a → aj	3+4
requerer	xekeɾ'eɾ	11	3	2	15	e→ɛ / e→ej	1b/3
medir	medʒ'iɾ	11	5	0	15	e→ɛ/dʒ→s/ dʒ→d	1/3/2b
cobrir	kobɾ'iɾ	12	4	1	16	o→u / ɔ (*)	3/1b
rir	x'iɾ	13	3	5	20	x→x'i /'idʒɪ	7/8
ler	l'eɾ	13	4	5	21	e→ej/l→l'e/l'edʒɪ	3/1b/8
prover	pɾov'eɾ	13	4	5	21	eʒ / e /edʒɪ	3/1b/8
perder	peɾd'eɾ	19	5	0	23	e→ɛ/d→k/d→dʒ	1b/3/6

This simple method is used to identify and count the irregularities in the pronunciation of Portuguese inflected verb forms. Tables 2, 3 and 4 show verb paradigms sorted by this irregularity index.

Any change occurring in the orthography, which maintains phonetic regularity, such as the one between <g> and <j> in the example <fingir>/<finjo> ('to pretend/I pretend') [fĩʒ'iɾ]/[fĩʒʊ], is not considered irregular.

4   Characterization of Irregularity in Verbs

Based on our definition of verb regularity, by analyzing irregular inflections it was possible to identify patterns of irregularity in both EP and BP. Using the irregularity index, it is also possible to separate the verbs that are near regular from the strongly irregular ones. In fact, there is regularity in the irregular forms for most verbs, which we call "quasi-regular" verbs. Table 1 identifies the templates where the irregularities occur for different tenses, moods and persons. This analysis of irregularity, aided by our linguistic knowledge and by the list of most common verbs, led us to identify groups of verbs with the same irregularity patterns. For each group a verb is elected as a paradigm. Table 2 and Table 3 show the identified paradigms of regular and quasi-regular verbs, for EP and BP respectively, indicating the type of pronunciation change, the template of the changes as well as the number of changes as defined in the variables of the equation (1). In terms of the irregularity index, there is a big gap between the quasi-regular verbs and the other irregular verbs, as can be observed in Table 4, which lists strongly irregular verbs. It can also be observed that these verbs are the same for EP and BP, as expected, however not in the same ranking order of irregularity.

Table 4. Irregular verbs in EP and BP with the highest index of irregularity

EP					BP				
Verb	*Pron.*	*IP*	*Rad*	*S*	*Ind*	*Verb*	*Pron.*	*IP*	*Rad*	*S*	*Ind*
aprazer	ɐpɾɐz'eɾ	33	5	26	63	aprazer	apɾaz'eɾ	26	5	25	55
caber	kɐb'eɾ	35	5	24	63	caber	kab'eɾ	31	5	23	58
querer	kəɾ'eɾ	35	5	25	64	saber	sab'eɾ	31	6	24	60
saber	sɐb'eɾ	35	6	25	65	querer	keɾ'eɾ	35	6	25	65
poder	pud'eɾ	35	7	24	65	dar	d'aɾ	31	5	30	65
dar	d'aɾ	31	6	32	68	haver	av'eɾ	35	9	28	71
haver	ɐv'eɾ	35	8	29	71	ver	v'eɾ	37	6	29	71
ver	v'eɾ	37	7	29	72	poder	pod'eɾ	42	10	23	74
estar	əʃt'aɾ	34	10	38	81	estar	ɪst'aɾ	34	12	34	79
ter	t'eɾ	43	9	35	86	ir	'iɾ	38	12	36	85
ir	'iɾ	38	12	38	87	ter	t'eɾ	43	11	34	87
vir	v'iɾ	44	9	36	88	trazer	tɾaz'eɾ	45	7	37	88
dizer	diz'eɾ	46	7	39	91	dizer	dʒiz'eɾ	46	7	38	90
trazer	tɾɐz'eɾ	47	8	38	92	fazer	faz'eɾ	46	8	38	91
ser	s'eɾ	44	12	38	93	vir	v'iɾ	43	11	39	92
fazer	fɐz'eɾ	48	10	39	96	ser	s'eɾ	44	12	38	93

A rule-based algorithm for paradigm attribution, given a verb in the infinitive form (orthographic and phonetic) has been implemented using common patterns learned from examples in lexical databases, such as in [19]. For example, verbs ending in *<air>* follow the paradigm *<trair>* 'to betray', while verbs ending in *<zir>* follow that of *<induzir>* 'to induce'. In most cases the algorithm takes the last vowel or consonant of the radical to choose the correct paradigm. In other cases, it uses a pattern from the orthographic form, or checks a list of verbs. For example, the verbs ending in *<ver>*, could be derivations of *<ver>* 'to see' as in *<prever>* 'to preview', *<antever>* 'to foresee' or *<rever>* 'to review', but not *<escrever>* 'to write', although it shares the same ending pattern with *<rever>*; or *<reaver>* 'to recover', which is pronounced as the irregular *<haver>* 'to have'; or *<prover>* 'to provide', which has its own paradigm. The rules have been tested using the verb corpora described in section 2, followed by manual verification, so it is almost certain that the paradigms presented in Tables 2, 3 and 4 are representative of all the verbs in the Portuguese language.

4.1   Comparing European with Brazilian Portuguese

Tables 2 and 3 present the verb paradigms sorted by index of pronunciation irregularity for EP and BP, respectively. BP contains 56 paradigms (two less than EP). For BP the number of paradigms can even be reduced by about 10 more, if a post-processing rule is applied to the pronunciation forms: for the alveolar-plosives [t] and [d] in the right sided context of [i], [ĩ] and [ɪ], a change to [ʧ] and [ʤ] occurs, respectively. This is a common phonological event in BP stated in the literature [22, 23]. The reverse also applies as in template 5 in Table 1.

The paradigms for the regular pronunciations are the same, as indicated in the first four rows of Tables 2 and 3. As expected, the most irregular verbs, listed in Table 4, are the same for EP and BP; though not in the same ranking order of irregularity.

Observing the 56 (BP)/58 (EP) verb paradigms, we can establish an irregularity threshold or, in other words, establish a level at which almost irregular and irregular verbs can be divided. Therefore, above this threshold, we can observe a great change in the level of irregularity. This can also be seen through a large number of changes in the pronunciation of certain verbs, either by changes in their radicals or suffixes. This threshold can be established at the irregularity index between 24 and 54, which corresponds to the division between Tables 2/3 and Table 4. If we see verbs organized by rising order (with respect to irregularity from least to greatest) we can note that from the point of the verb *<perder>* 'to lose', in BP, onwards and the verb *<prover>* 'to provide', in EP, with respective irregularity indices 23 e 22, the next verb is *<aprazer>* 'to please' which is presented in the irregularity index of 55 (BP) and 63 (EP), respectively.

In EP, the verb *<fazer>* 'to do' is the most irregular, with 48 irregularities in pronunciation, 39 of which are in the suffixes and 10 radicals: [fɐz, fas, faz, faʃ, fiʃ, fiz'ɛ, feʃ, fɐ, fɐs, fɐjt], while in BP this verb presents 46 irregularities, 38 in suffixes (the 3rd person plural of perfect past, [fiz'emʊs], has a regular suffix) and 8 radicals: [faz, fas, fis, fiz'ɛ, fes, fiz, fa, fɐjt]. In BP the most irregular verb is *<ser>* 'to be', with 44 irregularities, 38 in suffixes and 11 radicals: [s, s'o, 'ɛ, s'ẽw̃, 'ɛɾ, fuj, fo, foj, s'eʒ, s'e, s'edʒɪ]. Most of the differences between BP and EP derive from different unstressed vowels, which are more closed in EP than in BP.

The verb *<apoiar>* 'to support' follows the conjugation paradigm of the verb *<tocar>* 'to touch' but is included because it is the only verb with the diphthong *<oi>* which is not regular. The verb *<temer>* 'to fear' is regular in BP, but follows the same paradigm as *<beber>* 'to drink' in EP ([16]). Also, the verb *<comer>* 'to eat' is regular in BP, but follows the paradigm of *<mover>* 'to move' in EP ([16]).

5   Conclusion

The automatic decision as to which pronunciation of a word is the correct one is a nontrivial problem, even for Portuguese verbs, and has not been systematically solved yet, if verbal irregularity is taken into account. With this study we intended to enhance a grapheme-to-phoneme system, generating a verb pronunciation system for both varieties of Portuguese, Brazilian and European, given only its infinitive form. Two obvious applications for this system are in Portuguese language teaching/learning and in speech synthesis.

Systematic regular and non-regular pronunciations of inflected verb forms were evaluated for European and Brazilian Portuguese, which result in the definition of a set of paradigms that are, with a high level of certainty, representative of all the verbs for Portuguese. An index of verb irregularity was also proposed, based on the overall number of changes observed in verb pronunciation forms relative to a regular situation. Most of the pronunciation irregularities derive from vowel alternations, as pointed out in the "Phone Change" column in Tables 2 and 3. Finally, a rule-based algorithm to classify any verb according to the defined paradigms was implemented and tested using the most common verbs for Brazilian Portuguese and European Portuguese. The analysis resulted from this study allowed a comparison between the two varieties of Portuguese, showing a similarity in the number of paradigms for pronunciation. As expected, most of the differences between BP and EP derive from the different unstressed vowels, which are more closed in EP than in BP. This study also contributed to the study of São Paulo standard pronunciation characteristics, supporting the classical studies in the field with a large corpus.

The proposed algorithms could be very useful for other varieties of Portuguese, namely for African Portuguese, for which a compatibility evaluation is forseen in the future.

Acknowledgment. This work was co-funded by FAPESP (2014/00613-5), FCT strategic project PEst-OE/EEI/LA0008/2013 and QREN project (PTDC/CLE-LIN/11 2411/2009;   TICE.Healty-13842).   Sara   Candeias   is   supported   by   the SFRH/BPD/36584/2007.

References

1. Veiga, A., Candeias, S., Perdigão, F.: Generating a pronunciation dictionary for European Portuguese using a joint-sequence model with embedded stress assignment. J. Braz. Comp. Society 19(2), 127–134 (2013)
2. Braga, D., Coelho, L.: A rule-based grapheme-to-phone converter for TTS systems in European Portuguese. In: Proceedings of 6th International Telecommunications Symposium, pp. 328–333 (2006)
3. Oliveira, L.C., Viana, M.C., Trancoso, I.M.: A rule-based text-to-speech system for Portuguese. In: Proceedings of ICASSP, San Francisco, vol. 2, pp. 73–76 (1992)
4. Teixeira, J.P.: A prosody model to TTS systems. PhD Thesis. Faculdade de Engenharia da Universidade do Porto (2004)
5. Albano, E.C., Moreira, A.A.: Archisegment-based letter-to-phone conversion for concatenative speech synthesis in Portuguese. In: Proceedings of the 4th International Conference, vol. 3 (1996)
6. Barbosa, F., Pinto, G., Resende, F.G., Gonçalves, C.A., Monserrat, R., Rosa, M.C.: Grapheme-Phone Transcription Algorithm for a Brazilian Portuguese TTS. In: Mamede, N.J., Baptista, J., Trancoso, I., Nunes, M.d.G.V. (eds.) PROPOR 2003. LNCS (LNAI), vol. 2721, pp. 23–30. Springer, Heidelberg (2003)
7. Neto, N., Patrick, C., Klautau, A., Trancoso, I.: Free tools and resources for Brazilian Portuguese speech recognition. Journal of the Brazilian Computer Society 17, 53–68 (2011)
8. Braga, D., Marques, M.A.: Desambiguação de Homógrafos para Sistemas de Conversão Texto-Fala em Português. Diacrítica 21(1), 25–50 (2007)
9. Braga, D.: Algoritmos de Processamento da Linguagem Natural para Sistemas de Conversão Texto-Fala em Português. Tese de Doutorado. Faculdade de Filoloxía da Universidade da Coruna (2008)
10. Mateus, M.H.M.: Fonologia. In: Mateus, M.H.M., et al. (eds.) Gramática da Língua Portuguesa. Editorial Caminho (2003)
11. Barbosa, J.M.: Fonologia e Morfologia do Português. Livraria Almedina (1994)
12. Villalva, A.: Estrutura morfológica básica. In: Mateus, M.H.M., et al. (eds.) Gramática da Língua Portuguesa. Editorial Caminho (2003)
13. Carmo, M.C.: As vogais médias pretônicas dos verbos na fala culta do interior paulista. Dissertação de mestrado. Universidade Estadual Paulista 'Julio de Mesquita Filho' (2009)
14. Collischonn, G., Schwindt, L.C.: Harmonia vocálica no sistema verbal do português do sul do Brasil. In: Estudos de Fonologia e de Morfologia, Porto Alegre, vol. 36, pp. 73–82 (2004)
15. Bechara, E.: Moderna gramática portuguesa. Nova Fronteira, Rio de Janeiro (2009)
16. Cunha, C., Cintra, L.: Nova Gramática do Português Contemporâneo. Sá da Costa (2002)
17. Lxconjugator, http://lxcenter.di.fc.ul.pt/services/en/LXServicesConjugator.html (accessed on March 28, 2014)
18. PRIBERAM dicionário, http://www.priberam.pt/dlpo/Conjugar/ (accessed on March 28, 2014)
19. VOP, Vocabulário Ortográfico do Português, http://www.portaldalinguaportuguesa.org/ (accessed on March 28, 2014)
20. Candeias, S., Veiga, A., Perdigão, F.: Sistema Automático de Pronunciação de Verbos. In: Anais da III Jornada de Descrição do Português, pp. 28–35 (2013)
21. Candeias, S., Veiga, A., Perdigão, F.: Pronunciação de verbos. Lidel-Edições Técnicas, Lisboa (forthcoming)

22. Santos, D., Rocha, P.: Evaluating CETEMPúblico, a free resource for Portuguese. In: Proceedings of the 39th Annual Meeting of the Association for Computational Linguistics (ACL 2001) ACL 2001, pp. 442–449 (2001)
23. Santos, D., Sarmento, L.: O projecto AC/DC: acesso a corpora/disponibilização de corpora. In: Mendes, A., Freitas, T. (eds.) Actas do XVIII Encontro Nacional da Associação Portuguesa de Linguística (APL 2002), pp. 705–717 (2002)
24. Pinheiro, G., Aluisio, S.: Corpus NILC: descrição e análise crítica com vistas ao projeto Lacio-Web. Série de Relatórios do Núcleo Interistitucional de Linguística Computacional. Technical report, NILC-TR-03-03 (2003)
25. grɐfɔnə, http://www.co.it.pt/~labfala/g2p/index.html (accessed on March 28, 2014)
26. Siravenha, A.C., Neto, N., Macedo, V., Klautau, A.: Uso de Regras Fonológicas com Determinação de Vogal Tônica para Conversão Grafema-Fone em Português Brasileiro. In: 7th International Information and Telecommunication Technologies Symposium (I2TS 2008), Foz do Iguaçu (2008)
27. Koch, I.G.V.: Souza e Silva, M.C.P.: Linguística aplicada ao português: morfologia. Cortez Editora (1983)
28. Cagliari, L.C.: Elementos de Fonética do Português Brasileiro. Paulistana, São Paulo (2009)
29. Cagliari, L.C.: Um modelo de transcrição fonética para um dicionário. In: Petrov, P., Sousa, P.Q., Samartin, R.L.I., Feijó, E.J.T. (eds.) Avanços em Ciências da Linguagem, vol. 1, pp. 475–490. Através Editora, Faro (2012)
30. Cristófaro-Silva, T.: Fonética e fonologia dos português: Roteiro de estudos e guia de exercícios, 3rd edn. Contexto, São Paulo (2000)
31. Barbosa, P.A.: Do grau de não perifericidade da vogal /a/ pós-tônica final. In: Diadorim, vol. 12 (2012)
32. Cunha, C., Cintra, L.F.L.: Nova gramática do português contemporâneo, vol. 2. Nova Fronteira, Rio de Janeiro (1985)
33. Atallah, M.J. (ed.): Algorithms and theory of computation handbook. CRC press (1998)
34. Shulby, C.D., Mendonça, G., Marquiafável, V.: Automatic Disambiguation of Homographic Heterophone Pairs Containing Open and Closed Mid Vowels. In: 9th Brazilian Symposium in Information and Human Language Technology, STIL 2013, Fortaleza, CE, Brazil, pp. 126–137 (2013)

Acoustic Similarity Scores for Keyword Spotting

Arlindo Veiga[1,2], Carla Lopes[1,3], Luís Sá[1,2], and Fernando Perdigão[1,2]

[1] Instituto de Telecomunicações, 3030-290 Coimbra, Portugal
[2] Universidade de Coimbra – DEEC, Polo II, 3030-290 Coimbra, Portugal
[3] Instituto Politécnico de Leiria – ESTG, 2411-901 Leiria, Portugal
{aveiga,calopes,luis,fp}@co.it.pt

Abstract. This paper presents a study on keyword spotting systems based on acoustic similarity between a filler model and keyword model. The ratio between the keyword model likelihood and the generic (filler) model likelihood is used by the classifier to detect relevant peaks values that indicate keyword occurrences. We have changed the standard scheme of keyword spotting system to allow keyword detection in a single forward step. We propose a new log-likelihood ratio normalization to minimize the effect of word length on the classifier performance. Tests show the effectiveness of our normalization method against two other methods. Experiments were performed on continuous speech utterances of the Portuguese TECNOVOZ database (read sentences) with keywords of several lengths.

Keywords: keyword spotting, spoken term detection, phonetic features, log-likelihood ratio normalization.

1 Introduction

Keyword spotting is a technique for searching words or phrase instances in a given speech signal. The problem is similar to automatic speech recognition, however the keyword spotting system goal is to detect a keyword in continuous speech instead of transcribe all input speech signal.

Keyword spotting techniques followed the evolution of automatic speech recognition techniques. The first approaches used Dynamic Time Warping (DTW) [1,2] and progressed to Hidden Markov Models (HMM) [3], where the HMM states contain parameters that represent a Gaussian Mixtures Model (GMM). Hybrid architecture approaches has been developed that combine HMM with neural networks and even more complex networks [4]. In hybrid approaches the neural network is used to estimate the a posteriori probabilities of HMM states instead a priori probabilities given by GMMs [5].

Currently, word spotting approaches can be classified into three main classes [6]:

1. based on large vocabulary continuous speech recognition system (LVCSR);
2. based on a phone lattice;

J. Baptista et al. (Eds.): PROPOR 2014, LNAI 8775, pp. 48–58, 2014.
© Springer International Publishing Switzerland 2014

3. based on acoustic models for fillers, background or anti-models.

The first two classes require a LVCSR, which is not a trivial and reliable resource for many languages (mainly due to the lack of good and large databases) and resulting in systems with demanding of high computational resources. The word spotting system based on the first class examines the LVCSR output results leading to a performance highly dependent of the LVCSR word accuracy. The second class uses a LVCSR as a pre-processing stage to create phone/word lattices. The word spotting system uses theses lattices to search for word occurrences.

The third class requires fewer resources and uses a filler model (Figure 1); however, it is difficult to build consistent models for fillers, background or anti-models for all words. Most of the word spotting systems based on this latter approach uses the filler model proposed in [7,8] that is composed internally by the states of all phone models arranged in parallel, as shown in Figure 2. The word spotting system presented in this paper follows this approach with a filler model. The keyword model is made by concatenating the models of the corresponding phone sequence, which is valid for a single or a sequence of words. The speech signal is represented by a sequence of feature vectors (observations) resulting from a short-term analysis of the signal in a frame by frame basis.

2   Keyword Spotting Decoder

A common word spotting system based on HMM acoustic models uses the Viterbi algorithm through a decoding network (task grammar) that allows the filler model to compete with the keyword model. The decoding process ends with a backtracking pass which determines where the word model "wins" the filler model along the speech utterance.

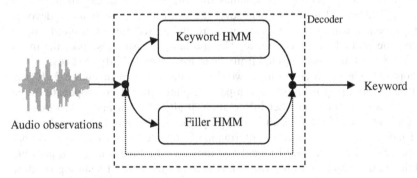

Fig. 1. Standard keyword spotting system

In this work we propose a new decoding process based only on model properties and on acoustic model's likelihoods. A score based on likelihood ratio and HMM information is proposed that allows decoding without the backtracking pass. In our case the filler model is not competing with the word model. Instead, it is used to compute the maximum likelihood of each observation frame given the phone models, because all the phone models are included in parallel. The decoding process adds a feedback arc (dotted arc in Figure 2), allowing us to obtain the best phone sequence, thus with maximum likelihood (free phone loop), for the given speech signal and phone models. It is possible to define a measure of similarity between the output free phone sequence and the predefined word's phone sequence to detect the word occurrence, but this implies the backtracking pass and the definition of a similarity score.

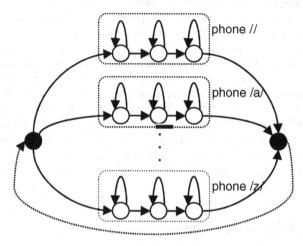

Fig. 2. Filler model composed by all HMM phone models arranged in parallel. Filled nodes in black are non-emitting states.

On the contrary, this work defines a similarity measure based only on the ratio between the keyword model likelihood and the filler model likelihood, avoiding the use of the phone sequence decoding. It assumes that the keyword model likelihood is similar to the filler model likelihood whenever the word occurs. Then, a fast decoder can be implemented using this similarity measure and a predefined threshold can be set to detect the spotted word occurrence. The similarity measure also uses the mean duration of the word models computed from the transition probability of states.

The proposed scheme to perform keyword spotting is shown in Figure 3. It performs a beam search step using the token-passing paradigm [9], to propagate frame likelihoods as well as frame number information. It allows the winner token that appears on the end state of the filler model to propagate to the initial state of the filler and word model. The frame number information is updated when the token propagates to the first state of the word model and indicates the time stamp of a possible keyword that is beginning. This information overcomes the need of a sliding window strategy to obtain the word boundaries.

At each signal frame a winning token appears at the end of the two models. The token leaving the word model has the time stamp of a possible keyword beginning and the likelihood of the keyword hypothesis. If this likelihood is similar to the filler's, then there is a high probability of being in the presence of the spotted word, otherwise the filler wins. Although this likelihood ratio performs well in word detection, there are other restrictions that must be applied. For instance, the likelihood ratio is highly dependent on the duration of the word and some kind of normalization must be taken into consideration.

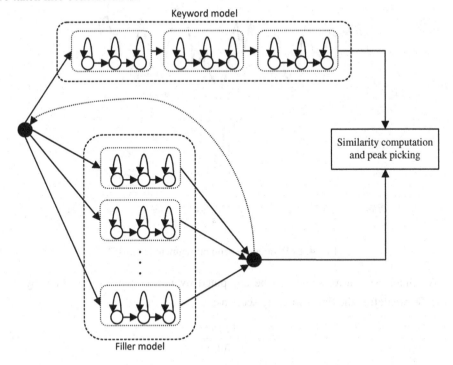

Fig. 3. Proposed keyword decoding scheme

3   Similarity Measure

We propose a similarity measure based on a likelihood ratio (or log-likelihood difference). Logarithm scale is used to avoid the precision issue on the likelihood computation. Considering that the token emitted by keyword model at frame t that lasts $N(t)$ frames to cross this model, then the log-likelihood ratio, $LLR(t)$, is defined as:

$$LLR(t) = \sum_{i=t-N(t)+1}^{t} LW(i) - \sum_{i=t-N(t)+1}^{t} LF(i) \tag{1}$$

where $LW(i)$ and $LF(i)$ are log-likelihoods of keyword and filler tokens at frame i, respectively. The $LLR(t)$ values are non-positive because the $LF(i)$ (best phone sequence) is always higher or equal to $LW(i)$. The LLR decrease with $N(t)$ so it is not an appropriate similarity measure as it is shown in Figure 4. It needs some sort of normalization, like those proposed in [10, 11], to make it independent of word length.

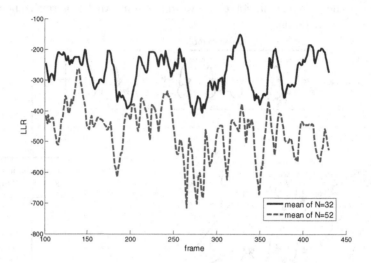

Fig. 4. LLR values depend on segment lengths

A reliable similarity score can be computed by normalizing the $LLR(t)$ by the $N(t)$. So we define the first similarity score as:

$$SS_1(t) = \frac{LLR(t)}{N(t)} . \tag{2}$$

The effect of this normalization is showed in Figure 5.

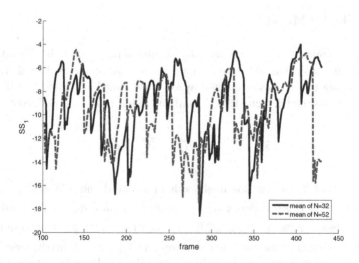

Fig. 5. LLR values normalized by the segment length N

Other normalization based on mean keyword model duration was used. The mean model duration with left-to-right topology is the sum of all its states' mean duration [12]. As is well known, considering a_{jj} the self-transition probability of a state j, the mean state duration, $\bar{d}_j$, is given by:

$$\bar{d}_j = \frac{1}{1-a_{jj}}.$$ (3)

As we use phones with 3 emitting states and 2 non-emitting states (state 1 and 5), the mean phone duration can be computed as:

$$\bar{d}_{phone} = \sum_{j=2}^{4} \frac{1}{1-a_{jj}}.$$ (4)

The mean word model duration, $\bar{D}_W$, is the sum of all its phones' mean durations. So, the second similarity score normalizes LLR by this mean duration:

$$SS_2(t) = \frac{LLR(t)}{\bar{D}_W}.$$ (5)

The $\bar{D}_W$ is computed once for each keyword, so the token can discard frame number information.

We propose a further normalization measure that enhances the occurrence of a peak in LLR, by accumulating the log-likelihood difference during the mean duration of the keyword. It is like an LLR area, as suggested by Figure 6. This normalization takes into account the number of frames that a token uses to cross a model and the shape of LLR during the token travel across the model. This similarity score is computed as:

$$SS_3(t) = \frac{LLR(t)}{A(t)}$$ (6)

where:

$$A(t) = \sum_{i=1}^{N(t)} \left(LLR(t) - LLR(t-i) \right)$$ (7)

A peak picking is then applied to the similarity score in order to obtain candidates of the spotted word. The normalization allows us to use a single threshold for all keywords.

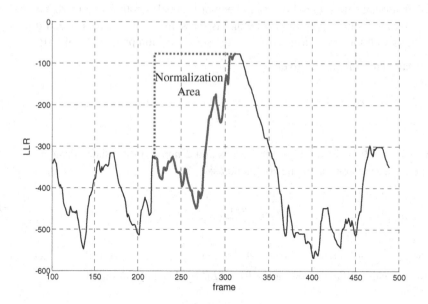

Fig. 6. Normalization area

4   Experiments and Results

4.1   Database

The word spotting system described in the previous section has been evaluated on the Portuguese TECNOVOZ [13] speech database. A total of 22,627 utterances (around 31.5 hours) were used which corresponds to 208 prompts of generic sentences.

A set of acoustic models for 37 context free phones were trained with the HTK toolkit [14]; 35 models correspond to the phonemes of the Portuguese language; one model is used for silence and one model for short pause. The number of Gaussians per state (mixture PDF) was incremented gradually until reaching 96 Gaussian in the training stage. The high number of Gaussian components is used in order to avoid the use of context-dependent models (triphones), which would complicate severely the filler model. These acoustic models are based on Mel-Frequency Cepstral Coefficients (MFCC) parameters which are commonly used on speech recognition front-end: 12 MFCCs plus energy and their first and second order time derivatives computed at a rate of 10ms and within a window of 25ms.

The database has 208 different prompts (phrases) and 1455 different words. It has a mean of 14 words per prompt with 4.4 words of standard deviation.

The reference marks of words (word boundaries) were taken by using forced alignment and some utterances were discarded because the inconsistences and hesitation events. A total of 1170 words were used as keywords composed by a number of phones between 1 and 16.

4.2   Results

It is common to use ROC (Receiver Operation Characteristics) curves [15] or DET (Detection Error Trade-off) curves [16] to evaluate binary decision systems. We use a DET curve that is a plot of false rejection error (type I error) rate (FRR) against false acceptance error (type II error) rate (FAR), by varying a decision threshold. We can then define a threshold corresponding to a particular operation point of the keyword spotting system from the DET curve. We consider the best peak between word boundaries as the score for a correct decision and the best peak off word boundaries as the score for an incorrect decision. The DET curve can be used to set the operation point that determines the threshold value for keyword detection.

The Figure 7 shows several DET curves of our similarity measure with different number of phones.

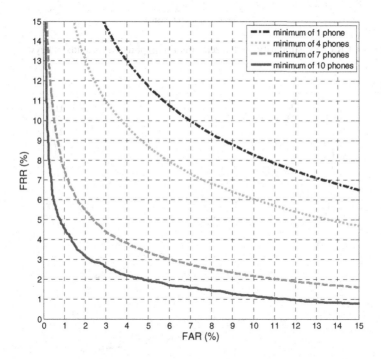

Fig. 7. DET curves of SS_3 similarity with several minimum phone thresholds in keywords

We observe that with limit of 1 phone, the equal error rate (EER – where the FRR is equal to the FAR) is equal to 8.84 % but with limit of 10 phones, the EER decreases to 2.78 % (Figure 8).

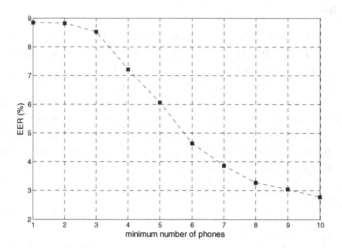

Fig. 8. EER of *SS₃* similarity with several minimum phone thresholds in keywords

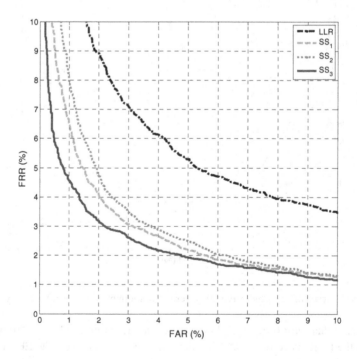

Fig. 9. DET curves of all similarities for keywords with minimum of 10 phones

Figure 9 shows the performances of the 3 normalization measures for keywords with a minimum of 10 phones. It is clear that the similarity score SS_3 has a better discrimination performance than the other ones.

The obtained results show that *LLR* can be used as similarity measure but the normalization has a dramatic impact on the detection performance. The simplest normalization method, SS_2, uses a constant value (mean keyword duration) and achieves EER equal to 3.27 %, a relative decrease of 36.26 % comparing to *LLR* alone. More elaborated normalization methods lead to better performance. The SS_1 has slightly better performance than SS_2 and has an EER equal to 3.05 %. SS_3 corresponds to the better normalized score with an EER equal to 2.78 %.

We can decrease FAR at the cost of FRR. For example, if an application imposes maximum FAR equal to 1 %, with the SS_3 measure the FRR increases to 4.54 %.

The results of this keyword spotting scheme is comparable to the reported performance of other keyword spotting scheme for the similar tasks. For example, in the [4] several keyword spotting approaches based on Dynamic Bayesian Networks (DBN) were tested on TIMIT corpus, and for which for a FAR equal to 1 % the FRR values are between 2.5 % to 12 %.

5   Conclusion

This paper proposes a fast and efficient decoding algorithm for keyword spotting system using only the Viterbi forward step. The decoding uses log-likelihood difference and model information to compute a similarity measure. The measure presented, achieved a good trade-off between the false alarm error and false rejection error and do not require a large vocabulary continuous speech recognition system. The results of our approach are comparable to others approaches in a similar task. As a future work we intend to use a deep neural network as a phone likelihood estimator in order to avoid using HMM context-dependent phone models in the filler model.

Acknowledgements. Arlindo Veiga would like to thank the Instituto de Telecomunicações for the Research Scholarship.

References

1. Bridle, J.S.: An Efficient Elastic-Template Method for Detecting Given Words in Running Speech. In: Proc. of the British Acoustical Society Meeting (1973)
2. Higgins, A., Wohlford, R.: Keyword Recognition Using Template Concatenation. In: Proc. of the International Conference on Acoustics, Speech, and Signal Processing, vol. 10, pp. 1233–1236 (1985)
3. Rabiner, L.R.: A Tutorial on Hidden Markov Models and Selected Applications in Speech Recognition. Proceedings of the IEEE 77(2), 257–286 (1989)
4. Wöllmer, M., Schuller, B., Rigoll, G.: Keyword Spotting Exploiting Long Short-Term Memory. Speech Communication 55, 252–265 (2013)

5. Zhu, Q., Chen, B., Morgan, N., Stolcke, A.: Tandem Connectionist Feature Extraction for Conversational Speech Recognition. In: Bengio, S., Bourlard, H. (eds.) MLMI 2004. LNCS, vol. 3361, pp. 223–231. Springer, Heidelberg (2005)
6. Szoke, I., Schwarz, P., Matejka, P., Burget, L., Karafiát, M., Fapso, M., Cernocky, J.: Comparison of Keyword Spotting Approaches for Informal Continuous Speech. In: Proc. of the 9th European Conference on Speech Communication and Technology, Lisbon, Portugal (2005)
7. Rohlicek, J.R., Russell, W., Roukos, S., Gish, H.: Continuous Hidden Markov Modeling for Speaker-Independent Word Spotting. In: Proc. of the International Conference on Acoustics, Speech, and Signal Processing, pp. 627–630 (1989)
8. Rose, R.C., Paul, D.B.: A Hidden Markov Model Based Keyword Recognition System. In: Proc. of the International Conference on Acoustics, Speech, and Signal Processing, vol. 1, pp. 129–132 (1990)
9. Young, S., Russell, N.H., Thornton, M.: Token Passing: A Simple Conceptual Model for Connected Speech Recognition Systems. Cambridge University Engineering Department, Cambridge (1989)
10. Junkawitsch, J., Ruske, G., Höge, H.: Efficient Methods for Detecting Keywords in Continuous Speech. In: Proc. of the 5th European Conference on Speech Communication and Technology, Rhodes, Greece, pp. 259–262 (1997)
11. Weintraub, M.: Keyword-Spotting Using SRI's DECIPHER Large-Vocabulary Speech-Recognition System. In: Proc. of the International Conference on Acoustics, Speech, and Signal Processing, vol. 2, pp. 463–466 (1993)
12. Papoulis, A.: Probability, Random Variables and Stochastic Processes, 3rd edn. McGraw-Hill Companies (1991)
13. Lopes, J., Neves, C., Veiga, A., Maciel, A., Lopes, C., Perdigão, F., Sá, L.: Development of a Speech Recognizer with the Tecnovoz Database. In: Teixeira, A., de Lima, V.L.S., de Oliveira, L.C., Quaresma, P. (eds.) PROPOR 2008. LNCS (LNAI), vol. 5190, pp. 260–263. Springer, Heidelberg (2008)
14. Young, S., Evermann, G., Gales, M., Hain, T., Kershaw, D., Liu, X., Moore, G., Odell, J., Ollason, D., Povey, D., Valtchev, V., Woodland, P.: The HTK Book (for HTK Version 3.4). Cambridge University Engineering Department, Cambridge (2006)
15. Egan, J.P.: Signal Detection Theory and ROC Analysis. Academic Press, New York (1975)
16. Martin, A., Doddington, G., Kamm, T., Ordowski, M., Przybocki, M.: The DET Curve in Assessment of Detection Task Performance. DTIC Document (1997)

JMorpher: A Finite-State Morphological Parser in Java for Android

Leonel F. de Alencar, Mardonio J.C. França, Katiuscia M. Andrade,
Philipp B. Costa, Henrique S. Vasconcelos, and Francinaldo P. Madeira

Universidade Federal do Ceará, Group of Computer Networks, Software Engineering,
and Systems (GREat)
Campus do Pici, Bloco 942-A, CEP 60455-760 Fortaleza, Brazil
{leonelararipe,mardoniofranca,katiusciaandrade,philippcosta,
henriquevasconcelos}@great.ufc.br,
naldomadeira@gmail.com
http://www.great.ufc.br

Abstract. This paper presents JMorpher, a morphological parsing utility that is implemented in pure Java. It is apparently the first tool of this type that natively runs on Android mobile devices. JMorpher compiles a lexical transducer definition in the AT&T raw text format, of the type generated by Foma and other open source finite-state packages, into an internal Java representation which is drawn upon to parse input strings. Besides the API, JMorpher comprises of a simple graphical interface that allows the user to load a transducer file, type in some text and parse it. Results of an evaluation based on large Portuguese lexical transducers of different complexity degrees are provided. The implementation was shown to be very efficient on a desktop PC. Although, on an Android smartphone, JMorpher's performance is much lower, it is still suited to the needs of NLP tasks in this environment.

Keywords: NLP. Finite-State Morphology. Morphological Analysis. Morphological Parsing. Lexical Transducer. Android Technology.

1 Introduction

Morphological analysis is a key module in the syntactic parsing pipeline, fulfilling the job of mapping word forms occurring in texts to sets of abstract representations, typically consisting of lemma, POS-tag and subcategorial features [1],[9]. Independent of sentence structure processing, the output of a morphological parser may be useful in a wide range of applications. Two examples from the growing field of mobile computing include the use of lemma information for dictionary lookup in an e-book reader application and the use of morphosyntactic features to improve performance of classification algorithms for organizing the user's messages.

Over the last two decades or so, finite-state transducers, due to compact storage and fast processing, have been a preferred implementation of morphological

J. Baptista et al. (Eds.): PROPOR 2014, LNAI 8775, pp. 59–69, 2014.

analyzers. Many industrial-scale lexical transducers were compiled for typologically diverse languages, using a variety of finite-state toolkits [1],[8],[12],[15].

Xerox Finite State Tools (XFST) [1] are one of the most efficient implementations of finite-state morphology, from the linguist's point of view it is the best documented and is a user-friendly toolkit. XFST has the caveat of being proprietary software, although it is freely distributed for non-commercial research purposes. Since 2010, computational morphology could be carried out with an open source, free software alternative to XFST, the Foma finite compiler and C library [8]. In general Foma is as efficient as XFST, Foma also has the advantage of being almost completely compatible with the XFST syntax, including the Lexc formalism.

As far as we know, Portuguese has no freely available large lexical transducer distributed under a free software/open source license. For the development of free, wide-coverage lexical resources for Portuguese using finite-state techniques, Foma is a very attractive option, especially when targeting lesser powered mobile devices with low storage capacity. As shown in Table 1, a lexical transducer based on the DELAF_PB computational lexicon [13], for example, takes up less than one tenth of the original entries in raw text format. However, in the mobile setting, Foma has the disadvantage of restricted portability. On the Android platform, for example, compiling Foma proved to be a difficult task due to library dependencies. It is worth noting that different mobile processors require different binaries, complicating the distribution of language-aware apps.

Summing up, Foma would be very useful in the computational processing of Portuguese on mobile devices, but, it is seemingly impractical in this environment. In this paper, we present JMorpher, a pure Java solution developed in order to solve this problem, JMorpher requires to no external libraries and is natively able to run on Android. JMorpher compiles finite-state transducers in the AT&T raw text format, of the type generated by Foma, HFST [12] etc., into an internal Java representation. This representation is drawn upon by the lookup algorithm, which outputs the parses for the input strings contained in the surface language of the transducer. JMorpher is not a port of Foma's flookup utility. Instead, JMorpher's internal transducer representation as well as its parsing algorithms were developed from scratch, aiming at an efficient Java implementation.

The main motivation for developing JMorpher was that HFST's Java lookup utility [6], [12], the only tool of this type we know about implemented in this language, is not open source. Their implementation has not been maintained since 2009 and was found to be unstable in simple tests we carried out.

The paper is structured as follows. In the next section, we describe the different types of transducers of Portuguese and test sets used to evaluate JMorpher. In Section 3, we outline the general concept of the package as well as its API and graphical interface. Then, in Section 4, the evaluation results are presented. Finally, the last section summarizes the main characteristics of the tool and points out possible directions for further research.

2   Test Data

In order to implement and evaluate JMorpher, we used Foma to compile the transducers listed in 1.[1] These finite-state networks encode different aspects of Portuguese morphology, presenting different degrees of complexity for a parsing algorithm to deal with. Unknown words are not handled. Networks FST01, INFL12, FST02, and FST03 are final lexical transducers, representing lexicon fragments of varied sizes, while RULES1C and RULES2C result from cascades of conditional replacement rules modeling orthographic alternations. Transducer FST01 was compiled from a Lexc grammar automatically generated from DE-LAF_PB [13].[2] The original tagset of DELAF_PB was converted to a more mnemonic notation, mostly based on the Leipzig Glossing Rules [10], complemented with abbreviations from a project on French morphology of the Laboratoire de Linguistique Formele [5].

Table 1. Evaluation transducers

Transducer	Source	Size	Characteristics
FST01	Raw text version of DELAF_PB comprising 878450 entries taking up 23.9 MB on disk	2 MB on disk, 43506 states, 138188 arcs, 1156736 paths	Non-sequential. Arcs labeled with epsilons on the surface side only after consumption of input string.
RULES1C	Orthographic alternation rules for the 1st conjugation	4 KB on disk, 9 states, 83 arcs, cyclic	Arcs labeled with the *other* symbol
RULES2C	Orthographic alternation rules for the 2nd conjugation	4 KB on disk, 6 states, 66 arcs, cyclic	Same as RULES1C
INFL12	Composition of a lexicon transducer with the rule transducers above	64 KB on disk, 2618 states, 4738 arcs, 140560 paths	Non-sequential. Arcs labeled with epsilons on the surface side before consumption of input string.
FST02	Union of FST01 and INFL12	2MB on disk, 44974 states, 142410 arcs, 1297296 paths	Same as INFL12.
FST03	Word-parse pairs from FST02	2.2 MB on disk, 44148 states, 143020 arcs, 1297296 paths	Same as FST01.

Three characteristics of the morphology of Portuguese (and natural language in general) lead to more complexity of the transducers generated by Foma than

---

[1] This is part of the development of a large lexical transducer of Brazilian Portuguese which will be the subject of another paper in the near future.

[2] The original DELAF_PB.dic file takes up 48.9 MB on disk. INFL12 encodes the morphotactics of 2008 verb lemmas from [11] which are not contained in DELAF_PB.

is desirable. The implementation of a parsing algorithm needs to be as resource inexpensive as possible, thus making it suitable for the limited storage and processing capacity of mobile devices. The ambiguity pervading the lexicon and the length mismatch between surface forms and analysis strings result in non-sequential transducers (i.e. non-deterministic on their input [9]) with epsilon-transitions. This is the case, for example, of FST01 and FST02 in Table 1. As for the epsilon-transitions, there is an important difference between these two transducers. While the former has arcs labeled with epsilons on the surface side only *after* consumption of the input string, the latter also contains such arcs *before* the input string has been consumed. In the following, we will refer to these two types as restricted and unrestricted transducers. This terminology reflects the existence or absence of this constraint on the position of epsilon-transitions.

Transducers INFL12 and FST02 are unrestricted because they incorporate RULES1C and RULES2C in their compiling process. By contrast, FST01 results from directly compiling the mappings of full forms to lemmas and features. One interesting point to note about these two types of transducers is that a non-cyclic unrestricted transducer can be transformed into an equivalent restricted transducer by means of a simple four step procedure: (i) execute Foma's `lower-words` to generate the surface language of the transducer, (ii) apply the transducer to this language, (iii) convert all word-parse pairs into Foma's spaced-text format, and (iv) use Foma's `read spaced-text` to compile a new transducer. FST03 in Table 1 was constructed from FST02 through this procedure.

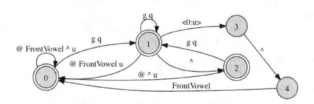

Fig. 1. Transducer modeling an orthographic alternation in the verb conjugation of Portuguese: insertion of **u** after stems ending in **g** or **q** before a front vowel

On the other hand, modeling an orthographic alternation such as *blogo* 'I blog' versus *bloguei* 'I blogged' by means of a conditional replacement rule, as exemplified in (1), introduces another complexity layer due to arcs labeled with the @ symbol in the transducer the rule is compiled into, as depicted in Fig. 1.[3] This special symbol, called ANY symbol in [1] and *other* symbol in [9], stands for any character not contained in the transducer's alphabet.

$$[..]->u||q|g\_"^"FrontVowel \tag{1}$$

---

[3] In the network in Fig. 1, the caret represents a morpheme boundary. FrontVowel is a placeholder for an actual transducer encoding non-posterior vowels in Portuguese.

While it is part of the definition of transducers like Fig. 1, @ is substituted for actual symbols of the alphabet when composing a lexicon transducer without it, modeling morphotactics, with one or more replacement rules. Therefore, INFL12 in Table 1 (as well as FST02) lacks arcs with this symbol, which would otherwise burden the parsing algorithm with an extra processing step.

To test the correctness of JMorpher's parsing algorithms, we used Foma's `lower-words` command to extract the lower projection of FST01, FST02, and FST03 in order to check JMorpher's output against the one produced by Foma's flookup utility. Since the relation between word forms and morphological analyses is often one-to-many, we eliminated the many repetitions in these three sets by means of the Unix sort and uniq utilities. The resulting sets corresponding to FST01 and FST02 will be referred to in the evaluation section as TESTSET1 and TESTSET2, respectively. As explained above, FST02 and FST03 are different, but equivalent transducers, encoding the same relation. To evaluate performance of the algorithm when applying FST01 and FST02 on real text, we chose the version of Mac-Morpho distributed with NLTK [2], from which we extracted our TESTSET3, composed of the first half million tokens of this corpus. This corpus suits our testing purposes because it is already tokenized and contractions are split up.[4] Table 2 shows the data sets used in the evaluation in section 4.

Table 2. Evaluation data

Test Set	Description	Size
TESTSET1	Lower projection of FST01	837755 word forms
TESTSET2	Lower projection of FST02	941454 word forms
MACMORPHO	Mac-Morpho converted to lower case	1170095 tokens
TESTSET3	Sample from MACMORPHO	500000 tokens

3   General Architecture

Since the main focus was morphological analysis in lesser-powered mobile devices with low RAM, limited storage capacity and limited processing abilities, JMorpher was tailored to obtain fast lookup times at low processing costs. A compromise had to be met concerning coverage. Therefore, JMorpher restricts itself to two subtypes of the possible transducers that integrate the repertoire of finite-state morphology, namely restricted and unrestricted transducers, as defined in the previous section. The special symbol @ is not supported. However, the experiments summed up in Table 1 suggest it is not necessary for a lookup algorithm dealing with final lexical transducers to process this symbol.

---

[4] Following [1], [4], [7], [9], and [12], among others, we do not consider tokenization as part of morphological parsing (or lexical analysis). Contractions could be straightforwardly handled by a lexical transducer. However, we prefer to split contractions during tokenization, adopting the solution proposed by [3]. Their approach interpolates statistical tagging into a two step tokenization. Therefore, input to JMorpher should be preprocessed by such a tokenizer.

JMorpher's core is the AbstractFST superclass with the two subclasses UnrestrictedFST and RestrictedFST, which model unrestricted and restricted transducers, respectively. The parent class provides an instance method buildFSTFromFile which converts a transducer specification in the AT&T raw text format into a Java data structure. This internal representation is drawn upon by the instance method parseWord of the child classes. This method takes a String parameter and returns a list of the parses for the word according to the transducer specification the object is instantiated from.

Distinguishing between the two subclasses UnrestrictedFST and RestrictededFST allows for faster processing of restricted transducers, since the parseWord method of the latter subclass only needs to check for epsilon-transitions after the input string has been consumed. By contrast, the parseWord method of the former class has to check for every input letter if there is also a path leading from an epsilon-transition to explore. As we will see in the next section, avoiding this additional processing step has a significant impact on performance. This is especially interesting in view of the fact that a non-cyclic unrestricted transducer can be converted into an equivalent restricted transducer by means of the simple procedure we sketched in the previous section.

It should be stressed that JMorpher is not a Java port of Foma's flookup, since this utility does not allow the user to choose between different parsing algorithms according to transducer type. The source code of Foma's flookup utility was not consulted. We developed our own algorithms from scratch in order to attain the same basic functionality as Foma's flookup, but limiting our scope to restricted and unrestricted transducers without the @ symbol.

Apart from the special symbol @, transducers compiled by Foma still pose some challenges for a parsing algorithm, due to non-sequentiality and epsilon-

Fig. 2. JMorher's Android app exemplified with an experimental transducer handling Internetese

transitons. Despite these complications, the parseWord methods of both subclasses process such transducers (the RestrictedFST class being limited to restricted transducers), handling non-sequentiality by means of recursion. More detailed evaluation results are given in the next section.

JMorpher was mainly designed to function as an API, providing morphological parsing e.g. in a text analysis pipeline. For demonstration purposes, however, a simple GUI was implemented which allows non-programmers to load a transducer file in the AT&T raw text format, type in some text and apply the transducer on the text (Fig. 2).[5]

4   Evaluation

In evaluating JMorpher, the following criteria were considered: a) correctness, b) lookup time, i.e. corpus processing time in terms of words per second, and c) load time, i.e. how long it takes to convert the transducer specification stored in a text file in the AT&T format into JMorpher's internal representation.

Since our tool provides two different parsing algorithms, we had to evaluate them separately. As the parseWord method of the RestricedFST class only applies to restricted transducers, the experiments carried out with the FST01, FST02, and FST03 transducers constitute these two scenarios: (i) FST01 and FST03 with the RestricedFST and UnrestricedFST classes, (ii) FST02 with the UnrestricedFST class.

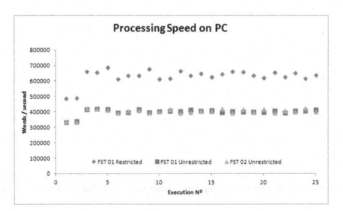

Fig. 3. Lookup times of the restricted and unrestricted parse methods on PC

To test correctness, we assumed Foma's flookup utility to generate the correct results when performing the experiments summarized in Table 3. In all of these experiments, JMorpher's output was identical to the one produced by Foma's

---

[5] The text typed in Fig. 2 is a made up example of Brazilian Portuguese Internetese. The transducer used is based on FST03 from Table 1, including about 100 extra items frequently used in informal computer-mediated discourse.

Table 3. Testing correctness of the transducer algorithms

Transducer	Test Set	Transducer Classes	Coverage
FST01	TESTSET1	RestrictedFST and UnrestrictedFST	100%
FST03	TESTSET2	RestrictedFST and UnrestrictedFST	100%
FST02	TESTSET2	UnrestrictedFST	100%
FST02	MACMORPHO	UnrestrictedFST	83.56%
FST03	MACMORPHO	RestrictedFST and UnrestrictedFST	83.56%

flookup utility, i.e. each word form received the same set of parses. It should be noted that these transducers are highly ambiguous.[6] A total of 40073 different word forms from MACMORPHO were analysed with FST02 and FST03. Each of these forms was assigned 1.8 parses on average.

Fig. 3 and Fig. 4 display corpus processing times of 25 executions obtained from Java SE on a PC and from Android running on a smartphone, respectively, by applying transducers FST01 and FST02 to TESTSET3. These experiments were carried out on a 64 bit Intel Core i7-4770 CPU clocked at 3.40 GHz with 8 GB RAM running Windows 7 Professional and on a Quad-core Krait 300 clocked at 1.7 GHz with 2 GB RAM running Android v4.1.2.

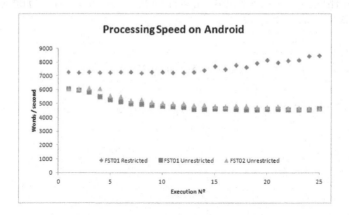

Fig. 4. Lookup times of the restricted and unrestricted parse methods on Android

The data in Fig. 3 show the parseWord methods of both the RestrictedFST and UnrestrictedFST class to be efficient. Lookup performance reaches 631193 words per second by applying the former method on TESTSET3. It is about

---

[6] This feature was inherited from DELAF_PB [13], see Table 1. The word *que* (roughly corresponding to *that* and *what*), for example, receives 14 different analyses. Many other similar examples could be mentioned. It is questionable if all these analyses are really necessary. In a future work, we will investigate ways to reduce the ambiguity in the morphological parsing of Portuguese.

57% faster than the latter method, which reaches an average of 400070 and 401229 words per second with the transducers FST01 and FST02, respectively. All methods were tested 25 times to derive these average numbers.

On the smartphone running Android, JMorpher's performance on the same data showed to be much more modest (see Fig. 4). It slowed down to an average of 7612 words per second when applying FST01 on TESTSET3 with the RestrictedFST class. As expected, performance is about 51% faster in this case than with the UnrestrictedFST class.

The last evaluation criterion is the time required to instantiate the RestrictedFST and UnrestrictedFST classes. On the PC, JMorpher needs just 60 milliseconds to create a transducer instance of FST01 or FST02. On the Android smartphone, instantiating these transducers takes longer, load time ranging between 4.5 and 5.5 seconds.

5   Conclusion

In this paper JMorpher, a morphological parsing tool intended to contribute to the computational processing of Portuguese on lesser-powered Android mobile devices was presented. JMorpher is implemented in Java and is totally portable. JMorpher is also capable of running on standard desktop computers benefitting from the additional performance present when using desktop hardware.

As far as we know, JMorpher is the first tool that performs morphological analysis on Android with lexical transducers generated by the free software, open source finite-state package Foma, avoiding the complications of porting this library to this platform. Therefore, JMorpher brings to this environment the advantages of finite-state transducers, which allow for very compact storage and fast processing of lexical resources. This is especially relevant in the light of the processing and memory limitations of mobile devices.

In order to achieve reasonably high lookup performance on Android devices with modest processing capacity, JMorpher provides two distinct parsing methods, tailored to two subtypes of the possible lexical transducers generated by Foma. These two types of transducers differ in respect to the position in the input string epsilon-transitions are allowed, restricted transducers containing these transitions only after consumption of the input string. By contrast, in unrestricted transducers, epsilon-transitions also occur before the input string has been consumed, demanding an extra processing step.

To test correctness and performance of JMorpher's parsing algorithms, large lexical transducers of Portuguese were constructed, including one which integrates conditional replacement rules modeling orthographic alternations in the verb conjugation. Experiments conducted with different test sets showed the implementation to produce the same output as the one generated by Foma's flookup utility.

In tests with Java SE on a desktop PC, JMorpher processed the first half million tokens of the Mac-Morpho corpus at a rate of more than 630000 words per second using a restricted lexical transducer based on the computational lexicon DELAF-PB from the Unitex-PB project. With the unrestricted parsing algorithm, JMorpher's lookup performance decreases about 36%.

Lookup performance strongly varies with the complexity of the lexical transducer, among other factors. Abstracting away from these factors, JMorpher compares well with HFST Optimized Lookup [6],[14]. The best performance of this tool was 408000 words per second on an 64 bit Intel Xeon E5450 CPU clocked at 3.00 GHz with 64 GB RAM, as reported in [14]. This speed was reached with a French transducer that, like FST01, results from a Lexc grammar based on a full-form lexicon. It should be noted, this transducer recognizes 34% fewer word forms than FST01.

Although JMorpher's lookup is slower running on an Android smartphone in comparison to the performance achieved on the desktop PC, the performance is still very good. Currently it would take less than one second to parse an average newspaper article or chapter from a novel, with the largest of the Portuguese transducers constructed for testing. It should also be noted that JMorpher works flawlessly on this platform without any code adaptations.

Finally, evaluation took into account the time it takes to create an instance of the two transducer classes implemented. On the desktop PC, loading a large transducer is also very fast, taking up only 60 milliseconds. On the Android smartphone, this process takes between 4.5 and 5.5 seconds.

JMorpher represents a contribution to the development of language-aware Android apps targeted at Portuguese speaking users of mobile devices, particularly in countries like Brazil where such devices are very expensive. Further research is needed to improve the performance on Android devices, another interesting direction of investigation would be to evaluate the tool with large lexical transducers for typologically diverse languages, systematically comparing performance across different finite-state tools.

Acknowledgments. This work was financially supported by Fundação Cearense de Pesquisa e Cultura and LG Electronics do Brasil Ltda. We are grateful to Miguel Franklin de Castro, Alexander Ewart, Ericson Macedo, José Wellington Silva, Italo Wanderley, Clayson Celes, and Lívio Freire for valuable suggestions and contributions. Thanks are also due to Diana Santos and the other two anonymous reviewers for their detailed comments. As usual, all remaining errors are our own responsibility.

References

1. Beesley, K.R., Karttunen, L.: Finite State Morphology. CSLI, Stanford (2003)
2. Bird, S., Klein, E., Loper, E.: Natural Language Processing with Python: Analyzing Text with the Natural Language Toolkit. O'Reilly, Sebastopol (2009)
3. Branco, A., Silva, J.: Evaluating Solutions for the Rapid Development of State-of-the-Art POS Taggers for Portuguese. In: Lino, M.T., Xavier, M.F., Ferreira, F., Costa, R., Silva, R. (eds.) Proceedings of the 4th International Conference on Language Resources and Evaluation (LREC 2004), pp. 507–510. ELRA, Paris (2004)

4. Dale, R.: Classical Approaches to Natural Language Processing. In: Indurkhya, N., Damerau, F.J. (eds.) Handbook of Natural Language Processing, 2nd edn., pp. 3–7. Chapman & Hall/CRC, Boca Raton (2009)
5. Fradin, B.: Abbréviation des gloses morphologiques. Laboratoire de Linguistique Formelle, Paris (2013), http://www.llf.cnrs.fr/gloses-fr.php
6. Hardwick, S.: HFST: Optimized Lookup Format (2009), https://kitwiki.csc.fi/twiki/bin/view/KitWiki/HfstOptimizedLookupFormat
7. Hippisley, A.: Lexical Analysis. In: Indurkhya, N., Damerau, F.J. (eds.) Handbook of Natural Language Processing, 2nd edn., pp. 31–58. Chapman & Hall/CRC, Boca Raton (2009)
8. Hulden, M.: Foma: A Finite-State Compiler and Library. In: EACL (Demos), pp. 29–32 (2009)
9. Jurafsky, D., Martin, J.H.: Speech and Language Processing: An Introduction to Natural Language Processing, Computational Linguistics, and Speech Recognition. Pearson, London (2009)
10. Leipzig Glossing Rules, http://www.eva.mpg.de/lingua/resources/glossing-rules.php
11. Michaelis Moderno Dicionário da Língua Portuguesa. Melhoramentos, São Paulo (2009), http://michaelis.uol.com.br/moderno/portugues/index.php
12. Lindén, K., Silfverberg, M., Pirinen, T.: HFST Tools for Morphology: An Efficient Open-Source Package for Construction of Morphological Analyzers. In: Mahlow, C., Piotrowski, M. (eds.) SFCM 2009. CCIS, vol. 41, pp. 28–47. Springer, Heidelberg (2009)
13. Muniz, M.C.M.: Projeto Unitex-PB. NILC, São Paulo (2004), http://www.nilc.icmc.usp.br/nilc/projects/unitex-pb/web/
14. Silfverberg, M., Lindén, K.: HFST Runtime Format: A Compacted Transducer Format Allowing for Fast Lookup. In: FSMNLP (2009), http://www.ling.helsinki.fi/klinden/pubs/fsmnlp2009runtime.pdf
15. Xerox: Linguistic Tools: Morphological Analysis. Morphology, http://open.xerox.com/Services/fst-nlp-tools/Pages/

Tagging and Labelling Portuguese Modal Verbs

Paulo Quaresma[1,4], Amália Mendes[2], Iris Hendrickx[2,3], and Teresa Gonçalves[1]

[1] Department of Informatics, University of Évora, Portugal
[2] Center for Linguistics of the University of Lisbon, Portugal
[3] Center for Language Studies, Radboud University Nijmegen, The Netherlands
[4] L2F – Spoken Language Systems Laboratory, INESC-ID, Portugal

Abstract. We present in this paper an experiment in automatically tagging a set of Portuguese modal verbs with modal information. Modality is the expression of the speaker's (or the subject's) attitude towards the content of the sentences and may be marked with lexical clues such as verbs, adverbs, adjectives, but also by mood and tense. Here we focus exclusively on 9 verbal clues that are frequent in Portuguese and that may have more than one modal meaning. We use as our gold data set a corpus of 160.000 tokens manually annotated, according to a modality annotation scheme for Portuguese. We apply a machine learning approach to predict the modal meaning of a verb in context. This modality tagger takes into consideration all the features available from the parsed data (pos, syntactic and semantic). The results show that the tagger improved the baseline for all verbs, and reached macro-average F-measures between 35 and 81% depending on the modal verb and on the modal value.

1 Introduction

Nowadays we observe a growing interest in text mining applications that can automatically detect opinions, facts and sentiments in texts. Many of the current opinion or sentiment mining applications use a crude division between negative, neutral and positive sentiments. Modality (defined as the speaker's attitude towards the proposition in the text [19]) that has been studied for many decades in Linguistics, offers a theoretical framework to make a rather fine-grained distinction between different attitudes. For example, the speaker expresses his or someone else's commitment, hope, belief or knowledge about a certain proposition. We believe that such more detailed tagging of modal information will be helpful to improve opinion mining applications. Furthermore, a systematically tagged corpus with modal information will also be a rich resource for linguists who are interested in modality and how this influences the meaning of a text.

Modality can be expressed by different grammatical categories like modal adjectives, adverbs, nouns and verbs which we will denote as *modal trigger* in the rest of this paper. Here we focus on modality tagging for the Portuguese language and we concentrate on high frequent modal verbs. For our experiments we deliberately selected the highly ambiguous verbs that can express multiple modal meanings like for example the Portuguese verb *poder* that can express

J. Baptista et al. (Eds.): PROPOR 2014, LNAI 8775, pp. 70–81, 2014.

"to be possible", "to be able" or "to give permission". The modality tagger that we developed has two objectives: the identification of modal triggers and the attribution of a modal value to this trigger.

Constructing an automatic modality tagger requires a data set with labelled examples to train and evaluate the tagger. As we are currently in need of a suitable data set, one of the goals of the current experiments is to develop an automatic modality tagger on a small manually labelled sample that can later be applied (semi-automatically) to generate a larger data set.

In this paper, we restrict our experiment to 9 modal verbs taken from a small sample of 160K tokens, manually annotated with a modality scheme for Portuguese [11]. The selected verbs are ambiguous and have at least two modal meanings. This polysemy increases the level of difficulty of the automatic annotation task. To create the modality tagger, we first automatically assign lemmas, POS and syntactic tags, we then automatically identify modal triggers and apply a machine learning approach to attribute a modal value to the triggers, comparing the results with our gold (manually annotated) labelling.

Next we first discuss related work on modality annotation and tagging in section 2. In section 3 we describe the modality scheme and manually annotated dataset for Portuguese that we used in the experiments. The modality tagging approach is presented in section 4 and the results of automatic attribution of modal value in 5. We conclude in section 6.

2   Related Work

The annotation schemes covering modality differ greatly in their objectives and in the nature of the concepts that are labelled. Modality may be one aspect of the semantic information encoded in the properties of events, such as Time and Condition (Matsuyoshi et al. [14], Baker et al. [2] and Nirenburg and McShane [17]). Modal values may also be included in annotation schemes that cover both factuality and modality, as in Sauri et al. [24], or that are concerned with subjectivity, beliefs and opinions. Contrary to many of these approaches, Rubinstein et al. [22] use a restricted notion of modality and establish conditions for an expression to be considered modal, such as the requirement for a propositional argument.

The options regarding what textual elements to annotate also differ: the modal value may be attributed globally to the sentence/event or it can be encoded on specific triggers (and other components such as target and source). The nature of the data over which to apply the annotation also varies: some work has been developed specifically on spoken data, as the scheme described for Brazilian Portuguese in Avila and Melo (2013) [26], while modality has enriched the semantic annotation of events in biomedical texts. For more contrastive information on the existing annotation schemes, see an overview in Nissim et al. [18].

Some of these modality schemes have led to experiments in the automatic annotation of modal values, mainly for English. For instance, in BioNLP, Miwa et al. [16] annotated pre-recognised events with the epistemic value "level of certainty" and attain F-measures of 74,9 for "low confidence" and 66,5 for "high but not complete confidence". Another system is described in Battistelli et al. [3] and plans to attribute enunciative and modal features (E_M) to textual segments. Modal triggers are to be identified through semantic clues and a syntactic parser will be used to calculate the length of each segment. The design of the system is well described but has not been fully implemented yet. The modal scheme presented in [24] focuses on factuality and has been applied to an experiment in the automatic identification of events in text, together with the characterization of their modality features. The authors report results with accuracy values of 97.04 with the EviTA tool. Also, a specific task for detecting uncertainty through the use of hedging clues was organized at CoNLL2010 [7].

Baker et al. [2] go a little further than our own experiment in this paper: they identify both the modal trigger and its target and report results of 86% precision with two rule-based modality taggers over a standard LDC data set. The approach of Diab et al. [6] covers modality but is essentially geared towards the identification of belief. Contrary to our own approach, the authors do not take into consideration the polysemy of the auxiliary verbs and encode a single value (epistemic), although they do report that the verbs may be deontic in some contexts. This experiment has been extended to other modality values (Ability, Effort, Intention, Success and Want), as reported in Prabhakaran et al [20]. The authors applied their system to their Gold data (containing emails, sentences from newswire, letters and blogs) and on a corpus of emails annotated on MTurk. The two experiments attained quite different overall measures: 79.1 F-measure for the MTurk experiment and 41.9 on the Gold data, that the authors attribute to the difference in the corpora design and to the higher complexity of the experts annotations.

Our objective to identify and label Portuguese modal verbs with modal values is closely related to the work of Ruppenhofer and Rehbein [23] on English modals. The five English modal verbs (*can/could, may/might, must, ought, shall/should*) were first identified in texts and their modal value was predicted by training a maximum entropy classifier on features extracted from the training set. The experiment achieved an improvement of the baseline for all verbs but *must*, and accuracy numbers between 68.7 and 93.5.

The diversity of the concept of modality and of the textual scope of the annotation makes it difficult to fully contrast the results described in the experiments, since no standard has yet emerged in this domain.

3   Data Set

Our experiment in the automatic annotation of modality is based on the annotation scheme for Portuguese presented in [12] and [11]. The annotation is based on the identification of lexical clues, called triggers, which include verbs, but also other POS classes, such as nouns, adjectives and adverbs. Each modal trigger is labelled with a modal value.

The scheme covers seven main modal values (epistemic, deontic, participant-internal, volition, evaluation, effort and success) and several sub-values. The first

five modal values follow linguistic typologies ([19][1]), while the last two are influenced by typologies used in annotation schemes for modality (e.g. [2]). Epistemic modality is the more complex value (and the more stable across typologies)
and is subdivided into five sub-values: knowledge, belief, doubt, possibility and
interrogative. Evidential modality (i.e., belief supported by evidence) is not considered an independent value and is rather annotated as epistemic belief. Deontic
modality is divided into obligation and permission and includes values that are
described as participant-external modality in van der Auwera and Plungian [1]
(obligation and possibility that are not dependent on the participants but rather
on external conditions that make something required or possible). However, the
scheme does include what these authors describe as Participant-internal modality and its two subvalues: necessity and capacity (an internal capacity or an
internal necessity of the participant, usually the subject). Finally, four other values are included: evaluation (of the proposition), volition (desires, wishes and
fears), effort and success.

The annotation scheme components comprise the trigger (which is the lexical clue conveying the modal value), its target, the source of the event mention
(speaker or writer) and the source of the modality (for instance, the entity that
considers the proposition to be possible, that establishes an obligation or that
has an internal capacity to do something). The source of the event mention and
the source of the modality are in many cases the same entity: the speaker/writer
produces a discourse/text unit where it states its belief or doubt or the possibility that something may happen. However, they may also be different entities
when the text presents the views of someone else than its producer. The trigger
receives an attribute *modal value*, while both trigger and target are marked for
polarity. Modal verbs may have more than one meaning and it is sometimes difficult to distinguish between those modal values, even when the annotator takes
into consideration a larger context. To address this issue, the scheme includes the
Ambiguity component, where the annotators can write down secondary meanings that would also be available in that specific context. This annotation scheme
was further enriched with Focus information to address the interaction betwen
exclusive adverbs and modal triggers [15], although we will not take these components into consideration in our current experiment.

The verbs that are the subject of our experiment are good examples of the
polysemy of modal verbs. Two of them may also have non modal meanings,
as it is the case with *dever* and *poder*. We give in (1) an example of a non
modal use of *dever* and in (2) and (3) examples of modal uses of the verb. In
(2), *dever* has an epistemic reading, stating that the proposition is probable,
while in (3), it has a Deontic obligation reading (the adverb *obrigatoriamente*
'obligatorily' is also a modal trigger). In (3) we also provide the total description
of the components of the sentence annotation. The obligation has scope over the
whole proposition [devem [faltas sucessivas aos julgamentos ser contrariadas]]
and there is no overt source of the modality (in these cases the verb itself is

marked as the source)[1]. Cases marked as ambiguous are illustrated in (4): the context is ambiguous between an Epistemic possibility and a Deontic obligation reading (it is probable vs it is required that the money be spent in such a way). The annotator selects what he considers to be the primary value and marks the ambiguity in the Ambiguity component.

(1) O Governo *deve* explicações muito mais claras e completas do que as que deu (...).
'The Government owes much more clear and complete explanations than the ones that it gave'

(2) Entre copos de vinho, muitos cigarros e piadas trocadas com os músicos, Chico Buarque de Hollanda ensaia o seu próximo "show", que *deve* estrear em Janeiro.
'Among wine, cigarettes and jokes with the musicians, Chico Buarque de Hollanda rehearses his next show, which should start in January.'

(3) Faltas sucessivas aos julgamentos contribuem para a morosidade da justiça e, obrigatoriamente, *devem* ser contrariadas.
'Repeated absence to trials contribute to the slowdown of justice and must obligatorily be opposed.'
Trigger: devem
    Modal value: deontic_obligation
    Polarity: positive
Target: Faltas sucessivas@ser contrariadas
Source of the modality: devem
Source of the event: writer
Ambiguity: none

(4) O seu orçamento global é de 12,5 milhões de euros, sendo que uma grande maquia deverá ser aplicada na construção de infraestruturas rodoviárias.
'Its global funding is 12,5 million euros, and a large part (might/has to) be applied in the construction of road infrastructures.'
Trigger: deverá
    Modal value: Epistemic_possibility
    Polarity: positive
Target: uma grande maquia@ser aplicada na construção de infraestruturas rodoviárias
Source of the modality: deverá
Source of the event: writer
Ambiguity: Deontic_obligation

This annotation scheme was applied to a corpus sample of 2000 sentences extracted from the written subpart of the Reference Corpus of Contemporary Portuguese (CRPC)[9]. The sentences were selected on the basis of a list of

---

[1] Notice that the discontinuity of the target is marked with the symbol @ in this example, but is encoded in XML in the data set.

potential modal verbs mentioned in the literature. In the annotation however, all modal triggers including nouns, adjectives and adverbs were annotated. In total the data set contains approximately 3200 modal triggers.

In this experiment however, we only evaluate on a set of highly ambiguous verbs that can have multiple modal values. Many of the verbs in the data set, such as high frequent modal verbs like *querer 'to want'* and *tentar 'to try'*, only have one modal value and for those assigning the correct modal value becomes a trivial task. We aim to only study the hard cases that involve true ambiguous verbs. We focus on modal verbs with multiple modal meanings that each occur at least 5 times in the small annotated corpus sample. Only a handful of verbs met this criteria, giving us a list of 9 verbs to work with. In Table 1 we show the 9 verbs and their distribution between the different main modal values. We see that all different modal values of the data set are covered but most verbs only have two different modal values resulting in a sparse matrix. Two of the verbs (*poder* and *dever*) are polysemous verbs that can be used as a semi-auxiliary verb where it has a modal meaning or as a main verb without a modal meaning.

The current experiment is a follow-up of our experience with tagging 3 Portuguese verbs (*poder, dever* and *conseguir*) with modality values, as reported in [21]. Here, we extend our work, so as to cover a larger list of verbs and provide a better testing set for our system.

Table 1. Data set used in these experiments. Abbreviations used: non(non-modal), EB (epistemic belief), EP (epistemic possibility), EF (effort), EV (evaluation), OB (deontic obligation), NE (participant-internal necessity), PC (participant-internal capacity), PE (deontic permission), SE (success) and VO (volition)

verb	English	total	non	EB	EP	EF	EV	OB	NE	PC	PE	SE	VO
arriscar	risk	44			25	19							
aspirar	aspire	50		19									31
considerar	consider	29		18			11						
conseguir	succeed	84								41		43	
dever	must/might	120	12		37			71					
esperar	hope	52		26									26
permitir	allow	78			60						18		
poder	can/may/be able	258	22		154					40	42		
precisar	need	54						9	45				
total		769	34	63	276	19	11	80	45	81	60	43	57

4   Modality Tagging

The modality tagging is done in three different steps. First we preprocess the data set with a syntactic parser. Next, the parser output is used to detect the modal triggers. And finally we label each modal trigger with the appropriate modal value in context.

We used the PALAVRAS parser [5] for the syntactic analysis, and rewrote the result to an XML format with logical terms using the tool Xtractor [8]. We use a list of the selected verbs described in section 3 to detect the modal triggers. Some of the verbs can be used as modal or non-modal like *dever* and *poder*. The parser labels these verbs as auxiliary verbs when they are used as modals and we exploit these parser predictions to detect the modal usage.

For the final step of labelling the modal triggers with their appropriate modal value we apply a machine learning approach. We adopted the "word expert" strategy that is often used in automatic word sense disambiguation: we train a specialised classifier for each verb. This strategy is ideal for classification tasks where every verb has a different type of meanings or modal values and a different distribution in usage [4].

After experimenting with several machine learning algorithms with Weka [10], we achieved the best results with SVM, Support Vector Machines [25] using the subsequence string kernel [13]. We performed two experiments with different feature representations (using default algorithm values). In the first one we only use the words in the original sentences and in the second experiment we used the POS tags and functional and syntactic information extracted from the sentence's parse tree, in a window of 70 characters around the verb. For the evaluation we used a 10-fold stratified cross-validation procedure. Remark that this is a challenging task as we only have a few examples for every modal verb word expert to train and test. In the next section we present the results of our experiments.

5   Experiments: Attribution of the Modal Value

In the context of this work, and as presented in table 1, we analysed 9 modal verbs and created automatic classifiers able to attribute a modal value to each occurrence of these verbs in sentences written in the Portuguese language.

Occurrences of these 9 verbs were automatically detected from the analysis of the output of the syntactical parser PALAVRAS [5], which associates a lemma to each word in a sentence. As verbs *poder* and *dever* may occur with non-modal uses, it was necessary, for these verbs, to distinguish the two kind of uses (modal and non-modal). We used the sentences' parse trees to identify the semi-auxiliary situations, which are modal. Based on the parsers' output we were able to detect the modal usage of *poder* and *dever* with an F-score of 98%.

To identify the modal value, we applied a machine learning approach to the sentences with occcurrences of the 9 modal verbs analysed in this work. Our system takes into consideration all the features available from the PALAVRAS output: lemma and POS of the trigger, left and right syntactic context, and semantic features: predicate argument structure, [±human] nature of arguments. The results for both experiments (using the word sentences and a text linearized format of the parse tree within a window around the verb) are presented in Table 2. We give results for a baseline and for both experiments (sentences and window parse tree), computing precision (P), recall (R) and F-value (F) and the macro-average over the different modal values. As baseline we used a system that always assigns the most frequent modal value for each verb.

As summary we can refer the following points:

- Verb *arriscar*: the classifier which uses as input the partial parse tree was able to improve the baseline F-measure from 36.2 to 63.0.
- Verb *aspirar*: the classifier which uses as input the partial parse tree was able to improve the baseline F-measure from 38.3 to 81.1.
- Verb *conseguir*: the classifier which uses as input the partial parse tree was able to improve the baseline F-measure from 33.9 to 76.2.
- Verb *considerar*: the classifier which uses as input the sentences was able to improve the baseline F-measure from 38.3 to 41.6.
- Verb *dever*: the classifier which uses as input the partial parse tree was able to improve the baseline F-measure from 39.7 to 63.8.
- Verb *esperar*: the classifier which uses as input the sentences was able to improve the baseline F-measure from 33.3 to 53.6.
- Verb *permitir*: the classifier which uses as input the partial parse tree was able to improve the baseline F-measure from 43.5 to 55.8.
- Verb *poder*: the classifier which uses as input the sentences was able to improve the baseline F-measure from 33.3 to 35.3.
- Verb *precisar*: the classifier which uses as input the sentences was able to improve the baseline F-measure from 45.7 to 62.5.

So, our system was able to clearly improve the F-measure baseline approach for all 9 modal verbs, with more than 40 points improvement for the verbs *aspirar* and *conseguir*, more than 20 points for the verb *arriscar*, and more than 10 points for the verbs *dever*, *esperar*, *permitir*, and *precisar*. The values for the verb *poder* are lower, probably because this verb has three modal meanings while the other verbs have two.

We also observed that for 5 modal verbs the partial parse tree input was able to produce a better classifier and for the other 4 verbs the list of words of the sentences were enough to obtain the best classifier.

We plan to follow up this analysis with a detailed study identifying the individual role of the syntactic and semantic features that are used for the automatic attribution of the modal value in our system. More specifically, for each parse tree we intend to evaluate the relevance of the following features in the modal value classification: word, lemma, POS tag, syntactic tag, semantic information, role label. The analysis will be performed over the partial parse trees that include the modal verbs and their parents and grand-parents and also over the nodes in the path from the root of the sentence parse trees to the modal verbs. We will also make a comparative study of the relevant features for each of the studied modal verbs. In fact, from the analysis of the already obtained results, we foresee the existence of great differences between the 9 modal verbs. The fact that, for some verbs, the parse tree input obtains better results and for others it is better to use the list of words of the sentences, suggests that syntactic and semantics features might not be equally relevant for all verbs.

Table 2. Results of all verbs

arriscar		baseline			sentences			tree		
		P	R	F	P	R	F	P	R	F
effort	19	0	0	0	47.4	47.4	47.4	57.9	57.9	57.9
possibility	25	56.8	100	72.5	60.0	60.0	60.0	68.0	68.0	68.0
average	44	28.4	50.0	36.2	53.7	53.7	53.7	63.0	63.0	**63.0**

aspirar		baseline			sentences			tree		
		P	R	F	P	R	F	P	R	F
volition	31	62.0	100	76.5	65.7	74.2	69.7	86.7	83.9	85.2
belief	19	0	0	0	46.7	36.8	41.2	75.0	78.9	76.9
average	50	31.0	50.0	38.3	56.2	55.5	55.5	80.9	81.4	**81.1**

conseguir		baseline			sentences			tree		
		P	R	F	P	R	F	P	R	F
capacity	41	0	0	0	57.1	48.8	52.6	76.9	73.2	75.0
success	43	51.2	100	67.7	57.1	65.1	60.9	75.6	79.1	77.3
average	84	25.6	50.0	33.9	57.1	57.0	56.8	76.3	76.2	**76.2**

considerar		baseline			sentences			tree		
		P	R	F	P	R	F	P	R	F
belief	18	62.1	100	76.6	60.0	83.3	69.8	55.0	61.1	57.9
evaluation	11	0	0	0	25.0	9.1	13.3	22.2	18.2	20.0
average	29	31.0	50.0	38.3	42.5	46.2	**41.6**	38.6	39.7	39.0

dever		baseline			sentences			tree		
		P	R	F	P	R	F	P	R	F
obligation	71	65.7	100	79.3	74.0	80.3	77.0	75.0	76.1	75.5
possibility	37	0	0	0	54.8	45.9	50.0	52.8	51.4	52.1
average	108	32.9	50.0	39.7	64.4	63.1	63.5	63.9	63.8	**63.8**

esperar		baseline			sentences			tree		
		P	R	F	P	R	F	P	R	F
belief	26	50.0	100	66.7	57.1	46.2	51.1	54.5	46.2	50.0
volition	26	0	0	0	54.8	65.4	59.6	53.3	61.5	57.1
average	52	25.0	50.0	33.3	56.0	55.8	**55.4**	53.9	53.9	53.6

permitir		baseline			sentences			tree		
		P	R	F	P	R	F	P	R	F
possibility	60	76.9	100	87.0	75.7	88.3	81.5	79.4	83.3	81.3
permission	18	0	0	0	12.5	5.6	7.7	33.3	27.8	30.3
average	78	38.5	50.0	43.5	44.1	47.0	44.6	56.4	55.6	**55.8**

poder		baseline			sentences		tree				
		P	R	F	P	R	F	P	R	F
permission	42	0	0	0	33.3	9.5	14.8	12.5	11.9	12.2
possibility	154	65.3	100	79.0	65.8	83.8	73.7	63.5	68.8	66.0
capacity	40	0	0	0	17.9	12.5	14.7	24.1	17.5	20.3
average	236	21.8	33.3	26.3	39.0	**35.3**	34.4	33.4	32.7	32.8

precisar		baseline			sentences			tree		
		P	R	F	P	R	F	P	R	F
necessity	45	84.0	100	91.3	86.3	97.8	91.7	83.3	88.9	86.0
obligation	9	0	0	0	66.7	22.2	33.3	16.7	11.1	13.3
average	54	42.0	50.0	45.7	76.5	60.0	**62.5**	50.0	50.0	49.7

6   Conclusion

We have presented our experiment to automatically annotate modality for 9 Portuguese modal verbs, using a corpus sample of 160K tokens, manually tagged with modal values. For this experiment, we selected verbs that had more than one modal meaning, which occurred at least 5 times in the corpus. We used SVM and performed two experiments: one uses the original sentences and another uses the information available from the parser's output. The results of the attribution of modal value improved the baseline for all verbs, with macro-average F-measures between 35.3 and 81.1%, depending on the modal verb and on the modal value. Considering that our training corpus was relatively small and that we selected challenging verbs in our experiment, we believe that our goal, of creating a larger corpus with modal information by a (semi) automatic tagging process, could lead to positive results in the future. We plan to study the role played by each feature in our system and to observe in more detail the reason why some verbs reach higher scores with the experiment that uses sentences instead of the parse tree. We also aim to compute a learning curve to estimate the amount of manually annotated examples that are needed to get a good performance from the modality tagger. Also, for our system to be able to label new verbs that didn't occur in the initial data set, we plan to train a general modal trigger classifier that is not dependent on the verb itself.

Acknowledgements. This work was partially supported by national funds through FCT - Fundação para a Ciência e Tecnologia, under project Pest-OE/EEI/LA0021/2013 and project PEst-OE/LIN/UI0214/2013.

References

1. der Auwera, J.V., Plungian, V.A.: Modality's semantic map. Linguistic Typology 1(2), 79–124 (1998)
2. Baker, K., Bloodgood, M., Dorr, B., Filardo, N.W., Levin, L., Piatko, C.: A modality lexicon and its use in automatic tagging. In: Chair, N.C.C., Choukri, K., Maegaard, B., Mariani, J., Odijk, J., Piperidis, S., Rosner, M., Tapias, D. (eds.) Proceedings of the Seventh International Conference on Language Resources and Evaluation (LREC 2010). European Language Resources Association (ELRA), Valletta (2010)
3. Battistelli, D., Damiani, M.: Analyzing modal and enunciative discursive heterogeneity: How to combine semantic resources and a syntactic parser analysis. In: Proceedings of the IWCS 2013 Workshop on Annotation of Modal Meanings in Natural Language (WAMM), pp. 7–15. Association for Computational Linguistics, Potsdam (2013)
4. Berleant, D.: Engineering "word experts" for word disambiguation. Natural Language Engineering 1(4), 339–362 (1995)
5. Bick, E.: The parsing system PALAVRAS. Aarhus University Press (1999)
6. Diab, M.T., Levin, L.S., Mitamura, T., Rambow, O., Prabhakaran, V., Guo, W.: Committed belief annotation and tagging. In: Third Linguistic Annotation Workshop, pp. 68–73. The Association for Computer Linguistics, Singapore (2009)

7. Farkas, R., Vincze, V., Móra, G., Csirik, J., Szarvas, G.: The conll-2010 shared task: Learning to detect hedges and their scope in natural language text. In: Proceedings of the Fourteenth Conference on Computational Natural Language Learning, pp. 1–12. Association for Computational Linguistics, Uppsala (2010)

8. Gasperin, C., Vieira, R., Goulart, R., Quaresma, P.: Extracting xml syntactic chunks from portuguese corpora. In: Proc. of the TALN Workshop on Natural Language Processing of Minority Languages and Small Languages, pp. 223–232 (2003)

9. Généreux, M., Hendrickx, I., Mendes, A.: Introducing the reference corpus of contemporary portuguese online. In: Calzolari, N., Choukri, K., Declerck, T., Dogan, M.U., Maegaard, B., Mariani, J., Odijk, J., Piperidis, S. (eds.) LREC 2012 – Eighth International Conference on Language Resources and Evaluation, pp. 2237–2244. European Language Resources Association (ELRA), Istanbul (2012)

10. Hall, M., Frank, E., Holmes, G., Pfahringer, B., Reutemann, P., Witten, I.H.: The weka data mining software: An update. SIGKDD Explor. Newsl. 11(1), 10–18 (2009)

11. Hendrickx, I., Mendes, A., Mencarelli, S.: Modality in text: A proposal for corpus annotation. In: Calzolari, N., Choukri, K., Declerck, T., Dogan, M.U., Maegaard, B., Mariani, J., Odijk, J., Piperidis, S. (eds.) LREC 2012 – Eighth International Conference on Language Resources and Evaluation, pp. 1805–1812. European Language Resources Association (ELRA), Istanbul (2012)

12. Hendrickx, I., Mendes, A., Mencarelli, S., Salgueiro, A.: Modality Annotation Manual, vol. 1. Centro de Linguística da Universidade de Lisboa, Lisboa

13. Lodhi, H., Saunders, C., Shawe-Taylor, J., Cristianini, N., Watkins, C.: Text classification using string kernels. Journal of Machine Learning Research 2, 419–444 (2002)

14. Matsuyoshi, S., Eguchi, M., Sao, C., Murakami, K., Inui, K., Matsumoto, Y.: Annotating event mentions in text with modality, focus, and source information. In: Chair, N.C.C., Choukri, K., Maegaard, B., Mariani, J., Odijk, J., Piperidis, S., Rosner, M., Tapias, D. (eds.) Proceedings of the Seventh International Conference on Language Resources and Evaluation (LREC 2010). European Language Resources Association (ELRA), Valletta (2010)

15. Mendes, A., Hendrickx, I., Salgueiro, A., Ávila, L.: Annotating the interaction between focus and modality: The case of exclusive particles. In: Proceedings of the 7th Linguistic Annotation Workshop and Interoperability with Discourse, pp. 228–237. Association for Computational Linguistics, Sofia (2013)

16. Miwa, M., Thompson, P., McNaught, J., Kell, D.B., Ananiadou, S.: Extracting semantically enriched events from biomedical literature. BMC Bioinformatics 13, 108 (2012)

17. Nirenburg, S., McShane, M.: Annotating modality. Tech. rep. University of Maryland, Baltimore County (2008)

18. Nissim, M., Pietrandrea, P., Sanso, A., Mauri, C.: Cross-linguistic annotation of modality: A data-driven hierarchical model. In: Proceedings of IWCS 2013 WAMM Workshop on the Annotation of Modal Meaning in Natural Language, pp. 7–14. Association for Computational Linguistics, Postam (2013)

19. Palmer, F.R.: Mood and Modality. Cambridge textbooks in linguistics. Cambridge University Press (1986)

20. Prabhakaran, V., Bloodgood, M., Diab, M., Dorr, B., Levin, L., Piatko, C.D., Rambow, O., Van Durme, B.: Statistical modality tagging from rule-based annotations and crowdsourcing. In: Proceedings of the Workshop on Extra-Propositional Aspects of Meaning in Computational Linguistics, ExProM 2012, pp. 57–64. Association for Computational Linguistics, Stroudsburg (2012)

21. Quaresma, P., Mendes, A., Hendrickx, I., Gonalves, T.: Automatic tagging of modality: Identifying triggers and modal values. In: Proceedings of ISA-10 - 10th Joint ACL-ISO Workshop on Interoperable Semantic Annotation. European Language Resources Association (ELRA), Reykjavik (2014)

22. Rubinstein, A., Harner, H., Krawczyk, E., Simonson, D., Katz, G., Portner, P.: Toward fine-grained annotation of modality in text. In: Proceedings of the Tenth International Conference for Computational Semantics, IWCS 2013 (2013)

23. Ruppenhofer, J., Rehbein, I.: Yes we can!? annotating english modal verbs. In: Chair, N.C.C., Choukri, K., Declerck, T., Doğan, M.U., Maegaard, B., Mariani, J., Odijk, J. (eds.) Proceedings of the Eight International Conference on Language Resources and Evaluation (LREC 2012). European Language Resources Association (ELRA), Istanbul (2012)

24. Sauri, R., Verhagen, M., Pustejovsky, J.: Annotating and recognizing event modality in text. In: FLAIRS Conference, pp. 333–339 (2006)

25. Vapnik, V.N.: Statistical Learning Theory. Wiley-Interscience (1998)

26. Ávila, L., Melo, H.: Challenges in modality annotation in a brazilian portuguese spontaneous speech corpus. In: Proceedings of IWCS 2013 WAMM Workshop on the Annotation of Modal Meaning in Natural Language. Association for Computational Linguistics, Potsdam (2013)

Training State-of-the-Art Portuguese POS Taggers without Handcrafted Features

Cícero Nogueira dos Santos and Bianca Zadrozny

IBM Research - Brazil
Av. Pasteur, 138/146 Botafogo, Rio de Janeiro, 22296-903, Brazil
{cicerons,biancaz}@br.ibm.com

Abstract. Part-of-speech (POS) tagging for morphologically rich languages normally requires the use of handcrafted features that encapsulate clues about the language's morphology. In this work, we tackle Portuguese POS tagging using a deep neural network that employs a convolutional layer to learn character-level representation of words. We apply the network to three different corpora: the original Mac-Morpho corpus; a revised version of the Mac-Morpho corpus; and the Tycho Brahe corpus. Using the proposed approach, while avoiding the use of any handcrafted feature, we produce state-of-the-art POS taggers for the three corpora: 97.47% accuracy on the Mac-Morpho corpus; 97.31% accuracy on the revised Mac-Morpho corpus; and 97.17% accuracy on the Tycho Brahe corpus. These results represent an error reduction of 12.2%, 23.6% and 15.8%, respectively, on the best previous known result for each corpus.

Keywords: Portuguese Part-of-Speech Tagging, Deep Learning, Convolutional Neural Networks.

1 Introduction

Part-of-speech tagging consists in labeling each word in a text with a unique POS tag, e.g. noun, verb, pronoun and preposition. POS tagging for morphologically rich languages normally requires the use of handcrafted features that encapsulate clues about the language's morphology. For Portuguese POS tagging, features like suffixes, prefixes and capitalization have been used in previous work [1,2,3,4]. Some recent approaches for POS tagging have been using distributed word representations [5,6], also known as word embeddings, which are normally learned using neural networks and capture syntactic and semantic information about words. However, with few exceptions [7,8,9], work on learning of representations for natural language processing (NLP) has focused exclusively on the word level and does not capture morphological and word shape information.

In this paper we approach Portuguese POS tagging using CharWNN, a recently proposed deep neural network (DNN) architecture that jointly uses word-level and character-level representations [9]. CharWNN employs a convolutional layer that allows effective feature extraction from words of any size. At tagging time, the convolutional layer generates character-level embeddings for each

J. Baptista et al. (Eds.): PROPOR 2014, LNAI 8775, pp. 82–93, 2014.

word, even for the ones that are outside the vocabulary. We perform experiments that show the effectiveness of CharWNN for Portuguese POS tagging. Using the same set of hyper-parameters, our approach achieves state-of-the-art results for three different corpora. Additionally, in our experiments we provide information about the impact of character-level representations for POS tagging of unknown words and show evidence of CharWNN's effectiveness to learn features that carry morphological information.

This work is organized as follows. In Section 2, we briefly describe the deep neural network architecture. In Section 3, we discuss some related work. Section 4 details our experimental setup and results. Section 5 presents our final remarks.

2   CharWNN Architecture

CharWNN extends Collobert et al.'s [10] neural network architecture by adding a convolutional layer to extract character-level representations [9]. Given a sentence, the network gives for each word a score for each tag $\tau \in T$. In order to score a word, the network takes as input a fixed-sized window of words centralized in the target word. The input is passed through a sequence of layers where features with increasing levels of complexity are extracted. The output for the whole sentence is then processed using the Viterbi algorithm to perform structured prediction. A detailed description of CharWNN architecture can be found in [9].

2.1   Character- and Word-Level Representations

The first layer of the network transforms words into real-valued feature vectors (embeddings) that capture morphological, syntactic and semantic information about the words. We use a fixed-sized word vocabulary V^{wrd}, and we consider that words are composed of characters from a fixed-sized character vocabulary V^{chr}. Given a sentence consisting of N words $\{w_1, w_2, ..., w_N\}$, every word w_n is converted into a vector $u_n = [r^{wrd}; r^{wch}]$, which is composed of two subvectors: the *word-level embedding* $r^{wrd} \in \mathbb{R}^{d^{wrd}}$ and the *character-level embedding* $r^{wch} \in \mathbb{R}^{cl_u}$ of w_n. While word-level embeddings are meant to capture syntactic and semantic information, character-level embeddings capture morphological and shape information.

Word-Level Embeddings. Word-level embeddings are encoded by column vectors in an embedding matrix $W^{wrd} \in \mathbb{R}^{d^{wrd} \times |V^{wrd}|}$. Each column $W_i^{wrd} \in \mathbb{R}^{d^{wrd}}$ corresponds to the word-level embedding of the i-th word in the vocabulary. We transform a word w into its word-level embedding r^{wrd} by using the matrix-vector product: $r^{wrd} = W^{wrd}v^w$, where v^w is a vector of size $|V^{wrd}|$ which has value 1 at index w and zero in all other positions. The matrix W^{wrd} is a parameter to be learned, and the size of the word-level embedding d^{wrd} is a hyper-parameter to be chosen by the user.

Character-Level Embeddings. In the POS tagging task, informative features may appear in the beginning (like the prefix "*un*" in "*unfortunate*"), in the middle (like the hyphen in "*self-sufficient*" and the "*h*" in "*10h30*"), or at the end (like suffix "*ly*" in "*constantly*"). Therefore, robust methods to extract morphological and shape information from words must take into consideration all characters of the word. In CharWNN this problem is approached using a convolutional layer [11,12] that produces local features around each character of the word, and then combines them using a max operation to create a fixed-sized character-level embedding of the word.

Given a word w composed of M characters $\{c_1, c_2, ..., c_M\}$, we first transform each character c_m into a character embedding r_m^{chr}. Character embeddings are encoded by column vectors in the embedding matrix $W^{chr} \in \mathbb{R}^{d^{chr} \times |V^{chr}|}$. Given a character c, its embedding r^{chr} is obtained by the matrix-vector product: $r^{chr} = W^{chr} v^c$, where v^c is a vector of size $|V^{chr}|$ which has value 1 at index c and zero in all other positions. The input for the convolutional layer is the sequence of character embeddings $\{r_1^{chr}, r_2^{chr}, ..., r_M^{chr}\}$.

The convolutional layer applies a matrix-vector operation to each window of size k^{chr} of successive windows in the sequence $\{r_1^{chr}, r_2^{chr}, ..., r_M^{chr}\}$. Let us define the vector $z_m \in \mathbb{R}^{d^{chr} k^{chr}}$ as the concatenation of the character embedding m, its $(k^{chr} - 1)/2$ left neighbors, and its $(k^{chr} - 1)/2$ right neighbors:

$$z_m = \left(r_{m-(k^{chr}-1)/2}^{chr}, ..., r_{m+(k^{chr}-1)/2}^{chr} \right)^T$$

The convolutional layer computes the j-th element of the vector r^{wch}, which is the character-level embedding of w, as follows:

$$[r^{wch}]_j = \max_{1 < m < M} \left[W^0 z_m + b^0 \right]_j \tag{1}$$

where $W^0 \in \mathbb{R}^{cl_u \times d^{chr} k^{chr}}$ is the weight matrix of the convolutional layer. The same matrix is used to extract local features around each character window of the given word. Using the max over all character windows of the word, we extract a "global" fixed-sized feature vector for the word.

Matrices W^{chr} and W^0, and vector b^0 are parameters to be learned. The size of the character vector d^{chr}, the number of convolutional units cl_u (which corresponds to the size of the character-level embedding of a word), and the size of the character context window k^{chr} are hyper-parameters.

2.2 Scoring and Structured Inference

We follow Collobert et al.'s [10] window approach to score all tags T for each word in a sentence. This approach follows the assumption that the tag of a word depends mainly on its neighboring words, which is true for various NLP tasks, including POS tagging. Given a sentence with N words $\{w_1, w_2, ..., w_N\}$, which have been converted to joint word-level and character-level embedding $\{u_1, u_2, ..., u_N\}$, to compute tag scores for the n-th word in the sentence, we first create a vector x_n resulting from the concatenation of a sequence of k^{wrd} embeddings, centralized in the n-th word, $x_n = \left(u_{n-(k^{wrd}-1)/2}, ..., u_{n+(k^{wrd}-1)/2} \right)^T$. We use a special *padding token* for the words with indices outside of the sentence boundaries. Next, the vector x_n is processed by two usual neural network layers, which extract one more level of representation and compute the scores:

$$s(x_n) = W^2 h(W^1 x_n + b^1) + b^2 \qquad (2)$$

where matrices $W^1 \in \mathbb{R}^{hl_u \times k^{wrd}(d^{wrd}+cl_u)}$ and $W^2 \in \mathbb{R}^{|T| \times hl_u}$, and vectors $b^1 \in \mathbb{R}^{hl_u}$ and $b^2 \in \mathbb{R}^{|T|}$ are parameters to be learned. The transfer function $h(.)$ is the hyperbolic tangent. The size of the context window k^{wrd} and the number of hidden units hl_u are hyper-parameters to be chosen by the user.

In POS tagging, the tags of neighboring words are strongly dependent. Some tags are arranged in chunks (e.g, proper names with two or more words), and some tags are very unlikely to be followed by other tags (e.g. verbs are very unlikely to follow determiners). Therefore, a sentence-wise tag inference that captures structural information from the sentence can deal better with tag dependencies. Like in [10], we use a prediction scheme that takes into account the sentence structure. The method uses a transition score A_{tu} for jumping from tag $t \in T$ to $u \in T$ in successive words, and a score A_{0t} for starting from the t-th tag. Given the sentence $[w]_1^N = \{w_1, w_2, ..., w_N\}$, the score for tag path $[t]_1^N = \{t_1, t_2, ..., t_N\}$ is computed as follows:

$$S\left([w]_1^N, [t]_1^N, \theta\right) = \sum_{n=1}^{N} \left(A_{t_{n-1} t_n} + s(x_n)_{t_n} \right) \qquad (3)$$

where $s(x_n)_{t_n}$ is the score given for tag t_n at word w_n and θ is the set of all trainable network parameters $\left(W^{wrd}, W^{chr}, W^0, b^0, W^1, b^1, W^2, b^2, A\right)$. After scoring each word in the sentence, the predicted tag sequence is inferred using the Viterbi [13] algorithm.

2.3   Network Training

We train CharWNN by minimizing a negative likelihood over the training set D. In the same way as in [10], we interpret the sentence score (3) as a conditional probability over a path. For this purpose, we exponentiate the score (3) and normalize it with respect to all possible paths. Taking the log, we arrive at the following conditional log-probability:

$$\log p\left([t]_1^N | [w]_1^N, \theta\right) = S\left([w]_1^N, [t]_1^N, \theta\right) - \log\left(\sum_{\forall [u]_1^N \in T^N} e^{S([w]_1^N, [u]_1^N, \theta)} \right) \qquad (4)$$

The log-likelihood in Equation 4 can be computed efficiently using dynamic programming [5]. We use stochastic gradient descent (SGD) to minimize the negative log-likelihood with respect to θ:

$$\theta \mapsto \sum_{([w]_1^N, [y]_1^N) \in D} -\log p([y]_1^N | [w]_1^N, \theta) \qquad (5)$$

where $[w]_1^N$ corresponds to a sentence in the training corpus D and $[y]_1^N$ represents its respective tag labeling.

We use the backpropagation algorithm to efficiently compute gradients of the neural network. We implemented CharWNN using the *Theano* library [14].

3   Related Work

CharWNN architecture is mainly inspired by the work of Collobert et al. [10] which has the specific goal of avoiding task-specific engineering of features for standard natural language processing tasks, while still achieving good results. The main difference between our approach and Collobert et al.'s work is the automatic extraction of character-level features employed in our method. Moreover, they apply their method for English language, while we deal with Portuguese.

The importance of taking into consideration the morphological structure of words for natural language processing appears in other related work. Alexandrescu and Kirchhoff [15] present a factored neural language model where each word is represented as a vector of features such as stems, morphological tags and cases. In [7], the authors also choose to operate at the morpheme level, assuming access to a dictionary of morphemic analyses of words. Then, they use a recursive neural network (RNN) to explicitly model the morphological structures of words and learn morphologically-aware embeddings. Lazaridou et al. [16] use compositional distributional semantic models, originally designed to learn meanings of phrases, to derive representations for complex words, in which the base unit is the morpheme. The main advantage of CharWNN over the cited approaches is its ability to extract morphological and shape information directly from the words. The convolutional layer employed in CharWNN avoids the use of hancrafted features and morpheme dictionaries.

4   Experimental Setup and Results

4.1   Unsupervised Learning of Word-Level Embeddings

Recent work has showed that large improvements in terms of model accuracy can be obtained by performing unsupervised pre-training of word embeddings [10,17,18]. In our experiments, we perform unsupervised learning of word-level embeddings using the *word2vec* tool[1], which implements the *continuous bag-of-words* and *skip-gram* architectures for computing word vector representations[19].

We use three sources of unlabeled text: the Portuguese Wikipedia; the CETENFolha[2] corpus; and the CETEMPublico[3] corpus. We process the Portuguese Wikipedia corpus using the following steps: (1) remove paragraphs that are not in Portuguese; (2) substitute non-roman characters for a special character; (3)

---

[1] https://code.google.com/p/word2vec/
[2] http://www.linguateca.pt/cetenfolha/
[3] http://www.linguateca.pt/cetempublico/

tokenize the text using the tokenizer that we implemented. The CETENFolha and CETEMPublico corpora are distributed in a tokenized format. They also contain tags indicating sentences, lists, author of the text, etc. We only include the parts of the corpora tagged as sentences. Additionally, for the three corpora, we remove sentences that are less than 20 characters long (including white spaces) or have less than 5 tokens. Like in [10] and [7], we lowercase all words and substitute each numerical digit by a 0 (for instance, *1905* becomes *0000*). When concatenating the three corpora: Portuguese Wikipedia, CETENFolha and CETEMPublico, the resulting corpus contains around 401 million words.

When running the *word2vec* tool, we set that words must occur at least 5 times in order to be included in the vocabulary, which resulted in a vocabulary of 453,990 entries. To train our word-level embeddings we use *word2vec*'s skip-gram method with a context window of size five.

We do not perform unsupervised learning of character-level embeddings. The character vocabulary is quite small if compared to the word vocabulary. Therefore, we assume that the amount of data in the labeled POS tagging training corpora is enough to effectively train character-level embeddings, which are initialized by randomly sampling each value from an uniform distribution: $\mathcal{U}(-r, r)$, where $r = \sqrt{\dfrac{6}{|V^{chr}| + d^{chr}}}$. The raw (not lowercased) words are used in the character-level embeddings, which allows the network to capture relevant information about capitalization.

4.2   Portuguese POS Tagging Copora

We use three labeled Portuguese POS tagging corpora to perform our experiments: (1) the (original) Mac-Morpho corpus [20], which contains around 1.2 million manually tagged words; (2) a revised version of the Mac-Morpho corpus [6], where errors were corrected in order to create a more reliable POS tagged corpus; (3) and the Tycho Brahe corpus [21], which contains historical Portuguese texts. We use the same training/test partitions used by Fernandes [4], Fonseca and Rosa [6] and Milidiú et al.[3], hence we can compare our results with the ones reported by these authors. Additionally, for each corpus we created a correspondent development set by randomly selecting 5% of the training set sentences. In Table 1, we present detailed information about the three corpora. The last two columns of Table 1 inform the number of out-of-the-supervised-vocabulary words (OOSV) and the number of out-of-the-unsupervised-vocabulary words (OOUV). OOSV words are the ones that do not appear in the training set, while OOUV words are the ones that do not appear in the vocabulary created using the unlabeled data (see Sec. 4.1), i.e, words for which we do not have word embeddings.

The proportion of OOUV in the Tycho Brahe corpus is larger than for the other two corpus. For instance, in the Tycho Brahe test corpus around 2.5% of the words are OOUV, while in the Mac-Morpho test corpus only around 0.47% of the words are OOUV. We explain this difference by the fact that Tycho Brahe corpus is composed by historical Portuguese texts, while the unlabeled corpus

Table 1. Portuguese POS Tagging Copora used in the experiments

Corpus	Partition	Sentences	Tokens	OOSV	OOUV
	Train	42,021	959,413	0	4155
Mac-Morpho	Dev.	2,212	48,258	1,360	202
	Test	9,141	213,794	9,523	1004
	Train	42,741	808,500	0	5039
Rev. Mac-Morpho	Dev.	2,250	42,450	1,481	294
	Test	4,999	94,993	3,290	609
	Train	29,163	734,922	0	19076
Tycho Brahe	Dev.	1,535	40,679	1,358	1103
	Test	10,234	259,991	9,191	6567

that we use, which is composed by Wikipedia, CETEMPublico and CETEN-Folha, contains mostly contemporary Portuguese.

4.3 Model Setup

We use the development sets to tune the neural network hyper-parameters. We spent more time tuning the learning rate than tuning other parameters, since this is the hyper-parameter that has the largest impact in the prediction. Like in [9], we use a learning rate schedule that decreases the learning rate λ according to the training epoch. The same set of hyper-parameters are used for the three datasets. The best results are achieved using around 9 training epochs. Using 4 threads in a Intel® Xeon® E5-2643 3.30GHz machine, our *Theano* based implementation of CharWNN takes around 2h30min to complete eight training epochs for the Mac-Morpho corpus. In Table 2, we show the selected hyper-parameter values.

Table 2. Neural Network Hyper-Parameters

Parameter	Parameter Name	Value
d^{wrd}	Word embedding dimensions	100
k^{wrd}	Word context window size	5
d^{chr}	Char. embedding dimensions	10
k^{chr}	Char. context window size	5
cl_u	Convolutinal units	50
hl_u	Hidden units	300
λ	Learning rate	0.0075

In order to assess the effectiveness of the character-level representation of words, we compare CharWNN with an architecture that uses only word embeddings and additional features instead of character-level embeddings of words. In our experiments, WNN represents a network which is fed with word representations only, i.e, for each word w_n its embedding is $u_n = r^{wrd}$. WNN is essentially

Collobert et al.'s [10] NN architecture. Where indicated, it also includes two additional handcrafted features: capitalization and suffix. The capitalization feature has five possible values: all lowercased, first uppercased, all uppercased, contains an uppercased letter, and all other cases. We use suffix of size three. In our experiments, both capitalization and suffix embeddings have dimension five. We use the same NN hyper-parameters values (when applicable) shown in Table 2.

4.4   Results for Portuguese POS Tagging

In Table 3, we report the performance of different NNs for POS tagging of the three test corpora. We show accuracies for whole test sets (*All words*), as well as for OOSV and OOUV words separately. In the column *All words*, we can see that the behavior of the different NN configurations is quite similar for the three corpora. The architecture WNN which does not use character-level embeddings, performs very poorly without the use of the capitalization feature. The main reason for that is the use of lowercased words in our word embeddings. For Mac-Morpho and Revised Mac-Morpho corpora, CharWNN achieves a slightly better result than the one of the architecture that uses the two handcrafted features. However, CharWNN has a larger impact for the Tycho Brahe corpus, producing an error reduction of 11.8% compared to WNN with two handcrafted features.

Regarding out-of-vocabulary words, we can see in Table 3 that for the Mac-Morpho and the Revised Mac-Morpho corpora, CharWNN is at least as effective as WNN with handcrafted features. For the Tycho Brahe corpus, CharWNN reduces the accuracy error of WNN by impressive 42.2% for OOUV words and by 30.8% for OOSV words. The impact of CharWNN is larger for Tycho Brahe because this corpus contains many (old) words that are not present in the corpora used for unsupervised pre-training. This result indicates that CharWNN is effective to improve the POS tagging accuracy of words for which the system does not have pre-trained word embeddings. Moreover, this result suggests that the features extracted by the convolutional layer contain additional information that is not encoded in the two handcrafted features.

In Table 4, we compare the result of CharWNN with reported state-of-the-art results. In [4], Fernandes uses a Structured Perceptron with Entropy Guided Feature Induction. In [3], Milidiú et al. use Entropy Guided Transformation Learning, which combines Decision Trees and Transformation-Based Learning. In both pieces of work, the authors use many handcrafted features, most of them to deal with unknown words. In [6], Fonseca and Rosa use a NN architecture similar to WNN with two handcrafted features, capitalization and suffixes. Our system reduces the error of the previously best systems by 12.2% for the Mac-Morpho corpus, 23.6% for the Revised Mac-Morpho corpus and 15.8% for the Tycho Brahe corpus. This is an impressive result, since we train our model from scratch, i.e., without the use of any handcrafted features.

Table 3. Comparison of different NNs for Portuguese POS Tagging

Corpus	System	Features	All Words	OOSV	OOUV
Mac-Morpho	CharWNN	–	**97.47**	92.49	**89.74**
	WNN	Caps + Suffix3	97.42	**92.64**	89.64
	WNN	Caps	97.27	90.41	86.35
	WNN	Suffix3	96.35	85.73	81.67
	WNN	–	96.19	83.08	75.40
Rev. Mac-Morpho	CharWNN	–	**97.31**	**93.43**	91.63
	WNN	Caps + Suffix3	97.24	92.64	**92.61**
	WNN	Caps	97.05	89.60	88.83
	WNN	Suffix3	96.18	86.96	84.56
	WNN	–	95.98	83.19	78.49
Tycho Brahe	CharWNN	–	**97.17**	**86.58**	**86.63**
	WNN	Caps + Suffix3	96.79	80.61	76.88
	WNN	Caps	95.66	60.87	50.59
	WNN	Suffix3	95.74	75.05	71.55
	WNN	–	94.63	56.04	44.86

Table 4. Comparison with state-of-the-art Portuguese POS taggers

Corpus	System	Accuracy
Mac-Morpho	This Work	**97.47**
	Fernandes [4]	97.12
Rev. Mac-Morpho	This Work	**97.31**
	Fonseca and Rosa [6]	96.48
Tycho Brahe	This Work	**97.17**
	Milidiú et al. [3]	96.64

We conduct an additional analysis to indirectly check the effectiveness of the morphological information encoded in character-level embeddings of words. The analysis consist in computing the similarity between the embeddings of rare words and words in the training set. In Table 5, we present five OOUV words from the Tycho Brahe test corpus and their respective most similar words in the training set. The similarity between two words w_i and w_j is computed as the cosine between the two vectors r_i^{wch} and r_j^{wch}. We can see in Table 5 that the character-level embeddings are very effective for learning affixes. These five words are correctly tagged by CharWNN, while WNN, even with the two handcrafted features is not able to correctly tag them.

In Fig.1 we present the behavior of CharWNN regarding the character-level features extracted for the word *"pré-cambrianas"*. This word does not appear in the Mac-Morpho corpus, which was used to train the respective model. Note that in the convolutional layer, 50 features are first extracted for each character. Then the max operator selects the 50 features which have the largest values among the characters to construct the character-level feature set r^{wch}. We can observe in Fig.1 that the extracted features concentrate mainly around the prefix *"pré-"* and the ending *"anas"*, which are in fact informative about the noun POS.

Table 5. Five OOUV words and their respective most similar words in the Tycho Brahe Corpus according to the character-level embeddings learned

rectissimamente	figurei	obrigadíssimas	debilita-se	lisonjeiam
fortissimamente	jurei	longuíssimas	deleita-se	lisongeiam
antiquissimamente	apurei	lindíssimas	facilita-se	reflectiam
exactissimamente	recobrei	diferentíssimas	respeita-se	revestiam
perfeitissimamente	procurei	alvíssimas	aproveita-se	receiam
discretissimamente	virei	dilatadíssimas	permita-se	repetiam

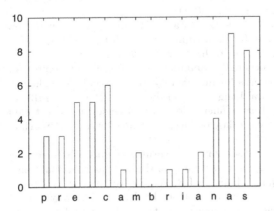

Fig. 1. Number of local features selected at each character of the word *"pré-cambrianas"*

5   Conclusions

In this work we approach Portuguese POS tagging using a deep neural network architecture that uses a convolutional layer to extract character-level features. The main contributions of the paper are: (1) the demonstration that it is feasible to train state-of-the-art Portuguese POS taggers for different corpora using the same model, with the same hyper-parameters, and without any handcrafted features; (2) the definition of new state-of-the-art results for three Portuguese POS tagging corpora; and (3) the demonstration that the proposed NN architecture is very suitable for Portuguese POS tagging.

As future work, we would like to analyze in more detail the contribution of character-level representations to POS tagging. For instance, it would be important to know which morphological structures are easier/harder to capture with this approach.

References

1. Branco, A., Silva, J.: Evaluating solutions for the rapid development of state-of-the-art pos taggers for portuguese. In: Proceedings of the Fourth International Conference on Language Resources and Evaluation (2004)

2. Nogueira dos Santos, C., Milidiú, R.L., Rentería, R.P.: Portuguese part-of-speech tagging using entropy guided transformation learning. In: Teixeira, A., de Lima, V.L.S., de Oliveira, L.C., Quaresma, P. (eds.) PROPOR 2008. LNCS (LNAI), vol. 5190, pp. 143–152. Springer, Heidelberg (2008)
3. Milidiú, R.L., dos Santos, C.N., Duarte, J.C.: Portuguese corpus-based learning using etl. J. Braz. Comp. Soc. 14(4), 17–27 (2008)
4. Fernandes, E.L.R.: Entropy Guided Feature Generation for Structure Learning. PhD thesis. Pontifícia Universidade Católica do Rio de Janeiro (2012)
5. Collobert, R.: Deep learning for efficient discriminative parsing. In: Proceedings of the Fourteenth International Conference on Artificial Intelligence and Statistics (AISTATS), pp. 224–232 (2011)
6. Fonseca, E.R., Ao Luís, G., Rosa, J.: Mac-morpho revisited: Towards robust part-of-speech tagging. In: Proceedings of the 9th Brazilian Symposium in Information and Human Language Technology, pp. 98–107 (2013)
7. Luong, M.T., Socher, R., Manning, C.D.: Better word representations with recursive neural networks for morphology. In: Proceedings of the Conference on Computational Natural Language Learning, Sofia, Bulgaria (2013)
8. Chrupala, G.: Text segmentation with character-level text embeddings. In: Proceedings of the Workshop on Deep Learning for Audio, Speech and Language Processing, ICML (2013)
9. dos Santos, C.N., Zadrozny, B.: Learning character-level representations for part-of-speech tagging. In: Proceedings of the 31st International Conference on Machine Learning, Beijing, China. JMLR: W&CP, vol. 32 (2014)
10. Collobert, R., Weston, J., Bottou, L., Karlen, M., Kavukcuoglu, K., Kuksa, P.: Natural language processing (almost) from scratch. Journal of Machine Learning Research 12, 2493–2537 (2011)
11. Waibel, A., Hanazawa, T., Hinton, G., Shikano, K., Lang, K.J.: Phoneme recognition using time-delay neural networks. IEEE Transactions on Acoustics, Speech and Signal Processing 37(3), 328–339 (1989)
12. Lecun, Y., Bottou, L., Bengio, Y., Haffner, P.: Gradient-based learning applied to document recognition. Proceedings of the IEEE, 2278–2324 (1998)
13. Viterbi, A.J.: Error bounds for convolutional codes and an asymptotically optimum decoding algorithm. IEEE Transactions on Information Theory 13(2), 260–269 (1967)
14. Bergstra, J., Breuleux, O., Bastien, F., Lamblin, P., Pascanu, R., Desjardins, G., Turian, J., Warde-Farley, D., Bengio, Y.: Theano: A CPU and GPU math expression compiler. In: Proceedings of the Python for Scientific Computing Conference, SciPy (2010)
15. Alexandrescu, A., Kirchhoff, K.: Factored neural language models. In: Proceedings of the Human Language Technology Conference of the NAACL, New York City, USA, pp. 1–4 (June 2006)
16. Lazaridou, A., Marelli, M., Zamparelli, R., Baroni, M.: Compositional–ly derived representations of morphologically complex words in distributional semantics. In: Proceedings of the 51st Annual Meeting of the Association for Computational Linguistics (ACL), pp. 1517–1526 (2013)
17. Zheng, X., Chen, H., Xu, T.: Deep learning for chinese word segmentation and pos tagging. In: Proceedings of the Conference on Empirical Methods in NLP, pp. 647–657 (2013)
18. Socher, R., Bauer, J., Manning, C.D., Ng, A.Y.: Parsing with compositional vector grammars. In: Proceedings of the Annual Meeting of the Association for Computational Linguistics (2013)

19. Mikolov, T., Chen, K., Corrado, G., Dean, J.: Efficient estimation of word represen-
    tations in vector space. In: Proceedings of Workshop at International Conference
    on Learning Representations (2013)
20. Aluísio, S.M., Pelizzoni, J.M., Marchi, A.R., de Oliveira, L., Manenti, R.,
    Marquiafável, V.: An account of the challenge of tagging a reference corpus for
    brazilian portuguese. In: Mamede, N.J., Baptista, J., Trancoso, I., Nunes, M.d.G.V.
    (eds.) PROPOR 2003. LNCS, vol. 2721, pp. 110–117. Springer, Heidelberg (2003)
21. Namiuti, C.: O corpus anotado do português histórico: um avanço para as
    pesquisas em lingüística histórica do português. Revista Virtual de Estudos da
    Linguagem 2(3) (2004)

General Purpose Word Sense Disambiguation Methods for Nouns in Portuguese*

Fernando Antônio Asevedo Nóbrega and Thiago Alexandre Salgueiro Pardo

Interinstitutional Center for Computational Linguistics (NILC)
Institute of Mathematical and Computer Sciences, University of São Paulo
{fasevedo,taspardo}@icmc.usp.br

Abstract. Word Sense Disambiguation (WSD) aims at determining the appropriate sense of a word in a particular context. Although it is a highly relevant task for Natural Language Processing, there are few works for Portuguese, which are tailored to specific applications, such as translation and information retrieval. In this work, we report our investigation of some general purpose WSD methods for nouns in Portuguese, tackling two additional challenges: using Princeton Wordnet (for English) as the sense repository and applying/customizing a WSD method for multi-document applications, which, to the best of our knowledge, has not been addressed before. In this paper, we also report our efforts on building a sense annotated corpus (for nouns, only), which was used for evaluating the investigated WSD methods.

1   Introduction

Word Sense Disambiguation (WSD) aims at determining the appropriate sense for a word in a particular context [1]. For example, in the sentence "the bank has borrowed money in a variety of ways", the word "bank" might be associated to multiple senses, as: sloping land; a financial institution (which is the correct sense); a supply held in reserve for future use; etc. For many applications, it is essential to know the correct word senses in order to produce good results. Consider, for instance, machine translation: it is necessary to know the sense for correctly translating the word to the target language. Therefore, WSD is an important area, dealing with a phenomenon of semantic nature and, as such, is as difficult as useful. However, for the Portuguese language, there are only a few WSD works, which are tailored to specific applications. For example, [14] shows a disambiguation method for ten highly ambiguous verbs in English, aiming at machine translation applications. [9], in turn, presents a WSD method for ambiguous geographical words, as "São Paulo", which, in Portuguese, may refer to a city, a state, a soccer team or even a saint.

In this work, we are interested in investigating general purpose WSD methods for Portuguese. As a departure point, we opted for dealing only with nouns, which

---

* The corresponding MSc dissertation is available at
  www.teses.usp.br/teses/disponiveis/55/55134/tde-28082013-145948

J. Baptista et al. (Eds.): PROPOR 2014, LNAI 8775, pp. 94–101, 2014.

are usually the most frequent open morphosyntactic class found in texts, and, as [13] claims, the disambiguation of nouns is enough to positively impact some applications.

In particular, we dealt with two specific challenges: a multilingual challenge, since we use Princeton Wordnet [12] as sense repository, whose synsets were considered as the possible senses that the words in Portuguese might be associated to; and a multi-document challenge, in that we are also interested on checking how multiple documents may contribute to the WSD task. The former decision of using Princeton Wordnet as sense repository is due to (i) its widespread use in the area for WSD and also for other applications, (ii) it has been manually produced , and (iii) the current partial development state of most of the similar resources for Portuguese. The decision of also dealing with multiple documents comes from the current high demand of multi-document processing tasks, as information retrieval and extraction and multi-document summarization.

Our main research hypotheses that guided our work are that it is possible to achieve good accuracy with general purpose WSD methods, that the English sense repository suffices for representing most of the senses found in Portuguese texts, and that multiple documents may present more context information for improving WSD results. We have used a corpus of news texts written in Brazilian Portuguese to conduct this investigation. We have investigated and made the necessary adaptations to three classical methods, namely, the heuristic method that always select the more frequent sense for the words, the traditional Lesk algorithm [8] and some of its variations, and the web-based proposal of Mihalcea and Moldovan [10]. For multi-document WSD, we explore a graph-based multi-document representation, inspired by the work of Agirre and Soroa [2].

In general, our methods adopt the following procedures: considering a sentence in Portuguese, we initially find the appropriate translation to English (it is a necessary step considering that we use Princeton Wordnet); and, then, given the possible synsets in Wordnet for the translations, the synset that represents the best sense for the word must be selected. In this work, we use the online bilingual dictionary WordReference® to automatically perform the translations (other tools have also been tested, as Google Translate®, but there were limitations of use and license). Our results show that the Mihalcea and Moldovan approach is better to disambiguate highly ambiguous words. This method also performs well for all the words, together with the classical (heuristic) method and our graph-based multi-document proposal.

This paper is organized in 6 sections. In section 2 we briefly present some related work. The corpus and the sense annotation task are described in Section 3. Section 4 shows the developed WSD methods, while their evaluation is described in Section 5. Section 6 presents some final remarks.

2   Basic Concepts and Related Work

WSD usually considers three main elements: 1) the target word to be disambiguated; 2) its context (usually, words surrounding the target word); and 3) a sense repository, as dictionaries, thesaurus, ontologies and wordnets. Some methods try to disambiguate all the words in a text ("all words" task), while some try to disambiguate only a group of words ("lexical sample" task). The methods may also be classified as corpus-based or knowledge-based. The formers use annotated corpora for producing machine learning classifiers. Generally, they perform the lexical sample task. The knowledge-based methods use linguistic resources and similarity measurements to be able to deal with a wider range of words. In this work, as we aim at general purpose methods, knowledge-based approaches are more appropriate. In what follows, we briefly introduce the main knowledge-based works on which we base our investigation.

Lesk [8] presents what is considered the most classical method for WSD. It is a simple method based on machine-readable dictionaries. It assumes that the best sense is the one whose definition in the dictionary is the most similar to the labels of the target word context (by comparing their words). Some authors, as [7], [3] and [15], present variations to the methods, considering wordnet synsets and their glosses and examples for performing the comparisons.

Mihalcea and Moldovan [10] propose an unrestricted method for WSD, based on the use of the web. The authors assume that the correct sense should be the one that occurs more frequently with the target word context in the web. In this method, the context is a single word, usually the closest word according to the observed syntactic pattern. For example, to disambiguate a noun, the context is the nearby verb.

Agirre and Soroa [2] propose a graph-based method. The authors use the PageRank algorithm [4] in a graph with words and synsets as nodes. The authors initially use the hierarchical structure of Wordnet to initialize the graph. In the next step, edges are created among words and their respective synsets. Finally, PageRank is applied in the graph to rank the synsets, with the best one being chosen as the correct sense.

3   The CSTNews Corpus

We have used the CSTNews[1] corpus [5] for testing the WSD methods. It has 140 news texts grouped by topic into 50 clusters. Each cluster has 2 or 3 texts. Since the corpus was not annotated with senses, this annotation was carried out as follows.

Given the difficulty of the sense annotation task and the limited time and human resources to annotate the full corpus, we opted to annotate the 10% most frequent nouns in each cluster. Each cluster was manually tagged by groups of two or three human annotators. For each new cluster to annotate, the annotation groups were mixed, in order to avoid any annotation bias. The task was carried out by 10 annotators in 1-hour daily meetings, during five weeks. To assist the participants, an easy-to-use annotation tool was built, called NASP[2] (in Portuguese, *NILC – Anotador de Sentidos para o Português*).

---

[1] Available at www.icmc.usp.br/pessoas/taspardo/sucinto/cstnews.html
[2] Available at www.icmc.usp.br/pessoas/taspardo/sucinto/files/NASP.zip

In general, 4366 words were annotated, with 519 different translations and 575 distinct synsets. The number of distinct synsets annotated per word in the corpus ranged from 1 to 5. However, when limiting the counting of different synsets for the clusters, the number ranged from 1 to 3. In general, 93% of the words happened with only one sense in their clusters; 6% had 2 senses; and 1% had 3 senses. This information suggests that to use multi-document information may assist the WSD in this context.

We analyzed the sense annotation task complexity by counting the number of available synsets for the annotation of each word. Figure 1 shows the results. One may see that it happens that some words have more than 50 possible synsets, which clearly shows how difficult the annotation task is. On the other extreme, some words have zero synsets. This happen for two reasons: our synset search approach could not find appropriate translations for some words; and/or there were no good synsets in Wordnet for the intended sense. For example, the word *licenciamento* ("licensing", in English) has no translation in WordReference® dictionary; on the other hand, the word *desabrigado* has two translation options ("unsheltered" and "unprotected") in the dictionary, but no corresponding synset in Wordnet (for noun category).

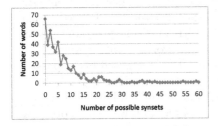

Fig. 1. Number of possible synsets per word in CSTNews noun annotation task

Table 1. Sense annotation task agreement in CSTNews

	Kappa	Percent agreement (%)		
		Total	Partial	Null
Translation	0.85	82.87	11.08	6.05
Synset	0.72	62.22	22.42	15.36
Translation + Synset	0.69	61.21	24.43	14.36

The annotation agreement was measured by both percent agreement counts and kappa [6]. For percent agreement, we computed the total agreement (when the annotators fully agreed), partial agreement (when the majority of the annotators agreed) and null agreement (for the remaining cases). Given the multilingual characteristic of our work, the agreement was measured for three elements: the chosen translations, the selected synsets, and the translation-synset pairs. Differently from percent agreement, the Kappa measure discounts the agreement by chance. Its values range from 0 to 1, with 1 indicating perfect agreement. Several researches consider that a kappa value above 0.6 is enough to have a reliable annotated corpus (although such value depends on the subjectivity and difficulty of the performed annotation). Table 1 shows the obtained agreement values. The rows show the evaluated elements, while the columns exhibit the agreement values. As expected, translation was easier than synset selection, but directly influences it. Once a bad translation is chosen, it is hard to find a good synset.

Again, as expected, the agreement is lower when we consider the perfect matching of translations and synsets among annotators. In general, the obtained values indicate that the annotation is reliable and may be used for the intended purpose.

4   Developed Methods

We tested four WSD methods. Besides translating and recovering synsets, all methods use the following pre-processing steps: (1) sentence splitting; (2) part-of-speech tagging; (3) removal of stopwords; (4) lemmatization of the remaining words; and (5) target words detection and context representation. Initially, we implemented the heuristic (baseline) method, frequently used in the literature, which assigns to the target word the most frequent sense in Wordnet (usually the first in the list), considering only the most frequent translation (the first one too) in the bilingual dictionary.

Our second method is an adaptation of Lesk algorithm [8] based on the work of [3]. Our algorithm has six variations: (G-T) using synset Glosses to compare with labels composed of possible word Translations in the context; (S-T) using synset sample Sentences to compare with labels composed of possible word Translations in the context; (GS-T) using synset Glosses and sample Sentences to compare with labels composed of possible word Translations in the context; (S-S) only synset sample Sentences to compare with labels composed of the sample Sentences of all possible synsets for the context words; (GS2) synset sample Sentences and synset Glosses to compare with labels composed of all possible synset sample Sentences and Glosses for the context words.

The third method is the WSD algorithm of Mihalcea and Moldovan [10] (which we will simple refer by Mihalcea method). In this method, we build word pairs for posting queries in the web. A word pair consists of the noun under focus and the nearest verb in the sentence. We use Microsoft Bing®for searching the web.

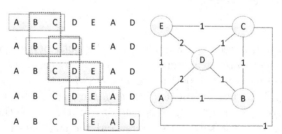

Fig. 2. Example of Multi-document Word Co-occurrence Network

The fourth method uses the best variation of the Lesk algorithm (G-T) for the multi-document WSD. This method uses a multi-document representation of context and assumes that all the occurrences of a word in a cluster have only one sense based in our corpus evidence. Furthermore, this method makes a second assumption: finding the most related words (which co-occurred the most) with the target word in its cluster helps selecting relevant context words and suggesting the best synset. We use a graph, which we call Multi-document Word

Co-occurrence Network (MWCN) (adapting from [11]) to represent the multi-document context. Thus, each node represents a single word in the cluster, and the edges indicate the frequency of occurrence of the corresponding words in a moving window of size N (which is a parameter of the algorithm), restricting that the same word pair in the same windown contributes only once to the edge weight. Figure 2 shows a MWCN for a sequence of hypothetical words A,B,C,D,E,A,D, using windows with three words (N=3). One may see that the words in the same windows (solid squares) are connected in the graph. The edge shows the number of windows in which the corresponding letters occurred together. For example, the edge A-D has weight 2 (since the words occurred in the fourth and fifth windows). It is important to note that window overlaps (indicated by dashed squares) contribute only once to the graph.

In the disambiguation process, the algorithm applies the G-T method using the N most related words with the target word in the MWCN. For example, for disambiguating the word D in the MWCN in Figure 2, the context words were A, E and either B or C. In our experiments, we test MWCN with windows sizes N=3 (MWCN3) and N=5 (MWCN5).

5   Evaluation

The WSD methods were evaluated on two tasks: the "all words" task; and the "lexical sample" task – in our case, the 20 most ambiguous words in the corpus. While the former task is necessary for measuring the coverage of the methods, the latter gives an idea of the robustness when dealing with difficult words.

In the all words task, the evaluation was measured by four metrics: (P) Precision – number of correct classifications over the number of words classified by the method; (R) Recall – number of correct classifications over the total number of tagged words in the corpus; (C) Coverage - number of words classified (correctly or not) by the method over the total number of tagged words in the corpus; and (A) Accuracy – equal to R, but using the heuristic method when no classification is found.

Table 2 shows the obtained results. The rows show the methods and the columns show the average values for the metrics. We show only the best configuration for the Lesk method (G-T). One may see that the MWCN3 method achieved 49.56% precision, and the Mihalcea method achieved 99.41% coverage. The heuristic method achieved the highest accuracy (51%). This may be explained by the fact that, in the CSTNews corpus, the majority of the words were annotated with the most frequent synset and this is a good scenario for this method. However, we did not found statistical difference in the precision and coverage values for the MWCN3 and heuristic methods.

Besides the heuristic method, Mihalcea method shows the best coverage. This is explained by the shorter context window (only one word). The MWCN3 e MWCN5 methods were better than Lesk [8] results for precision and coverage. It is important to note that the MWCN3 and MWCN5, for multi-document purposes, are better for WSD in this scenery. It is also interesting to notice that most of the accuracy values are similar to recall values, which indicates that the heuristic method do not find the correct senses for words that were not classified by others methods.

Table 2. All-word evaluation

Method	P(%)	R(%)	C(%)	A(%)
Heuristic	51.00	51.00	100	51.00
G-T	42.20	41.20	91.10	41.20
Mihalcea	39.71	39.47	**99.41**	39.59
MWCN3	**49.56**	**43.90**	88.59	43.90
MWCN5	46.87	41.80	87.65	41.80

Table 3. Lexical-sample evaluation

Word	Heuristic	S-T	Mihalcea	MWCN(3,5)
ano	90.50	86.30	**94.83**	47.22
hora	50.00	**50.00**	**50.00**	0.00
local	30.00	**30.00**	**33.33**	17.65
vez	0.00	**10.50**	**0.00**	**0.00**
>= Heuristic	-	12	**13**	8
Avg precision	27.88	28.46	**32.37**	19.10

In the lexical sample task, we used only the precision metric in order to evaluate the quality of the methods. Table 3 shows the evaluation results. The rows show some ambiguous words (with varying precision values) and the columns show each method (again, we show only the best configurations for Lesk method). For example, Mihalcea method achieved a 94.83% precision for the word *ano* ("year"). The bold values indicate cases that the methods performed as well as or better than the heuristic method. In general, Mihalcea method was the best method for dealing with highly ambiguous words. The MWCN3 and MWCN5 are presented together because they produced the same results. The last rows show the number of times that the methods were better or equal than the heuristic method. The last row shows the average precision of the methods for all the 20 words, with the Mihalcea method being the best one, in overall.

6 Final Remarks

As far as we know, this is the first investigation of general purpose WSD methods for nouns in Portuguese. We evaluated some classical methods and also proposed a new one that takes into account the available multi-document information. Another contribution of this work is the annotation of a corpus with noun senses, which is freely available for use.

Although Princeton Wordnet is a valuable resource (given its widespread use), our methods suffer with some lexical gaps. For instance, the word *caipirinha* (which is a typical drink in Brazil) has no specific synset in Wordnet. Instead, for cases like this, more generic synsets must be adopted (as the "drink" synset). The opposite – the specification – also happens. While in Portuguese we have the word *dedo* (does not mattering if it refers to hands or feet), in English it is necessary to decide among the specific words "finger" (if one talks about hands) or "toe" (about feet). For automatic WSD methods, this is a challenge that remains for future work.

Acknowledgements. To FAPESP and CNPq, for supporting this work.

References

1. Agirre, E., Edmonds, P.: Introduction. In: Word Sense Disambiguation: Algorithms and Applications. Springer (2006)
2. Agirre, E., Soroa, A.: Personalizing pagerank for word sense disambiguation. In: Proceedings of 12th Conference of the European Chapter of the ACL, pp. 33–41 (2009)
3. Banerjee, S.: Adapting the Lesk algorithm for word sense disambiguation to wordnet. Master's thesis. Department of Computer Science, University of Minnesota (2002)
4. Brin, S., Page, L.: The anatomy of a large-scale hypertextual web search engine. In: Proceedings of 17th International World-Wide Web Conference, WWW 1998 (1998)
5. Cardoso, P.C.F., Maziero, E.G., Jorge, M.L.R.C., Seno, E.M.R., Di Felippo, A., Rino, L.H.M., Nunes, M.D.G.V., Pardo, T.A.S.: CSTNews – a discourse-annotated corpus for single and multi-document summarization of news texts in brazilian portuguese. In: Anais do III Workshop "A RST e os Estudos do Texto", pp. 88–105. Sociedade Brasileira de Computação, Cuiabá (2011)
6. Carletta, J.: Assessing agreement on classification tasks: The kappa statistic. Computational Linguistics 22, 249–254 (1996)
7. Kilgarriff, A., England, B., Rosenzweig, J.: English senseval: Report and results. In: Proceedings of 2nd International Conference on Language Resources and Evaluation, pp. 1239–1244 (2000)
8. Lesk, M.: Automatic sense disambiguation using machine readable dictionaries: How to tell a pine cone from an ice cream cone. In: Proceedings of 5th Annual International Conference on Systems Documentation, pp. 24–26. Association for Computing Machinery, New York (1986)
9. Machado, I.M., de Alencar, R.O., de, O.C., Junior, R., Davis, C.A.: An ontological gazetteer and its application for place name disambiguation in text. Journal of the Brazilian Computer Society 17, 267–279 (2011)
10. Mihalcea, R., Moldovan, D.I.: A method for word sense disambiguation of unrestricted text. In: Proceedings of 37th Annual Meeting of the Association for Computational Linguistics, pp. 152–158. Association for Computational Linguistics, College Park (1999)
11. Mihalcea, R., Radev, D.: Graph-based natural language processing and information retrieval. Cambridge University Press (2011)
12. Miller, G.A.: Wordnet: A lexical database for english. Communications of the ACM 38, 39–41 (1995)
13. Plaza, L., Diaz, A.: Using semantic graphs and word sense disambiguation techniques to improve text summarization. Proceedings of Procesamiento del Lenguaje Natural 47, 97–105 (2011)
14. Specia, L.: Uma Abordagem Híbrida Relacional para a Desambiguação Lexical de Sentido na Tradução Automática. PhD thesis. Instituto de Ciências Matemáticas e de Computação–ICMC–USP (2007)
15. Vasilescu, F., Langlais, P., Lapalme, G.: Evaluating variants of the Lesk approach for disambiguating words. In: Proceedings of Language Resources and Evaluation (LREC 2004), Lisbon, Portugal, pp. 633–636 (2004)

Semi-supervised Parsing of Portuguese

Pablo Botton da Costa[1] and Fabio Natanael Kepler[2]

[1] Federal University of Sao Carlos (UFSCar)
[2] Federal University of Pampa (UNIPAMPA)

Abstract. Supervised methods for grammar induction depend on large manually annotated corpora, which is time consuming and needs linguistic expertise. Many languages do not have the necessary resources, and so have a lack of largely available hand-annotated data. This includes Portuguese. So good methods for unsupervised grammar induction are desired. Some models based on constituency are achieving promising results for English. We explore one of them, implementing a semi-supervised model and showing the results we get with the scarce resources we have at hand. With an f-score of 64.32% on around 2400 sentences of Historical Portuguese, it is the best result we know for unsupervised Portuguese parsing. We hope that new works on unsupervised grammar induction of Portuguese will be triggered, and more annotated corpora developed.

Keywords: Syntactic parsing; constituency parsing; semi-supervision.

1 Introduction

Syntax is a set of grammar rules that govern the way words are arranged to form phrases and sentences on a language. It also means the study of the structure and word order of phrases and sentences of a language. Automatic syntactic analysis, usually called parsing, involves making a parser program try to learn a language's syntax and then assign the correct tree structure to new sentences.

The most successful approaches are based on *supervised learning* methods, usually using some kind of probabilistic model. These methods use hand annotated corpora to learn structure, and can currently reach above 90% of accuracy [14,13] on English newspaper text [11]. However, two important drawbacks are that these approaches require a large corpus manually annotated with syntactic structure and are too restricted to its domain, application, or genre.

The task of building a large corpus demands high linguistic expertise and time. As an example, the annotation of the first 100K words of the Chinese Treebank took almost two years [18] (including the development of the guidelines for segmentation, part-of-speech (POS) tagging and syntactic bracketing). This problem is further complicated when dealing with languages other than English, which usually lack the amount of necessary resources to bootstrap the annotation task. In the case of Portuguese, hand-crafted treebanks are still hard to find. There has been increasing efforts in building Portuguese corpora, but they are still not very large, and are mainly annotated with just POS tags [5,4,9,1].

J. Baptista et al. (Eds.): PROPOR 2014, LNAI 8775, pp. 102–107, 2014.

Another learning method, called *unsupervised learning*, is getting some promising results in the past decade. It tries to induce structure from raw or tagged data, without the need of bracketed text. Since it does not require too many annotated human resources, it could therefore be used for bootstraping Portuguese corpora development. One of the most promising classes of unsupervised learning algorithms uses distributional evidences to identify constituent structure. [7]'s constituent-based model leads to good results when applied to English.

These results, however, are still relatively lower than those of supervised models. Although unsupervised methods are still compelling given a non resourceful language or domain, *semi-supervised learning* could be used in order to try to achieve better results while demanding fewer resources than supervised methods.

In this work we describe how we applied [7]'s model to Historical Portuguese data [5] and how a simple semi-supervised model can produce promising results. All results for Portuguese are lower than those for English, but yet well above the baseline.

In the next section we explain the model of [7], and in Section 3 we show other works on unsupervised induction and Portuguese parsing. In Section 4 we explain our proposed model and, in Section 5, the experiments performed and the results achieved. Finally, we draw some conclusions and future work ideas.

2   Constituent-Based Model

Certain words and groups of words function as a single unit within a sentence structure. They behave as *constituents*, and can be detected, for example, if they can be substituted by a pronoun or expanded, or if they can occur in various positions inside the clause.

[7] show a generative parsing model, called *Constituent Context Model* (CCM), where the fundamental assumption is that constituents appear in constituent contexts. The model exploits the constituency tests that long constituents often have shorter and more common equivalents which appear in similar contexts, and so are easier to be tested. Besides constituents, the model also describes distituents, which are groups of words that are not constituents. And by also modelling the contexts the model tries to transfer the constituency of a sequence to its context, this way making new sequences occurring in the same context to be parsed with similar constituency later.

A *span* is a contiguous subsequence of a sentence enclosing a *yield y* (a sequence of terminals such as D ADJ N), and occurring in a *context x*, such as N_VB, where x is the ordered pair of preceding and following terminals. A *bracketing b* of a sentence is a matrix indicating which spans are constituents and which are not. b is *tree-equivalent* if it has no two constituent spans crossing, if the size-one terminal spans and the full sentence span are constituents, and if the size-zero spans are distituents. It is *binary* if it corresponds to a binary tree.

Given a sentence s, the model chooses a bracketing b according to a distribution $P_T(b)$, and then generates s given b: $P(s, b) = P_T(b)P(s|b)$. Each span is filled independently, and each yield and context of each span are independent of each other:

$$P(s|b) = \prod_{\langle i,j \rangle} P(y_{ij}|b_{ij})P(x_{ij}|b_{ij}). \tag{1}$$

The P_T distribution is the distribution obtained by a uniform splitting process of generating trees described by Klein: choose a split point at random, then recursively build trees on each side of the split. This distribution puts relatively more weight on unbalanced trees than the uniform distribution over binary trees.

To induce structure, the EM algorithm is ran over the model, with the set S of sentences being treated as observed and the set B of bracketings as unobserved. The parameters Θ are the multinomial distributions over yields and contexts, $P(y|b)$ and $P(x|b)$.

E-Step: According to the current Θ, find the expectations of the bracketings B, $P(B|S,\Theta)$. The expected count of there being a bracket around the span y_{ij} of s, that is, the fraction of trees over s that contain y_{ij} as a constituent, is given by $P_\mathcal{B}(y_{ij}|s) = \frac{\sum_{b \in B:b_{ij}=C} P(s,b)}{\sum_{b \in B} P(s,b)}$. This quantity is calculated using a cubic dynamic program very similar to the *Inside-Outside* algorithm. The expected count equation for distituents is a little more complicated, and is not specified in Klein's text. Intuitively, though, it corresponds to the sum of the expected number of brackets crossing the span y_{ij}: $P_\mathcal{C}(y_{ij}|s) = \sum_{p<i,i<q<j} P_\mathcal{B}(y_{pq}|s) + \sum_{i<p<j,q>j} P_\mathcal{B}(y_{pq}|s)$, normalized by the number of crossing brackets.

M-Step: Compute the new set of parameters Θ', re-estimating the yields and contexts distributions. Re-estimation is computed by Maximum Likelihood estimates. The constituents distributions are computed by $\bar{P}(y^a|b = C) = \frac{\sum_{s \in S} \sum_{y_{ij}=y^a} P_\mathcal{B}(y_{ij}|s)}{\sum_{s \in S} \sum_y P_\mathcal{B}(y|s)}$ and $\bar{P}(x^a|b = C) = \frac{\sum_{s \in S} \sum_{x_{ij}=x^a} P_\mathcal{B}(x_{ij}|s)}{\sum_{s \in S} \sum_x P_\mathcal{B}(x|s)}$. The distituents distributions are computed similarly, using the $P_\mathcal{C}$ expectations.

The process begins at the M-Step, using P_T as the initial distribution. For smoothing, 2 counts are added for each yield and context as being constituents, and 8 counts as being distituents. This reflects the relative skew that random spans are more likely to be distituents.

3   Previous Work

The main recent works on unsupervised grammar induction using constituency-based models are the ones from Clark [3] and Klein and Manning [6,8,7]. Clark uses distributional clustering to group sequences and then a mutual information criterion to filter constituents from distituents. Klein and Manning use a similar idea, as we explained above. [15] show a slightly better CCM version by making some adaptations, specially regarding the EM algorithm and its local maximum problem.

All these works are mainly built and tested against English. For Portuguese, there is no such similar work that we could find. The syntactic analysis problem in Portuguese is approached only by supervised models [2,12,17].

4   Semi-supervised CCM

Our proposed model for semi-supervised learning is based on the original CCM model as implemented by [10], and is able to use fractions of supervision during the stage of grammar induction. Gold trees can be used for feeding the model with information about constituent labels, that is, syntactic labels occurring in internal nodes of the trees. During training, instead of computing $P(y_{ij}|b_{ij})$ and $P(x_{ij}|b_{ij})$, the model computes $P(y_{ij,l}|b_{ij})$ and $P(x_{ij,l}|b_{i,j})$, where l is the syntactic label of the span in case the given sentence has a gold tree and the span is a constituent, `true` if the span is a constituent without a label, and `false` if the span is a distituent.

During induction the model tries to find the best label for each span, computing $P(s|b)$ by using an updated version of Equation 1:

$$P(s|b) = \prod_{\langle i,j \rangle} \max_{l \in L} P(y_{ij,l}|b_{ij})P(x_{ij,l}|b_{ij}), \tag{2}$$

where L is the set of syntactic labels observed during training plus the generic labels for constituents, `true`, and for distituents, `false`.

5   Experiments and Results

The corpus used for training and testing is based on the texts from the *Tycho Brahe Corpus of Historical Portuguese* [5], which are freely available. The texts with syntactic annotation were concatenated, all punctuation was removed, and every sentence longer than 10 words was discarded. The resulting corpus, which we will be calling TB10, has 3938 sentences and trees[1].

We run a ten-cross validation using the TB10 corpus with our model described in Section 4, which we call SSCCM, and the original CCM model as implemented by [10], which we call CCM-L. Each cross validation of the SSCCM model was run ten times using different levels of supervision: from 0 to 100% of the training corpus. Results were then averaged (detailed results are not shown due to space restrictions).

Since all cited models only produce binary trees, they are bounded at the maximum they can achieve over n-ary trees. UBOUND compares the binarized corpus against the normal corpus. For the baseline we use the right branching strategy (RBRANCH).

In order to allow reproducible evaluation, we measured efficiency using unlabelled PARSEVAL metrics computed by the EVALB program[2]. UP, UR and UF refer to unlabelled precision, recall, and F-measure, respectively.

Table 1 shows the average results for CCM-L, SSCCM, and UBOUND over the TB10 corpus, and, for comparison, also for CCM-L over the English corpus,

---

[1] We will gladly provide a copy while we arrange with the Tycho Brahe Project for hosting it.

[2] Freely available at `http://nlp.cs.nyu.edu/evalb/`

WSJ10, as reported by [10]. Since this is the first result on unsupervised parsing of Portuguese that we know of, there are no other results to compare to. The original CCM model achieves much lower results for Portuguese than for English, not even surpassing the baseline, probably due to the small size of the training corpus. However, the SSCCM greatly improves the performance, reducing the error by about 28%, and surpassing the baseline by far.

Table 1. Average unlabelled results of the unsupervised and semi-supervised models over the TB10 corpus

Model	UF (%)	UP (%)	UR (%)	Corpus
CCM-L	71.90	64.20	81.60	WSJ10
RBranch	55.0	45.3	70.0	TB10
CCM-L	50.50	41.60	64.20	TB10
SSCCM	**64.32**	**58.25**	**71.85**	TB10
UBound	81.62	72.04	96.56	TB10

6   Conclusions

Given the scarce amount of manually structure-annotated corpora in Portuguese, and the domain adaptation problem of supervised parsers, the (successful) use of unsupervised parsers is highly desired. Surprisingly, it appears that not much work has been done focusing Portuguese, contrary to the increasing interest shown by works on other languages.

We have implemented a semi-supervised model based on one of the current best unsupervised models for grammar induction. [7] get 71.9% of f-measure on English, on the Wall Street Journal (WSJ10) treebank, but just 50.50% for Portuguese. Our implementation gets 64.32% on Portuguese, with a smaller corpus than the WSJ10. This is the best known published semi-supervised parsing score on any Portuguese data set.

Future work will focus on the weaknesses of the model, which is its ability to only handle binary trees, and the limited sentence length of ten words. Another task is to experiment with different parameter search methods, like Contrastive Estimation and annealing techniques for EM [16,15]. Our hope is that this work will trigger more research on Portuguese grammar induction, and also effectively help annotation of new Portuguese data.

References

1. Barreto, F., Branco, A., Ferreira, E., Mendes, A., Nascimento, M.F., Nunes, F., Silva, J.: Open resources and tools for the shallow processing of portuguese: The tagshare project. In: Proceedings of LREC 2006 (2006)
2. Bonfante, A.G.: Parsing Probabilístico para o Português do Brasil. Tese de doutorado, Programa de Pós-Graduação em Ciência da Computação, Instituto de Ciências Matemáticas e Computação, Universidade de São Paulo (Junho 2003)

3. Clark, A.: Unsupervised Language Acquisition: Theory and Practice. Phd thesis. University of Sussex (2001)
4. ICMC-USP: NILC's Corpora (2008), http://www.nilc.icmc.usp.br
5. IEL-UNICAMP and IME-USP: Corpus Anotado do Português Histórico Tycho Brahe (2008), http://www.ime.usp.br/~tycho/corpus
6. Klein, D.: The Unsupervised Learning of Natural Language Structure. Ph.d. thesis. Stanford University (2005), http://www.cs.berkeley.edu/~klein/
7. Klein, D., Manning, C.D.: A generative constituent-context model for improved grammar induction. In: ACL 2002: Proceedings of the 40th Annual Meeting of the Association for Computational Linguistics, pp. 128–135. Association for Computational Linguistics, Morristown (2002)
8. Klein, D., Manning, C.D.: Corpus-based induction of syntactic structure: Models of dependency and constituency. In: Proceedings of the 42nd Annual Meeting of the ACL 2004 (2004)
9. Linguateca.pt, The Floresta Sintá(c)tica project (2008), http://linguateca.dei.uc.pt/Floresta/
10. Luque, F.M.: Una implementacion del modelo dmv+ccm para parsing no supervisado. In: 2do Workshop Argentino en Procesamiento de Lenguaje Natural (2011)
11. Marcus, M., Santorini, B., Marcinkiewicz, M.: Building a large annotated corpus of english: the penn treebank. Computational Linguistics 19(2), 313–330 (1994)
12. Martins, R., Nunes, G., Hasegawa, R.: Curupira: A functional parser for brazilian portuguese. In: Mamede, N.J., Baptista, J., Trancoso, I., Nunes, M.d.G.V. (eds.) PROPOR 2003. LNCS (LNAI), vol. 2721, pp. 179–183. Springer, Heidelberg (2003)
13. McClosky, D., Charniak, E., Johnson, M.: Effective self-training for parsing. In: Proceedings of HLT/NAACL, pp. 152–159. Association for Computational Linguistics (June 2006)
14. Petrov, S., Klein, D.: Improved inference for unlexicalized parsing. In: Human Language Technologies 2007: The Conference of the North American Chapter of the Association for Computational Linguistics. Proceedings of the Main Conference, pp. 404–411. Association for Computational Linguistics, Rochester (2007), http://www.aclweb.org/anthology/N/N07/N07-051
15. Smith, N.A., Eisner, J.: Annealing techniques for unsupervised statistical language learning. In: Proceedings of the 42nd Annual Meeting of the Association for Computational Linguistics (ACL 2004), Barcelona, Spain (July 2004)
16. Smith, N.A., Eisner, J.: Guiding unsupervised grammar induction using contrastive estimation. In: Proceedings of IJCAI Workshop on Grammatical Inference Applications (2005)
17. de Carvalho e Sousa, F.: Analisador Sintático Estatístico Orientado ao Núcleo-Léxico para a Língua Portuguesa. Dissertação de mestrado, Programa de Pós-Graduação em Ciência da Computação, Instituto de Matemática e Estatística, Universidade de São Paulo (Outubro 2003)
18. Xia, F., Palmer, M., Xue, N., Ocurowski, M.E., Kovarik, J., Chiou, F.-D., Huang, S., Kroch, T., Marcus, M.: Developing guidelines and ensuring consistency for chinese text annotation. In: Proceedings of the Second Language Resources and Evaluation Conference LREC-00, Athens, Greece (June 2000), http://www.cis.upenn.edu/~chinese/ctb.html

What We Can Learn from Looking at Profanity

Gustavo Laboreiro and Eugénio Oliveira

LIACC, Universidade do Porto, Faculdade de Engenharia
{gustavo.laboreiro,eco}@fe.up.pt

Abstract. Profanity is a common occurrence in online text. Recent studies found swearing words in over 7% of English tweets and 9% of Yahoo! Buzz messages. However, efforts in recognizing, understanding and dealing with profanity do not share resources, namely, their dataset, which imposes duplication of effort and non-comparable results.

We here present a freely available dataset of 2500 messages from a popular Portuguese sports website. About 20% of the messages had profanity, thus we annotated 726 swear words, 510 of which were obfuscated by the authors. We also identified the most frequent profanities, and what methods, and combination of methods, people used to disguise their cursing.

1 Introduction and Related Work

In the context of this work we define profanity, curse, swear or taboo words, as words used with offensive or vulgar intentions. Although swearing can be studied in the context of multiple disciplines, from the computational perspective, it is commonly associated with the automatic identification of abusive comments.

Most often the intent of profanity identification lays on censoring these words or posts, but profanity is also tightly related with *sentiment analysis* and *opinion mining* tasks [1], since it can adequately express certain emotions [2,3,4], mostly negative. Its use seems to depend on several factors, such as gender, age and social class [5,4].

How common is cursing on-line? Most pages of 16 year olds on MySpace, and about 15% of pages of middle-aged people contained strong swearing [5]. 9.28% of comments in Yahoo! Buzz showed profanity [6]. Out of 51 million tweets in English, at least 7.73% of messages contained cursing [4], where swear words represented 1.15% of all words seen — as frequent as first person plural pronouns (we, us, our) [7].

While profanity *is* a common occurrence on-line, correct spelling is not. Curse words are not always written in the same way, a consequence of their use and spread being more oral than written. Graphical diversity is also augmented by accidental misspellings or intended obfuscations. Sood, Antin and Churchill found 76% of the top profane words not being written correctly, and describe why this variability is a hurdle for list-based profanity detection systems [6].

To study *how* users obfuscate their texts, we required a dataset annotated for profanity study. As we were unaware of any, we proceeded with the tiresome

J. Baptista et al. (Eds.): PROPOR 2014, LNAI 8775, pp. 108–113, 2014.

process of creating our own. Our goals were: *i)* the messages should relate to a swearing-prone subject; *ii)* the dataset needs to be of adequate size; *iii)* the annotation needs to address individual *words*; and *iv)* the dataset should be distributable, to avoid the duplication of effort and promote result comparison.

We annotated 2500 comments from a large sports news website in Portugal, and made it available, with extra information, at `http://labs.sapo.pt/2014/05/obfuscation-dataset/`.

We will next elaborate on the nature of our data, and on the annotation process. Then we will present a number of methods used for obfuscation, and look at their presence in the dataset. Finally we present our conclusions and how we expect to follow-up on our work.

2   Description of the Dataset

Our dataset was based on 2 years of text messages published on SAPO Desporto (`http://desporto.sapo.pt/`), a sport news website, with a strong emphasis in soccer, a sport known as important for the social identity in several countries [8]. We randomly selected 2500 messages, written in Portuguese.

The website checks all posts against a small list of forbidden "words" (the *blacklist*), and rejects any message that contains one. Users can choose to *not* use those taboo words, or they can attempt to bypass the filter in some way. Many took up the challenge, and the filter did not end cursing — it just pushed it into disguise. Hence, this data is appropriate for the study of obfuscation.

The blacklist contained 252 entries, including the most common profanities in the Portuguese language. Of interest to us were 44 curse words, 85 curse word variants (plural, diminutive, etc.), 30 curse words written in a graphical variant (e.g. no diacritical sign, "k" replacing "c", ...), 41 curse word variants written with graphical variants (e.g. "put@s"), and 10 foreign curse words and variants. The remainder of the list contained entries used for *spam* control. This list is distributed with our dataset.

In order to find how users used their creativity to overcome the obstacle of censorship, we had to analyse the messages. Three annotators used their sensibility on what constituted profanity. Once a word was considered swearing, it was tagged in the entire corpus. If misspelled, we add its canonical form to the annotation. In the end we identified 521 messages with profanities (1 in every 5), and 726 individual instances of profanity use (we ignored graphical duplicates in the same message), of which 510 were obfuscated.

We can summarise our profanity dictionary in 40 different base profanities, totalling 103 when counting variants. Despite the possibility of profanity variants being used as a kind of obfuscation (e.g., employing "shitful" instead of "shit"), we decided to consider them as distinct profanities.

Of the 103 profanities we identified, 29 were present in the blacklist, and represented half of the cursing instances that we found. Therefore, SAPO targeted the most frequent swearing terms, but failed before obfuscation. Let us take a look at the methods that were used to bypass the filter.

3 Obfuscation Methods

We were able to identify a total of 17 different ways in which the words we found deviated from their canonical spelling. They are described below, next to the symbols we assigned to represent them.

^-Ac **Accent removed** A diacritical mark is removed from the word. For example, "cabrão" becomes "cabrao", or "piço" becomes "pico".

^-C **Characters removed** Letters and/or symbols are removed from the word.

^+Ac **Accent added** A diacritical mark is added where it is not supposed to exist. For example, we see "cócó" instead of "cocó". This alteration seldomly had any phonetic impact on the words.

^+L **Letters added** Extra letters are added to the word, but is not a repetition of the preceding letter, as in "pandeleiro" instead of "paneleiro".

^+N **Number added** A seemingly random number is added to the word.

^+P **Punctuation added** A punctuation sign (".", ",", "-" or "_") is inserted into the word. These characters are chosen because they are easy to distinguish from letters. Two examples: "f-o-d-a-m" and "me-r.da".

^+S **Symbols added** A symbol not from our punctuation set is inserted in a word. One example is "fu£der", meaning "foder".

^+Sp **Spaces added** A space in employed to break the word into two or more segments. E.g., "co rnos", "p u t a".

$^=Ac$ **Change accent** One letter with an accent is replaced by the same letter bearing a different accent. We saw "cù" instead of "cú" many times.

$^=L$ **Letters substituted** One letter is replaced by one other letter. Usually this change does not alter the pronunciation of the word.

$^=N$ **Number as a letter** A number takes the place of a letter. Often the digit resembles the letter somewhat. As an example, "foda" becomes "f0da".

$^=P$ **Punctuation as a letter** One of the characters of our punctuation set are used as a placeholder for one or more letters. For example, "p..." for "puta".

$^=S$ **Symbol as a letter** A symbol from outside our punctuation set is used as a letter. A common occurrence was "@" instead of "a", as in "put@".

Ag **Word aggregation** Two words are combined into just one. For example, "americuzinho" combining "Américo" and "cuzinho" ("cu" and "co" sounding similar in this case).

Cp **Complex alteration** Forms of obfuscation that are too complex to be described with the other methods. A common occurrence is "fdp", that are the initials for "son of a bitch" ("sob") in Portuguese.

Ph **Phonetic-driven substitution** The words sound similar, but differ beyond a simple letter substitution. E.g., "fodassse" instead of "foda-se".

Pun **Pun** The author relies on the context to explain the obfuscation.

R **Repetition** One letter is repeated, as in "merddddddddda".

These operations were selected to provide a descriptive view, rather than to provide the smallest set of operations that could transform a word from its canonical representation. We focus on the way the reader *perceives* the obfuscation method, since multiple combinations can lead to the same result (e.g. substitution vs. insertion and removal). We will see that authors tend to choose methods that are easy to understand.

4 Analysis of the Dataset

In this work our main concern was on method *choice*. The number of times each method is used on each word strongly depends on word length, and provides little insight on how to reverse it. Also, some methods are more prone to overuse (e.g. *Repetition* and *Punctuation added*) than others (e.g. *Letter substituted* or *Accent removed*). Thus, if the author uses two letter substitutions in the same word, we count it as *one*.

We divided our analysis into two types of obfuscation: those that *maintain word length* and those that *alter word length*. We then look at how many operations were combined to obfuscate each word, as an indicator of complexity.

In general, the length of a word provides an additional clue that helps the reader in recognizing it, even when disguised. We found 261 obfuscations keeping word length (out of 510).

Many obfuscation methods cannot be used in order to achieve this. Tables 1a and 1b show the absolute frequency of the methods we saw. Letter substitution was the most popular choice, which can be explained with "c" and "k" usually being phonetically similar, and 1/3 of the curse words starting with a "c".

Obfuscation through only one method (Table 1a) is achieved mostly by substitutions ($=L$, $=N$ and $=S$) or accents manipulation ($=Ac$ and ^-Ac). When two methods are used (Table 1b), the clear preference lies in the combination of $=L$ and ^-Ac, mostly by writing "cú" (ass) as "ku".

When word length no longer constrains their efforts, authors show different preferences. In Table 1c and 1d we can see the method choice distribution across the remaining 189 obfuscation instances.

If no other method is used, Table 1c shows that *Repetition* (R) is the preferred choice, possibly because it calls attention to the word itself, and makes the modification obvious. The same characteristics we can claim to be shared by the insertion of easy-to-ignore noise (^+P and ^+Sp). Making puns was the fourth most popular method, something that is difficult to address automatically.

When two methods are used, there is a lack of clear predominance, as shown in Table 1d. The use of symbols as letters ($=S$) and repeating letters (R) are seen more frequently than the other methods, even if they are not combined often, which we found curious.

We also accounted for the rare concurrent use of three obfuscation methods. We saw the combination $=N\ Ph\ R$ three times (e.g., "f000daseeeeeee" for "foda-se"), while $^-Ac\ =L\ ^+Sp$ were seen together once ("ka brao" instead of "cabrão").

Table 1. How often each method was employed in our dataset. Maintaining word length: (a) and (b), changing word length: (c) and (d).

(a)

Method	Count	Method	Count
^-Ac	19	$=P$	3
^+Ac	7	$=S$	22
$=Ac$	22	$=Sp$	1
$=L$	102	Ph	2
$=N$	36	Pun	7

(b)

	$=L$	$=N$	$=S$	R
^-Ac	27		1	
^+Ac			1	
$=Ac$	1			
$=N$	1		5	
Ph		1		1
Pun	2			

(c)

Method	Count
^-C	8
^+L	4
^+P	38
^+S	3
^+Sp	21
Ag	10
Cp	4
Ph	4
Pun	20
R	77

(d)

	^+P	^+S	^+Sp	$=L$	$=P$	$=S$	Ag	Cp	Ph	Pun	R
^-Ac	1										5
^-C		1	1			1					
^+Ac								3			
^+L	1	1				1					
^+N						1					
^+P				1		6					
^+S				2				1		1	1
^+Sp				1	1	6					
$=Ac$									1		
$=L$											3
$=N$									1		4
$=P$											3
$=S$						1				1	2
Ag								1			
R								1	2		

5   Conclusion and Future Work

We hope that our work, while modest in size and scope, is a good first step towards greater research cooperation and validation of profanity identification.

We identified the most common swear words used in our corpus, and given that many were blacklisted, we inferred that the filter had no significant impact on the vocabulary of the users, as many circumvented it. We also surveyed a set of frequent obfuscation techniques we believe relevant when dealing with cursing.

We concluded that both graphical appearance and pronunciation are important when obfuscating profanity, but not necessarily both at the same time. Knowing that written "noise" can derive from personal choice [9], it could be interesting to study if this preference extends to obfuscation decisions.

By providing an annotation at a finer granularity (word level instead of message level), we believe that new techniques for word de-obfuscation can be enabled. This could be achieved by adapting the Levenshtein distance (new operations or statistics-derived costs), or through machine learning.

Acknowledgements. This project was funded by the UT Austin | Portugal International Collaboraboratory for Emerging Technologies, project UTA-Est/MAI/0006/2009, and SAPO Labs UP.

Also thanks to Luís Sarmento, Francisca Teixeira and Tó Jó for their help.

References

1. Constant, N., Davis, C., Potts, C., Schwarz, F.: The pragmatics of expressive content: Evidence from large corpora. Sprache und Datenverarbeitung: International Journal for Language Data Processing (33), 5–21 (2009)
2. Jay, T., Janschewitz, K.: Filling the emotional gap in linguistic theory: Commentary on Pot's expressive dimension (33), 215-221 (2007)
3. Jay, T.: The utility and ubiquity of taboo words. 4(2), 153-161 (2009)
4. Wang, W., Chen, L., Thirunarayan, K., Sheth, A.P.: Cursing in English on Twitter. In: Proceedings of the 17th ACM Conference on Computer Supported Cooperative Work & Social Computing, CSCW 2014 (February 2014)
5. Thelwall, M.: Fk yea I swear: Cursing and gender in MySpace. Corpora. 3(1), 83–107 (2008)
6. Sood, S.O., Antin, J., Churchill, E.: Profanity use in online communities. In: Proceedings of the SIGCHI Conference on Human Factors in Computing Systems, CHI 2012, pp. 1481–1490. ACM, New York (2012)
7. Mehl, M.R., Pennebaker, J.W.: The Sounds of Social Life: A Psychometric Analysis of Students Daily Social Environments and Natural Conversations. Journal of Personality and Social Psychology 84(4), 857–870 (2003)
8. Crisp, R.J., Heuston, S., Farr, M.J., Turner, R.N.: Seeing Red or Feeling Blue: Differentiated Intergroup Emotions and Ingroup Identification in Soccer Fans
9. Sousa Silva, R., Laboreiro, G., Sarmento, L., Grant, T., Oliveira, E., Maia, B.: 'twazn me!!!;(' Automatic Authorship Analysis of Micro-Blogging Messages. In: Muñoz, R., Montoyo, A., Métais, E. (eds.) NLDB 2011. LNCS, vol. 6716, pp. 161–168. Springer, Heidelberg (2011)

Extending a Lexicon of Portuguese Nominalizations with Data from Corpora

Cláudia Freitas[1], Valeria de Paiva[2], Alexandre Rademaker[3],
Gerard de Melo[4], Livy Real[5], and Anne Silva[1]

[1] PUC-Rio, Brazil
{anne.acsilva,maclaudia.freitas}@gmail.com
[2] Nuance Communications, USA
valeria.depaiva@gmail.com
[3] IBM Research and FGV/EMAp, Brazil
alexrad@br.ibm.com
[4] Tsinghua University, China
gdm@demelo.org
[5] UFPA, Brazil
livyreal@gmail.com

Abstract. We describe the extension of a lexicon of nominalizations in Portuguese, NomLex-PT, with nominals from a collection of corpora, the AC/DC corpora. The resulting lexicon of nominalizations is RDF-encoded and integrated with OpenWordNet-PT, a Portuguese WordNet freely available to download and consult. We discuss the reasons for this extension with corpus data, the methodology we followed, as well as our reasons for suggesting that the extended lexicon of nominalizations is a useful resource for researchers interested in Knowledge Representation of information extracted from Portuguese texts.

Keywords: Nominalizations, Portuguese, lexical resources, corpora.

1 Introduction

Lexical databases form an essential component of many modern Natural Language Processing (NLP) systems and frequently consist of large amounts of highly detailed and well-curated entries. Examples of such resources include the Princeton WordNet [1], a lexical-semantic network, NomLex [2], NomLexPlus, and NomBank [3], three resources that describe nouns, VerbNet [4], a verb lexicon, as well as FrameNet [5], a hierarchically structured database of prototypical situations, just to name a few. While each one of these resources is useful on its own, if we could unify them and relate them in a consistent manner, letting each resource complete the gaps in the others, we would have much more information than we have now.

In particular for languages where fewer resources exist, such as Portuguese, the ability to pool together the resources available and to complete them, using external resources and advanced natural language techniques to adapt these to the

J. Baptista et al. (Eds.): PROPOR 2014, LNAI 8775, pp. 114–124, 2014.

language at hand is a possible way forward.[1] We have been engaged in a project for producing lexical resources for Portuguese for a few years. So far, we have been able to share two resources with the community. The first is OpenWordNet-PT [7], an open-source version of WordNet for Portuguese. The second, NOMLEX-PT, is the result of our first steps towards the goal of creating a comprehensive lexicon of Portuguese nominalizations.

Continuing the overarching project of creating lexical resources for Portuguese, we started a smaller project, aimed at producing a lexicon of Brazilian Portuguese nominalizations similar to the English NomLex. In addition to NomLex, we also made use of the resources like French Nomage [8] and the collaborative dictionary Wiktionary in creating our lexicon. NomLex-PT is now a medium-sized resource that has been integrated with OpenWordNet-PT and is in the process of helping to improve the bigger resource.

Despite this growth, one may raise concerns that the multilingual resources used in bootstrapping our NomLex resource might have limited its scope to the common ground between the involved languages, without doing proper justice to the characteristic features of Portuguese. To address these concerns, we decided to make use of the rich linguistic data available from the AC/DC project [9] to increase our stock of nominalizations. In particular, the project provides publicly available Portuguese language corpora with morpho-syntactic annotations. While our original version focused on Brazilian Portuguese and was initially called NomLex-BR, we call the new resulting lexicon NomLex-PT, as the corpus has Portuguese texts from Portugal and Mozambique as well as from Brazil.

In this paper, we describe the extraction of data from the corpus and its hand curation, as well as the kind of evaluation we were able to perform and the applications we envisage for our resources. We also take steps to ensure that the new entries are aligned with OpenWordNet-PT, which, given its use of the RDF standard, enables an easier interchange of data with the large body of linguistic resources that have already been published as Linked Data in the Semantic Web.

2   Nominalizations in NomLex-PT and OpenWordNet-PT

Deverbal nouns, or nominalizations, can pose serious challenges for knowledge representation systems. A sentence like *Alexander destroyed the city in 332 BC* is easily parsed and its semantic arguments, such as the agent (Alexander), the thing destroyed (the city) and the time (332 BC), are readily obtained for a proposed logical representation of the sentence. By contrast, a sentence like *Alexander's destruction of the city took place in 332 BC* is much harder to interpret for many computational systems. It describes the same event of destruction, with the same semantic arguments, but these arguments are harder to obtain automatically from a syntactic parsing of the sentence. Hence it is advantageous to have a lexicon that marks such relatedness of meaning between the verbs and their nominalizations.

---

[1] The complementarity of lexical resources for Portuguese NLP is investigated in [6].

Deverbal nouns are well-studied in English, with the NomLex project [2] providing a well-established, open access baseline for corresponding results in other languages. Previous work has shown that NomLex is useful for natural language and knowledge representation tasks [10].

To construct a Portuguese lexicon of nominalizations, we initially started with a manual translation of the entries in the original NomLex. This gave us many of the most straightforward nominals in Portuguese, e.g. *construction/construção* and *writer/escritor*.

We then used cross-lingual projection methods to suggest additional candidates using resources like the French lexicon NOMAGE [8], the collaborative dictionary Wiktionary and its Portuguese version Wikcionário, as well as the FrameNet lexical resource [5]. These candidates were manually evaluated for correctness by two native speakers of Portuguese. Table 1 recalls the provenance of the nominals in our lexicon.

Table 1. Provenance of entries in NomLex-PT

Source	Number of Entries
NomLex Translations	1017
AC/DC Additions	1261
PAPEL Additions	458
Manual Additions	709
NOMAGE Translations	262
DHBB-Motivated Manual Additions	158
Wiktionary	152
FrameNet	142
OpenWordNet-PT	82
Total	4,241

Finally, as reported in our previous work [11] and [12], we integrated all of this data with our Portuguese wordnet. That resource, OpenWordNet-PT, started as a "projection" of the much more ambitious Universal WordNet (UWN) project, developed by Gerard de Melo and Gerhard Weikum [13]. The universal wordnet UWN extends the original Princeton WordNet with around 1.5 million meaning links for 800,000 words in over 200 languages, using data from Wiktionary, Wikipedia, and various other sources. OpenWordNet-PT is based on the Portuguese part of UWN but adds numerous manual revisions and new Portuguese-language glosses.

3   Corpus-Based Nominalization Detection

In this paper, we use a different technique to increase the stock of nominalizations in our database. We rely on text corpora, as explained below, to identify and compile lists of likely nominalization/verb pairs. These lists are then manually checked before being added to the inventory.

3.1   Gathering Data: The AC/DC Project

The AC/DC project was launched in 1998, with the goal of providing a place where it is possible to freely access Portuguese corpora. The acronym AC/DC stands for *Acesso a Corpora/Disponibilização de Corpora* (loosely translated as 'access and availability of corpora'), and its main purpose is 'to make a large number of corpus resources in Portuguese available on the web with a unified and simple interface that allows people to interact with the corpora without requiring physical access to institutions or software', as [14] reminds us.

The AC/DC cluster contains more than 1 billion words, distributed over the following genre parameters, listed in the order of their frequency distribution: general newspaper text, narrative fiction, specialized newspaper text, other or non-classified (which includes at least e-mail spam, EU calls, business letters, legal documents and web texts, especially blogs), informative, technical, and oral. In terms of language varieties, it contains Portuguese spoken in Portugal, Brazil, and Mozambique. For a detailed description of the Linguateca repository and the AC/DC project, the reader is referred to [15]. All AC/DC material was automatically parsed using the PALAVRAS parser [16]. The data thus comes with a range of linguistic information, including lemma, part-of-speech, morphology (gender, number, tense form, etc), and syntactical function.

3.2   Approach

Portuguese has specific suffixes that indicate the morphological relationship between nouns and verbs, such as *-ção, -agem* or *-or*[2]. Of these, some also convey, quite regularly, the semantic notion of agent or instrument, like the suffix *-or* or *-ura*. In light of these considerations, we query the corpora that make up the AC/DC cluster to find lists with instances of nominalization candidates that end with the chosen suffixes.

However, these suffix-based clues, despite being highly regular, are not infallible. Considering the suffix *-ção*, for example, while there is *relação-relacionar / relation-relate* or *produção-produzir/production-produce*, the pair *ignição-*ignar* does not exist, and many other nouns ending in *-ção* have no corresponding verb. Therefore a careful review of the lists generated via suffixes is imperative.

At this stage of NomLex-PT, we considered the following five suffixes, which we describe with examples both in Portuguese and English: *-ção* (and its allomorph *-ização*), for example *padronizar-padronização/standard-standardization*; *-ncia* (as in *dominar-dominância/dominate-dominance*); *-agem* an example that does not change form in English *(reciclar-reciclagem/recycle-recycle)*; and the suffixes for agentives *-or* (as in *trabalhar-trabalhador/work-worker*); and *-nte* as in *(estudar-estudante/study-student)*.

Through simple morphological rules, we produced the candidate verbs and then the noun-verb pairs were manually evaluated by two annotators. The

---

[2] The suffix *-or*, the most common agentive suffix, has two allomorphs: *-tor* (*construtor/constructor*) and *-dor* (*esfaqueador/*knifer, someone who attacks with a knife*).

annotators not only checked the pairs for correctness, but also added semantic markers to the nominalizations, a process we describe in great detail below
in Section 4.

3.3   Extension for Regressive and Other Nominalizations

In addition to the suffix-based search, we also considered regressive nominalizations, also known as zero-derivation nominalizations. It is well-known that
these make up a great number of Portuguese nominals (for example *atacar-
ataque/attack-attack*; *lutar-luta/fight-fight*), see for instance [17]. However, this
morphological process is not accompanied by an easily discernible morphological
trace, hence some linguists calling it a zero derivation. This hinders any corpus
search, as obviously the clue "ends with a vowel" is a very weak one and other
tools are required to bear on the problem.

Thus we turned our attention to PAPEL [18], a lexical resource for NLP
of Portuguese, based on the (semi-)automatic extraction of semantic relations
between the words from a general language dictionary. PAPEL makes available
lists of word and tuples grouped by semantic relation. Since PAPEL was created
semi-automatically, it is reasonable to assume that the lists contain mistakes.
However, there are no errors such as non-words. Obtained from a dictionary,
all listed words do exist in Portuguese. Given that we had already captured a
considerable amount of nominal-verb pairs from the corpus, we used the following
strategy: Only the verb-noun pairs that did not contain the already analyzed
suffixes went under revision by two linguists. The results were then reviewed,
just as the previous ones, by two linguists.

4   Linguistic Annotation

While links between nominalizations and the verbs they are derived from are
useful, natural language processing systems often require additional information
to interpret them correctly. While reviewing the noun-verb pairs, our annotators
thus also added additional annotations to assess the nominalizations in terms of
agentivity, animacy, and the possibility of lexicalization.

4.1   Agentives

Agentive nominals were defined following a grammatical criterion. Agentives are
nouns for which the paraphrase *one that Xs/ o que Xs* (where X is the verb) is
possible. So, *pintor (painter)* is *one that paints/o que pinta*; *estudante (student)*,
is *one who studies/o que estuda*; *computador (computer)* is *one that computes/o
que computa*[3]. The values of the feature 'agentive' are simply Y (yes) or N (no).

---

[3] Note that in Portuguese, as in most Romance languages, unlike English, the paraphrase does not have to decide a priori on the animacy (or not) of the subject:
computers/computadores and *fans/ventiladores* are treated as 'he', the same pronoun used for humans and we do not have to choose between 'who' and 'which',
'that' works for all.

The criterion for identification is only formal, so *democratização (democratisation)* (or *globalização–globalisation*), for example, in *A democratização da mídia permitiu a emergência do fenômeno (the democratisation of the media allowed the emergence of the phenomenon)*, although it may, from one point of view, be compatible with the idea of an agent, it does not match the paraphrase *"o que democratiza" (one that democratizes)*, therefore it is not considered agentive. For these cases, the agentive nominal would be *democratizador* (or *globalizador*). We should keep in mind that most of the nominalization suffixes are polysemous, so a nominal such as *democratização (democratisation)* has many meanings. Even if we can use it somehow as an agent nominal, its main use is still connected to the eventive reading, the process of *democratizar (democratize)*.

4.2   Animacy

The values for the feature animacy are: A (animate), I (inanimate), or Both (underspecified). For instance, *pintor/painter* is animate, *utilização/utilization* is inanimate, and the label 'Both' applies to nouns that can be used as either, see *recolhedor/collector* below. Animacy is known to be not as straightforward as it might appear, see the discussion in [19].

4.3   Lexicalization

Lexicalized nominals are those in which the actual meaning of the word no longer corresponds to the meaning resulting from the morphological process: *procuração* (the noun for 'power of attorney' in English) does not mean the act of *procurar/to search*, so we would say it is a lexicalized nominal; *coordenação (coordination)* corresponds both (a) to the act of *coordenar/to coordinate* as well as as to (b) the entity/group of people in charge of the coordination, the latter being a lexicalized noun. Examples:

a) A coordenação da pesquisa foi feita pelo departamento. The coordination of the research was done by the department.
b) A coordenação decidiu não aceitar a proposta. The coordination decided not to accept the proposal.

It is not always easy to decide whether we are facing a lexicalized nominal or not. The lexicalization process would be better described as a continuum, and some words can be more lexicalized than others. As an example, it can be difficult to decide if *educação/education* is lexicalized in *educação de jovens e adultos/youth and adult education*, since *jovens e adultos/youth and adults* can be either interpreted as arguments of the verb *educar/to educate* or as noun modifiers. On the other hand, *procuração/'power of attorney'* is clearly a noun that does not mean the act of *procurar/to search*. Besides, words can often have a lexicalized equivalent in a specific domain: *ocorrência/occurrence*, the event form of the verb *ocorrer/to occur*, is a deverbal nominal in *ocorrência de chuvas/occurrence of rain* but a lexicalized nominal in *ocorrência policial/police incident*. The values we assigned to the lexicalization feature are 0 (non-lexicalized), 1 (lexicalized), and B (both).

We assigned *procuração* 1; for *ocultação/concealment* we assigned 0; and for *declaração/declaration* we assigned B, since most speakers of Portuguese will know *declaração* both as an official document, as well as the action of the verb *declarar*. For nouns such as *civilização/civilization*, *organização/organization*, *coordenação/coordination*, and *vigilância/surveillance*, where the lexicalized version is considered equivalent to the group of people associated, we will consider the nominal *animate*, but they will not be agentive. The sentences below illustrate some cases as well as the importance of considering the sentence context:

(c) Até os profissionais da área de *vigilância* temem uma nova escalada de violência. Even the *surveillance* professionals fear a new climbing of the violence. (The nominal *vigilância/surveillance* is considered lexicalized and inanimate.)

(d) Em outro ponto do galpão, a *vigilância* encontrou alimentos como molho de tomate, arroz, bolacha, margarina e macarrão com prazo de validade vencido. In another area of the deposit, the *surveillance* (team) found provisions such as tomato sauce, rice, crackers, margerine and pasta with past sellby dates.(The same nominal is lexicalized and animate.)

4.4   Issues and Choices

By far, lexicalization was the most challenging point during our review process, but the *-or* suffix added some extra challenges. First of all, in order to generate the list of words ending by the suffix *-or*, we needed the syntactic annotation provided by the parser PALAVRAS applied to the AC/DC corpora. Since many *-or* nouns can also be used as adjectives, we refined the corpus search to search only for words that were classified as nouns. The search results still needed some revision, but the strategy saved us some time.

It was also crucial to check our *-or* candidates against their occurrences in corpus, in order to

(i) determine their use as animate/inanimate, and
(ii) verify any special context/domain in which the word might occur as a result of a lexicalization process.

Concerning the issue of animacy (i), we were informed by the corpora about the double nature (animate and inanimate) of nominals such as *recolhedor* and *desviador*, for example:

(e) A limpeza do córrego está sendo realizada com barreiras para evitar a disseminação do óleo, com equipamentos de sucção e com *recolhedores* de óleo. The cleaning of the creek is done with barriers to avoid dissemination of the oil, with suction equipment and with *collectors* of oil.(inanimate)

(f) Primeiro, trabalhou como *recolhedor* das apostas e há três anos teria ganho alguns pontos de jogo na zona norte. First (he) worked as a bet *collector* and three years ago (he) was supposed to have won some game spots in the North Zone.(animate)

(g) A armadura de protensão não aderente, disposta em elementos fictícios, paralelos aos elementos reais da estrutura, tendo em comum apenas os nós de extremidade da peça estrutural e os nós localizados nos *desviadores*. The shell of adhering protension, disposed in fictitious elements, parallel to the real elements of the structure, had in commun only the nodes of the extremities of the structural piece and the nodes located on the *routers*. (inanimate)

(h) Na presidência do Senado talvez o maior bandido do Brasil e o maior *desviador* de dinheiro público. In the Senate presidency, [we had] perhaps the biggest thief of Brazil and the biggest *swindler* of public funds. (animate)

Concerning the process of lexicalization (ii), the corpora showed us specialized uses we were not aware of initially. The observation of domain-specific uses was facilitated by the very nature of the AC/DC cluster. On the one hand, the Corpus Brasileiro (Brazilian Corpus) contains technical material; on the other hand, the presence of texts from both Brazilian and Portuguese varieties helped us with lexical differences.

(j) Valores médios de pH e acidez dos pães de forma produzidos com diferentes concentrações de *melhoradores* naturais (In, Mi e ML) e do melhorador comercial (MC). Average values of pH and acidity of sliced breads produced from different concentrations of natural and commercial *improvers*.

(k) Outro aliado de Gullit, Rijkaard, o *centralizador* dos lances de meio-campo. Another ally, Gullit, Rijkaard, the centralizer of midfield bids.

5   Results and Discussion

Table 2 shows the number of nominalizations (lemmas) after revision and their distribution by suffixes.

There is no other large lexicon of Portuguese nominalizations, freely available on the web, that we can compare our results with. As previously described, we started our project by first translating nominalizations from the English NomLex lexicon, followed by translating the French Nomage ones and gradually adding additional sources of data. Concerned that the translations might be missing the most used nominalizations in Portuguese, we had previously conducted a small

Table 2. Nominalizations in NomLex-PT by suffix

Suffix	Nominalizations
-ção	1345
-mento	334
-ida	24
-ura	107
-or	896
-nte	115
-ada	115
other	1302

coverage-focused experiment on a corpus of biographical stories. Subsequently, as reported in this paper, we decided to go for a full experiment on extracting nominalizations from the very large AC/DC corpora.

Lists of nominalization-verb candidate pairs were automatically produced and then manually checked by two linguists, who discussed difficult cases (some described in the paper) and reached a consensus. Since the process of reviewing and classifying the nominalizations was ultimately completely manual, traditional measures of precision and recall do not seem appropriate. We are considering other kinds of measures to test for coverage and accuracy of the classifications obtained.

For instance, we only became aware of the Spanish ANCORA lexicon recently and we intend to check the intersection of its coverage with ours. Another future work we are considering is extracting verbal subcategorization frames from the corpus, as also suggested by one of our reviewers. We are keenly aware that a sophisticated verb lexicon is a necessity for work on linguistic inference but this is a sizeable project, which would benefit substantially if other verbal lexica of Portuguese were freely available.

6   Conclusions

Using a combination of manual and automated methods and drawing on existing resources, we have created a representative lexicon of nominalizations in Portuguese that we hope is useful for linguists as well as for computational applications. Our original motivation to produce a lexicon of Portuguese nominalizations was to help with the knowledge representation of events in unstructured text, in a manner similar to the one described in [20]. The idea of a lexicon of nominalizations in Portuguese is part of a larger project of providing systems and resources to produce semantic analyses of Portuguese text, in a manner similar to what the Bridge Project at PARC [21] could accomplish for English. Many other components remain to be designed and implemented.

The approach taken here, populating a Portuguese nominalization lexicon with data extracted from a collection of corpora, adds a bottom-up dimension to a process that, so far, had been done in a top-down fashion. We do not start here from the (English or French-based) dictionaries, but from everyday Portuguese words. This has the good effect of making sure that we list roots and lemmas that are typically Portuguese (e.g. *esfaqueador*, *sambista*) and as a by-product, since it is performed by different linguists, re-checks the nominalizations already in place. We also retrieve words that are hardly present in dictionaries, such as slang and new constructions, which seem to have emerged in the language to describe specific new situations. Having both the top-down and the bottom-up approaches seems an appropriate way to obtain a balanced lexicon. Given that the lists of nominals we discussed have a wealth of information that we have not been able to fully operationalize, we are considering making these lists publicly downloadable as well as the lexicon. Our NomLex-PT data is available from https://github.com/arademaker/nomlex-pt, and we look forward to discussions on how to improve this resource as well as OpenWordNet-PT.

Much work remains to be done to increase the supply of freely available and robust semantic oriented lexical resources and tools for Portuguese. There is, however, a growing amount of available resources of which we have made heavy use, and to which we can now add NomLex-PT.

References

[1] Fellbaum, C.: WordNet: An electronic lexical database. The MIT press (1998)

[2] Macleod, C., Grishman, R., Meyers, A., Barrett, L., Reeves, R.: Nomlex: A lexicon of nominalizations. In: Proceedings of Eurale 1998, pp. 187–193 (1998)

[3] Meyers, A., Reeves, R., Macleod, C., Szekely, R., Zielinska, V., Young, B., Grishman, R.: The nombank project: An interim report. In: Meyers, A. (ed.) HLT-NAACL 2004 Workshop: Frontiers in Corpus Annotation, Boston, Massachusetts, USA, May 2-7, pp. 24–31. Association for Computational Linguistics (2004)

[4] Kipper, K., Korhonen, A., Ryant, N., Palmer, M.: Extending verbnet with novel verb classes. In: Proceedings of Fifth International Conference on Language Resources and Evaluation (LREC 2006), Genoa, Italy (June 2006)

[5] Baker, C.F., Fillmore, C.J., Lowe, J.B.: The Berkeley FrameNet project. In: Proc. COLING-ACL 1998, pp. 86–90 (1998)

[6] Santos, D., Barreiro, A., Freitas, C., Gonçalo Oliveira, H., Medeiros, J.C., Costa, L., Gomes, P., Silva, R.: Relações semânticas em português: Comparando o TeP, o MWN.PT, o Port4NooJ e o PAPEL. In: Textos seleccionados. XXV Encontro Nacional da Associação Portuguesa de Linguística, APL 2009, pp. 681–700 (2010)

[7] de Paiva, V., Rademaker, A., de Melo, G.: OpenWordNet-PT: An open Brazilian wordnet for reasoning (2012)

[8] Balvet, A., Barque, L., Condette, M.H., Haas, P., Huyghe, R., Marn, R., Merlo, A.: La ressource Nomage: Confronter les attentes thoriques aux observations du comportement linguistique des nominalisations en corpus. TAL 52(3), 124 (2011)

[9] Costa, L., Santos, D., Rocha, P.A.: Estudando o portugus tal como usado: o servio AC/DC. In: The 7th Brazilian Symposium in Information and Human Language Technology, STIL 2009, 8-11 de Setembro (2009)

[10] Gurevich, O., Crouch, R., King, T.H., de Paiva, V.: Deverbal nouns in knowledge representation. Journal of Logic and Computation (2008)

[11] de Paiva, V., Real, L., Rademaker, A., de Melo, G.: NomLex-BR: A lexicon of portuguese nominalizations. In: Proceedings of the 9th Language Resources and Evaluation Conference (LREC 2014). ELRA, Reykjavik (2014)

[12] Real, L.M., Rademaker, A., de Paiva, V., de Melo, G.: Embedding nomlex-br nominalizations into openwordnet-pt. In: Orav, H., Fellbaum, C., Vossen, P. (eds.) Proceedings of the 7th Global WordNet Conference, Tartu, Estonia, pp. 378–382 (January 2014)

[13] de Melo, G., Weikum, G.: Towards a universal wordnet by learning from combined evidence. In: Proceedings of the 18th ACM Conference on Information and Knowledge Management (CIKM 2009), pp. 513–522. ACM, New York (2009)

[14] Costa, L., Santos, D., Rocha, P.A.: Estudando o português tal como é usado: O serviço AC/DC. In: 7th Brazilian Symposium in Information and Human Language Technology – STIL 2009, São Carlos, Brazil (2009)

[15] Santos, D., Barreiro, A., Freitas, C., Oliveira, H.G., Medeiros, J.C., Costa, L., Gomes, P., Silva, R.: Relações semânticas em português: comparando o TeP, o MWN.PT, o Port4NooJ e o PAPEL. In: XXV Encontro Nacional da Associação Portuguesa de Linguística, pp. 681–700. APL, Porto (2009)

[16] Bick, E.: The Parsing System "Palavras": Automatic Grammatical Analysis of Portuguese in a Constraint Grammar Framework. PhD thesis. Aarhus University (2000)

[17] Basílio, M.: Estruturas lexicais do português; uma abordagem gerativa. Ática (1980)

[18] Oliveira, H.G., Santos, D., Gomes, P., Seco, N.: PAPEL: A dictionary-based lexical ontology for Portuguese. In: Teixeira, A., de Lima, V.L.S., de Oliveira, L.C., Quaresma, P. (eds.) PROPOR 2008. LNCS (LNAI), vol. 5190, pp. 31–40. Springer, Heidelberg (2008)

[19] Zaenen, A., Carletta, J., Garretson, G., Bresnan, J., Koontz-Garboden, A., Nikitina, T., O'Connor, M., Wasow, T.: Animacy encoding in english: Why and how. In: Proc. of the Association for Computational Linguistics Workshop on Discourse Annotation (2004)

[20] Gurevich, O., Crouch, D., King, T.H., de Paiva, V.: Deverbal nouns in knowledge representation. FLAIRS: The Florida Artificial Intelligence Research Society (May 2006)

[21] Bobrow, D.G., Cheslow, B., Condoravdi, C., Karttunen, L., King, T.H., Nairn, R., de Paiva, V., Price, C., Zaenen, A.: PARC's bridge and question answering system. In: Proceedings of Grammar Engineering Across Frameworks, pp. 26–45 (2007)

Body-Part Nouns and Whole-Part Relations in Portuguese

Ilia Markov[1,3], Nuno Mamede[2,3], and Jorge Baptista[1,3]

[1] Universidade do Algarve/FCHS and CECL,
Campus de Gambelas, 8005-139 Faro, Portugal
jbaptis@ualg.pt
[2] Instituto Superior Técnico, Universidade de Lisboa,
Av. Rovisco Pais, 1049-001 Lisboa, Portugal
Nuno.Mamede@tecnico.ulisboa.pt
[3] INESC-ID Lisboa/L2F – Spoken Language Lab,
R. Alves Redol, 9, 1000-029 Lisboa, Portugal
{Ilia.Markov,jbaptis}@l2f.inesc-id.pt

Abstract. In this paper, we target the extraction of *whole-part* relations involving human entities and *body-part nouns* occurrences in texts using STRING, a hybrid statistical and rule-based Natural Language Processing chain for Portuguese. Whole-part relation is a semantic relation between an entity that is perceived as a constituent part of another entity, or a member of a set.

Keywords: Whole-part relation, meronymy, body-part noun, disease noun, Portuguese.

1 Introduction

Automatic identification of semantic relations is an important step in extracting meaning out of texts, which may help several other Natural Language Processing (NLP) tasks, such as question answering, text summarization, machine translation, information extraction, information retrieval and others [13]. The whole-part relations acquired from a collection of documents are used in answering questions that normally cannot be handled based solely on keywords matching and proximity [14]. For automatic text summarization, where the most important information from a document or set of documents is extracted, semantic relations are useful for identifying related concepts and statements, so a document can be compressed [19].

The goal of this work is to improve the extraction of semantic relations between textual elements in STRING, a hybrid statistical and rule-based NLP chain for Portuguese [1][21]. This work will target whole-part relations (*meronymy*), that is, a semantic relation between an entity that is perceived as a constituent part of another entity, or a member of a set. In this case, we focus on the type of

---

[1] https://string.l2f.inesc-id.pt/ [last access: 04/05/2014].

J. Baptista et al. (Eds.): PROPOR 2014, LNAI 8775, pp. 125–136, 2014.
© Springer International Publishing Switzerland 2014

meronymy involving human entities and *body-part nouns* (henceforward, *Nbp*) when they co-occur in texts. Though STRING already extracts some types of semantic relations, meronymic relations are not yet being detected, in spite of the large set of *Nbp* that have already been semantically tagged in its lexicon. In other words, we expect to enhance the system's semantic relations extraction module by capturing meronymic relations. This paper is structured as follows: Section 2 briefly describes related work on whole-part dependencies extraction, while Section 3 explains with some detail how this task was implemented in STRING; Section 4 presents the evaluation procedure; and Section 5 draws the conclusions.

2   Related Work

Meronymy is a complex relation that "should be treated as a collection of relations, not as a single relation" [18]. In NLP, various information extraction techniques have been developed in order to capture whole-part relations from texts.

Hearst [16] tried to find lexical correlates to the *hyponymic* relations (type-of relations) by searching in unrestricted, domain-independent text for cases where known hyponyms appear in proximity. The author proposed six lexico-syntactic patterns; he then tested the patterns for validity, and used them to extract relations from a corpus. To validate his acquisition method, the author compared the results of the algorithm with information found in WordNet [9]. The author reports that when the set of 152 relations that fit the restrictions of the experiment (both the hyponyms and the hypernyms are unmodified) was looked up in WordNet:

> "180 out of the 226 unique words involved in the relations actually existed in the hierarchy, and 61 out of the 106 feasible relations (*i.e.*, relations in which both terms were already registered in WordNet) were found." [16, p. 544].

The author claims that he tried applying the same technique to meronymy, but without great success.

Berland and Charniak [5] addressed the acquisition of meronyms using manually-crafted patterns, similar to Hearst [16], in order to capture textual elements that denote whole objects (*e.g.*, *building*) and then to harvest possible part objects (*e.g.*, *room*). The authors used the North American News Corpus (NANC) - a compilation of the wire output of a certain number of newspapers; the corpus is about 1 million words. Their systems output was an ordered list of possible parts according to some statistical metrics. They report that their method finds parts with 55% accuracy for the top 50 words ranked by the system and a maximum accuracy of 70% over their top-20 results. The authors report that they came across various problems, such as tagger mistakes, idiomatic phrases, and sparse data – the source of most of the noise.

Girju *et al.* [13,14] present a supervised, domain independent approach for the automatic detection of whole-part relations in text. The algorithm identifies lexico-syntactic patterns that encode whole-part relations. The authors report an overall average precision of 80.95% and recall of 75.91%. The authors also state that they came across a large number of difficulties due to the highly ambiguous nature of syntactic constructions.

Van Hage *et al.* [15] developed a method for learning whole-part relations from vocabularies and text sources. The authors' method learns whole-part relations by

> "first learning phrase patterns that connect parts to wholes from a train-
> ing set of known part-whole pairs using a search engine, and then apply-
> ing the patterns to find new part-whole relations, again using a search
> engine." [15, p. 30].

The authors reported that they were able to acquire 503 whole-part pairs from the AGROVOC Thesaurus[2] to learn 91 reliable whole-part patterns. They changed the patterns' part arguments with known entities to introduce web-search queries. Corresponding whole entities were then extracted from documents in the query results, with a precision of 74%.

The Espresso algorithm [27] was developed in order to harvest semantic relations in a text. Espresso is based on the framework adopted in Hearst [16]:

> "It is a minimally supervised bootstrapping algorithm that takes as input
> a few seed instances of a particular relation and iteratively learns surface
> patterns to extract more instances." [27, § 3].

Thus, the algorithm extracts surface patterns by connecting the seeds (tuples) in a given corpus. The algorithm obtains a precision of 80% in learning whole-part relations from the Acquaint (TREC-9) newswire text collection, with almost 6 million words.

Thereby, for the English language, it appears that the acquisition of whole-part relation pairs by way of machine-learning techniques achieves fairly good results.

According to the very recent review of the literature on semantic relations extraction [1], no works on whole-part relations extraction for Portuguese have been identified. The current work also aims at extracting a specific type of whole-part relations, involving *Nbp*, but we adopt a rule-based approach, using the tools and resources available in STRING.

Next, in this work, we focus on state-of-the-art relations extraction in Portuguese, in the scope of ontology building. Some work has already been done on building *knowledge bases* for Portuguese, most of which include the concept of whole-part relations. These knowledge bases are often referred to as *lexical ontologies*, because they have properties of a lexicon as well as properties of an ontology [17,31]. Well-known, existing lexical ontologies for Portuguese are Portuguese WordNet.PT [22,23], later extended to WordNet.PT Global (Rede Léxico-Conceptual das Variedades do Português) [24]; MWN.PT-MultiWordNet

---

[2] http://www.fao.org/agrovoc [last access: 19.02.2014].

of Portuguese [30]; PAPEL (Palavras Associadas Porto Editora Linguateca) [26]; and Onto.PT [25]. Some of these ontologies are not freely available for the general public, while others just provide the definitions associated to each lexical entry without the information on whole-part relations. Furthermore, the type of whole-part relation targeted in this work, involving any human entity and its related *Nbp*, can not be adequately captured using those resources (or, at least, only those resources).

Attention was also paid to two well-known parsers of Portuguese, in order to discern how do they handle the whole-part relations extraction: the PALAVRAS parser [6], consulted using the Visual Interactive Syntax Learning (*VISL*) environment, and LX Semantic Role Labeller [7]. Judging from the available online versions/demos of these systems, apparently, none of these parsers extracts whole-part relations, at least explicitly.

In conclusion, the available resources identify some whole-part relations between lexical items in Portuguese, but they are not sufficient for the task of automatic extraction of whole-part relations as they occur between texts' instances of human entities and body-part nouns, as we here have targeted. Furthermore, we adopt a rule-based approach in order to extract this kind of relations from texts, considering the NLP system in which we intend to implement this module.

3   Whole-Part Dependency Extraction Module in STRING

3.1   STRING Overview

STRING performs all the basic steps of natural language processing (tokenization, sentence splitting, POS-tagging, POS-disambiguation and parsing) for Portuguese texts. The architecture of STRING is given in Fig. 1.

STRING has a modular, pipe-line structure, where: (i) the preprocessing stage (tokenization, sentence splitting, text normalization) and lexical analysis are performed by LexMan; (ii) followed by RuDriCo, which applies disambiguation rules, handles contractions and several special types of compound words; (iii) the

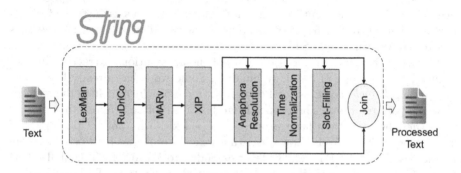

Fig. 1. STRING Architecture

MARv module then performs POS-disambiguation, using HMM and the Viterbi algorithm; and, finally, (iv) the XIP rule-based parser (Xerox Incremental Parser) [2] segments sentences into chunks and extracts dependency relations among chunks' heads. XIP also performs named entities recognition (NER). A set of post-parser modules have also been developed to handle certain NLP tasks such as anaphora resolution, temporal expressions' normalization, and slot-filling.

As part of the parsing process, XIP extracts *dependencies*. These dependencies correspond to different syntactic-semantic relations between the nodes of the sentence chunking tree, namely, the chunks' heads. Dependencies can thus be viewed as equivalent to (or representing) the syntactic relations holding between different elements in a sentence. Some of the dependencies extracted by XIP represent rather complex relations, such as the notion of *subject* (SUBJ) or *direct object* (CDIR), which imply a higher level of analysis of a given sentence. Other dependencies are much simpler and sometimes quite straightforward, like the determinative dependency DETD, holding between an article and the noun it determines, *e.g.*, *o livro* 'the book' – DETD(livro,o). Some dependencies can also be seen as auxiliary dependencies, and are required to build the more complex ones.

3.2   A Whole-Part Extraction Module in STRING

Next, we describe the way some of whole-part dependencies involving *Nbp* are extracted in the Portuguese grammar for XIP. To this end, a new module of the rule-based grammar was built, which is the first step towards a meronymy extraction module for Portuguese, and it contains most of the rules required for this work. In order to better present the different syntactic-semantic situations that the meronymy extraction module targets, some of the more simple cases are illustrated first and then some of the more complex situations follow. Example (1) is a simple case where there is a determinative PP, complement *de* 'of' N of the *Nbp*, so that the meronymy is overtly expressed in the text:

(1) *O Pedro partiu o braço do João* 'Pedro broke the arm of João'

The next rule captures the meronymy relation between *João* and *braço* 'arm':

```
IF( MOD[POST](#2[UMB-Anatomical-human],#1[human]) & PREPD(#1,?[lemma:de]) &
    CDIR[POST](#3,#2) & ~WHOLE-PART(#1,#2) )
    WHOLE-PART(#1,#2)
```

This rule is built using the XIP dependency rules' syntax, and it reads as follows: first, the parser determines the existence of a [MOD]ifier dependency, already calculated, between an *Nbp* (variable #2) and a human noun (variable #1); notice that, according to XIP conventions, the governor of the dependency is its first argument, hence *João* is said to be a modifier of *braço* 'arm'; this modifier must also be introduced by preposition *de* 'of', which is expressed by the dependency PREPD; then, a constraint is defined that the *Nbp* must be a direct object (CDIR) of a given verb (variable #3); and, finally, that there is still no previously calculated WHOLE-PART dependency between the *Nbp* and the human noun (variable #1); this last constraint is meant to ensure that there is only one meronymy relation between each *Nbp* and a given noun; if all these conditions are met, then, the parser builds the WHOLE-PART relation between the human determinative complement and the *Nbp*.

Next, in example (2), we present the (apparently) more simple case of a sentence with just a human subject and an *Nbp* direct object:

(2) *O Pedro partiu um braço* 'Pedro broke an arm'

In Portuguese, in the absence of a determinative complement, a possessive determiner or a dative complement (eventually reduced to a clitic dative pronoun), sentences like (2) are preferably interpreted as holding a whole-part relation between the human subject and the object *Nbp*. Thus, if there is a subject and a direct complement dependency holding between a verb and a human, on one side, and the verb and an *Nbp*, respectively; and if no WHOLE-PART dependency has yet been extracted for that *Nbp*, either for that human subject or another element in the same sentence, then the WHOLE-PART dependency is extracted.

There may be a relation within the same sentence between different *Nbp*, like in example (3). In this case, the WHOLE-PART relation should be established not only between the subject of the sentence and the *Nbp*, but also between *Nbp* in the sentence.

(3) *A Ana pinta as unhas dos pés* (lit: Ana paints the nails of the feet)
     'Ana paints the toenails'

In example (3), there is a meronymic relation between *Ana* and *unhas* 'nails', but also between *pés* 'feet' and *unhas* 'nails', so that two WHOLE-PART relations should be extracted.

There may be also a relation within the same sentence between an *Nbp* and a noun that designates a part of that same *Nbp*, and which we will call *npart* (*ponta da língua* 'tip of the tongue', *palma da mão* 'palm', etc.). This case differs from the previous one because, on the one hand, the whole-part relation should be established between the human noun and the *Nbp* and **not** the *npart* that precedes it; and, on the other hand, a second whole-part relation should also be established between the determinative *npart* and the *Nbp*, although this *npart* is not, by itself, an *Nbp*. Example (4) illustrates this situation:

(4) *O Pedro tocou com a ponta da língua no gelado da Ana*
     'Pedro touched with the tip of the tongue the ice cream of Ana'

WHOLE-PART(Pedro,língua) - correct; WHOLE-PART(língua,ponta) - correct; WHOLE-PART(Pedro,ponta) - incorrect.

The set of *npart* varies according to the *Nbp* and each set has to be established a priori. For example, for the *Nbp* *pé* 'foot' we can include the nouns *peito* 'instep', *alto* 'top', *cova* or *arco* 'arch', *dorso* 'instep', *planta* 'sole', and *ponta* 'tiptoe'. This is done by way of rules that add the feature *npart* to the nouns in the set associated to each *Nbp*, in the context of a determinative complement *de N* 'of N' of that *Nbp*. So far, 54 rules were built to associate the *Nbp* with their parts. All in all, 27 general rules have been built and implemented in STRING in order to extract whole-part relations involving *Nbp*.

We now turn to another type of meronymic relation. In some cases, a whole-part relation is only implicit, and though *Nbp* are involved, they are not mentioned directly (*gastritis*-'stomach'). In these cases, we decided that, nevertheless, a whole-part relation between the human entity and the "hidden" *Nbp* should be established. At this time, we focus on some predicative nouns designating specially localized *diseases* (*Nsick*). High lexical constraints apply in this relation: for each of these disease predicative nouns, the specific *Nbp* that is involved must be explicitly indicated in the lexicon. For example, the case where a disease noun is built with the support verb *ter* 'have', example (5):

(5) *O Pedro tem uma gastrite* 'Pedro has gastritis'

The rule that captures the meronymy relation between *Pedro* and *estômago* 'stomach' is given below:

```
IF( CDIR[POST](#1[lemma:ter],#2[lemma:gastrite]) & SUBJ(#1,#3) & ~WHOLE-PART(#3,?) )
    WHOLE-PART[hidden=+](#3,##noun#[surface:estômago,lemma:estômago])
```

The rule itself reads as follows: first, the system checks if the disease noun (in this case, *gastrite* 'gastritis') is the direct object (CDIR) of the verb *ter* 'have' (variable #1); secondly, the system verifies if there is an explicit subject (variable #3) for the verb; and if there is still no WHOLE-PART relation between that subject and the other node; in this case, the system builds the WHOLE-PART dependency between the subject of the verb and the 'hidden' *Nbp*, for which it creates a new (dummy) **noun** node. To express that a "hidden" noun is involved in this relation, a special tag '**hidden**' is also introduced in the dependency.

So far, 29 different pairs (*disease nouns, Nbp*) have been encoded in the lexicon.

To conclude, we have also addressed the issue of ambiguity raised by idioms involving *Nbp*. As it is well known, there are many frozen sentences (or idioms) that include *Nbp*. However, for the overall meaning of these expressions, the whole-part relation is often irrelevant, as in the next example:

(6) *O Pedro perdeu a cabeça* (lit: Pedro lost the [=his] head) 'Pedro got mad'

The overall meaning of this expression has nothing to do with the *Nbp*, so that, even though we may consider a whole-part relation between *Pedro* and *cabeça* 'head', this has no bearing on the semantic representation of the sentence, equivalent in (6) to 'get mad'. STRING's strategy to deal with this situation is, first, to capture frozen or fixed sentences, and then, after building all whole-part dependencies, exclude/remove only those containing elements that were also involved in fixed sentences' dependencies. In this way, two general modules, for fixed sentences and whole-part relations, can be independently built, while a simple "cleaning" rule removes the cases where meronymy relation is irrelevant.

Frozen sentences are initially parsed as any ordinary sentence, and then the idiomatic expression is captured by a special dependency (**FIXED**), which takes as its arguments the main lexical items of the idiom. The number of arguments varies according to the type of idiom. In the example (6) above, this corresponds to the dependency: **FIXED**(perdeu,cabeça), which is captured by the following rule:

```
IF (VDOMAIN(?,#2[lemma:perder]) & CDIR[post](#2,#3[surface:cabeça])) FIXED(#2,#3)
```

This rule captures any **VDOMAIN**, that is, a verbal chain of auxiliaries and the main verb whose lemma is *perder* 'loose', and a post-positioned direct complement whose head is the surface form *cabeça* 'head'.

In order to capture the idioms involving *Nbp*, we built about 400 of such rules[4], from 10 formal classes of idioms [3].

4   Evaluation

4.1   Evaluation Corpus

The 1st fragment of the CETEMPúblico corpus [32] was used in order to extract sentences that involve *Nbp*. This fragment of the corpus contains 14,715,055 tokens (147,567 types), 6,256,032 (147,511 different) simple words and 260,943 sentences. The existing STRING lexicon of *Nbp* and *Nsick* was adapted to to be used within the UNITEX corpus processor [28,29] along with the remaining available resources for European Portuguese, distributed with the system.

Using the *Nbp* (151 lemmas) and the *Nsick* (29 lemmas) dictionaries, 16,746 *Nbp* and 79 *Nsick* instances were extracted from the corpus (excluding the ambiguous noun *pelo* 'hair' or 'by-the', which did not appeared as an *Nbp* in this fragment). Some of these sentences were then excluded for they consist of incomplete utterances, or include more than one *Nbp* per sentence. A certain number of particularly ambiguous *Nbp*; *e.g.*, *arcada* 'arcade', *articulação* 'articulation', *etc.* that showed little or no occurrence at all in the *Nbp* sense were discarded from the extracted sentences. Finally, the sentences that lacked a full stop were corrected, in order to prevent errors from STRING's sentence splitting module. In the end, a set of 12,659 sentences with *Nbp* was retained for evaluation.

Based distribution of the remaining 103 *Nbp*, a random stratified sample of 1,000 sentences was selected, keeping the proportion of their total frequency in the corpus. This sample also includes a small number of disease nouns (6 lemmas, 17 sentences).

4.2   Inter-annotator Agreement

The output sentences were then divided into 4 subsets of 225 sentences each. Each subset was then given to a different annotator, and a common set of 100 sentences was added to each subset in order to assess inter-annotator agreement. For each sentence, the annotators were asked to append the whole-part dependency, as it was previously defined in a set of guidelines, using the XIP format.

Table 1. Average Pairwise Percent Agreement

Average pairwise percent agr.	Pairwise pct. agr. cols 1 & 4	Pairwise pct. agr. cols 1 & 3	Pairwise pct. agr. cols 1 & 2	Pairwise pct. agr. cols 2 & 4	Pairwise pct. agr. cols 2 & 3	Pairwise pct. agr. cols 3 & 4
85.031%	86.111%	**90.741%**	82.407%	81.481%	80.556%	88.889%

Table 2. Average Pairwise Cohen's Kappa

Average pairwise CK	Pairwise CK cols 1 & 4	Pairwise CK cols 1 & 3	Pairwise CK cols 1 & 2	Pairwise CK cols 2 & 4	Pairwise CK cols 2 & 3	Pairwise CK cols 3 & 4
0.629	0.65	**0.757**	0.59	0.558	0.518	0.699

For example, for (1) the annotators would produce WHOLE-PART(João,braço). Annotators could also mark a FIXED dependency (in the case of idioms) or no dependency at all (if no whole-part relation was present).

From the 100 sentences that were annotated by all the participants in this process, we calculated the Average Pairwise Percent Agreement, the Fleiss' Kappa [10], and the Cohen's Kappa coefficient for inter-annotator agreement [8] using ReCal3: Reliability Calculator [12], for 3 or more annotators [3].

The four annotators achieved the following results. First, the Average Pairwise Percent Agreement, that is, the percentage of cases each pair of annotators agreed with each other, is 85.031%, which is relatively high, as it is shown in Table 1. Next, the Fleiss' Kappa inter-annotator agreement coefficient was calculated, and it equals 0.625; the observed agreement of 0.85 is higher than expected agreement of 0.601, which we deem as a positive result. Finally, the Average Pairwise Cohen's Kappa is 0.629. as shown in Table 2.

According to Landis and Koch [20] this figures correspond to the lower bound of the "substantial" agreement; however, according to Fleiss [11], these results correspond to an inter-annotator agreement halfway between "fair" and "good".

In view of these results, we can assume as a reasonable expectation that the remaining, independent and non-overlapping annotation of the corpus by the four annotators is sufficiently consistent, so it will be used for the evaluation of the system output.

4.3   Evaluation of the System's Overall Performance

The results of the system performance are showed in Table 3. The number of instances (TP=*true-positives*; TN=*true-negatives*; FP=*false-positives*; FN=*false-negatives*) is higher than the number of sentences, as one sentence may involve several instances, and we count 5 partial TP as 0.5. The relative percentages of the TP, TN, FP and FN instances are similar between the 100 and the 900 set of sentences. This explains the similarity of the evaluation results and seems to

---

[3] http://dfreelon.org/utils/recalfront/recal3/ [last access: 08.02.2014].

Table 3. System's performance for *Nbp*.

Number of sentences	TP	TN	FP	FN	Precision	Recall	F-measure	Accuracy
100	8	73	7	14	0.53	0.36	0.43	0.79
900	73.5	673	55	118	0.57	0.38	0.46	0.81
Total:	81.5	746	62	132	0.57	0.38	0.46	0.81

confirm our decision to use the remaining 900 sentences' set as a golden standard for the evaluation of the system's output with enough confidence. The recall is relatively small (0.38), which can be explained by the fact that in many sentences, the *whole* and the *part* are not syntactically related. Precision is somewhat better (0.57). The accuracy is relatively high (0.81) since there is a large number of *true-negative* cases.

5   Conclusions

In this paper, we present a rule-based module for whole-part relations extraction involving human entities and body-part nouns (*Nbp*) in Portuguese, which has been implemented and integrated in the STRING NLP system. The most relevant syntactic constructions triggering this meronymic relations were described including the recovering of an implicit *Nbp* associated to several disease nouns. We also prevented idioms composed of *Nbp* to be captured, in view of their meaning non-compositionality, by capturing those idioms before the meronymy module. Around 17 thousand sentences with *Nbp* and disease nouns were extracted from a corpus. 4 Portuguese native speakers annotated a stratified random sample of 1,000 sentences and produced a golden standard, which was confronted against the system's output. The results show 0.57 precision, 0.38 recall, 0.46 F-measure, and 0.81 accuracy. In future work, we intent to improve recall by focusing on the *false-negative* cases already found, which shown that several syntactic patterns have not been paid enough attention, such as coordination.

Acknowledgments. This work was supported by national funds through FCT – Fundação para a Ciência e a Tecnologia, under project PEst-OE/EEI/LA0021/ 2013; and Erasmus Mundus Action 2 2011-2574 Triple I - Integration, Interaction and Institutions. We would like to thank the comments of the anonymous reviewers, which helped to improve this paper.

References

1. Abreu, S., Bonamigo, T., Vieira, R.: A review on Relation Extraction with an eye on Portuguese. J. Braz. Comp. Soc. 19(4), 553–571 (2013)
2. Ait-Mokhtar, S., Chanod, J., Roux, C.: Robustness beyond shallowness: Incremental dependency parsing. Natural Language Engineering 8(2/3), 121–144 (2002)

3. Baptista, J., Correia, A., Fernandes, G.: Frozen Sentences of Portuguese: Formal Descriptions for NLP. In: Workshop on Multiword Expressions: Integrating Processing. Intl. Conf. of the European Chapter of the ACL, Barcelona, Spain, pp. 72–79 (2004)

4. Baptista, J., Mamede, N., Markov, I.: Integrating Verbal Idioms into an NLP System. In: Baptista, J., Mamede, N., Candeias, S., Paraboni, I., Pardo, T., das Graça Nunes, M. (eds.) PROPOR 2014. LNCS(LNAI), vol. 8775, pp. 250–255. Springer, Heidelberg (2014)

5. Berland, M., Charniak, E.: Finding parts in very large corpora. In: Proceedings of the 37th Annual Meeting of the ACL on Computational Linguistics, pp. 57–64. ACL, Morristown (1999)

6. Bick, E.: The Parsing System "Palavras": Automatic Grammatical Analysis of Portuguese in a Constraint Grammar Framework. Ph.D. thesis, Aarhus Univ. Aarhus, Denmark: Aarhus Univ. Press (2000)

7. Costa, F., Branco, A.: LXGram: A deep linguistic processing grammar for portuguese. In: Pardo, T.A.S., Branco, A., Klautau, A., Vieira, R., de Lima, V.L.S. (eds.) PROPOR 2010. LNCS (LNAI), vol. 6001, pp. 86–89. Springer, Heidelberg (2010)

8. Cohen, J.: A coefficient of agreement for nominal scales. Educational and Psychological Measurement 20(1), 37–46 (1960)

9. Fellbaum, C.: WordNet: An Electronic Lexical Database. MIT, Cambridge (1998)

10. Fleiss, J.L.: Measuring nominal scale agreement among many raters. Psych. Bull. 76(5), 378–382 (1971)

11. Fleiss, J.: Statistical methods for rates and proportions, 2nd edn. John Wiley, New York (1981)

12. Freelon, D.: ReCal: Intercoder Reliability Calculation as a Web Service. Intl. J. of Internet Science 5(1), 20–33 (2010)

13. Girju, R., Badulescu, A., Moldovan, D.: Learning Semantic Constraints for the Automatic Discovery of Part-Whole Relations. In: Proceedings of HLT-NAACL, vol. 3, pp. 80–87 (2003)

14. Girju, R., Badulescu, A., Moldovan, D.: Automatic discovery of part-whole relations. Computational Linguistics 21(1), 83–135 (2006)

15. van Hage, W.R., Kolb, H., Schreiber, G.: A method for learning part-whole relations. In: Cruz, I., Decker, S., Allemang, D., Preist, C., Schwabe, D., Mika, P., Uschold, M., Aroyo, L.M. (eds.) ISWC 2006. LNCS (LNAI), vol. 4273, pp. 723–735. Springer, Heidelberg (2006)

16. Hearst, M.: Automatic acquisition of hyponyms from large text corpora. In: Proceedings of the 14th Conf. on Computational Linguistics, COLING 1992, vol. 2, pp. 539–545. ACL, Morristown (1992)

17. Hirst, G.: Ontology and the lexicon. In: Staab, S., Studer, R. (eds.) Handbook on Ontologies, pp. 209–230. Springer (2004)

18. Iris, M., Litowitz, B., Evens, M.: Problems of the Part-Whole Relation. In: Evens, M. (ed.) Relational Models of the Lexicon: Representing Knowledge in Semantic Networks, pp. 261–288. Cambridge Univ. Press (1988)

19. Khoo, C., Na, J.C.: Semantic Relations in Information Science. Annual Review of Information Science and Technology 40, 157–229 (2006)

20. Landis, J., Koch, G.: The measurement of observer agreement for categorical data. Biometrics 33(1), 159–174 (1977)

21. Mamede, N., Baptista, J., Diniz, C., Cabarrão, V.: STRING: An Hybrid Statistical and Rule-Based Natural Language Processing Chain for Portuguese. In: PROPOR 2012, vol. Demo Session. Paper available at http://www.propor2012.org/demos/DemoSTRING.pdf

22. Marrafa, P.: WordNet do Português: Uma base de dados de conhecimento linguístico. Instituto Camões (2001)

23. Marrafa, P.: Portuguese WordNet: general architecture and internal semantic relations. DELTA 18, 131–146 (2002)

24. Marrafa, P., Amaro, R., Mendes, S.: WordNet.PT Global – extending WordNet.PT to Portuguese varieties. In: Proceedings of the 1st Workshop on Algorithms and Resources for Modelling of Dialects and Language Varieties, pp. 70–74. ACL Press, Edinburgh (2011)

25. Oliveira, H.: Onto.PT: Towards the Automatic Construction of a Lexical Ontology for Portuguese. Ph.D. thesis, Univ. of Coimbra/FST (2012)

26. Oliveira, H.G., Santos, D., Gomes, P., Seco, N.: PAPEL: A Dictionary-based Lexical Ontology for Portuguese. In: Teixeira, A., de Lima, V.L.S., de Oliveira, L.C., Quaresma, P. (eds.) PROPOR 2008. LNCS (LNAI), vol. 5190, pp. 31–40. Springer, Heidelberg (2008)

27. Pantel, P., Pennacchiotti, M.: Espresso: Leveraging generic patterns for automatically harvesting semantic relations. In: Proceedings of Conf. on Computational Linguistics/ACL (COLING/ACL-06), Sydney, Australia, pp. 113–120 (2006)

28. Paumier, S.: De la reconnaissance de formes linguistiques à l'analyse syntaxique. Ph.D. thesis, Université de Marne-la-Vallée (2000)

29. Paumier, S.: Unitex 3.1beta, User Manual. Univ. Paris-Est Marne-la-Vallée (2014)

30. Pianta, E., Bentivogli, L., Girardi, C.: MultiWordNet: developing an aligned multilingual database. In: 1st Intl. Conf. on Global WordNet (2002)

31. Prévot, L., Huang, C., Calzolari, N., Gangemi, A., Lenci, A., Oltramari, A.: Ontology and the lexicon: A multi-disciplinary perspective (introduction). In: Huang, C., Calzolari, N., Gangemi, A., Lenci, A., Oltramari, A., Prévot, L. (eds.) Ontology and the Lexicon: A Natural Language Processing Perspective. Studies in Natural Language Processing, ch. 1, pp. 3–24. Cambridge Univ. Press (2010)

32. Rocha, P., Santos, D.: CETEMPúblico: Um corpus de grandes dimensões de linguagem jornalística portuguesa. In: Nunes, M. (ed.) V Encontro para o processamento computacional da língua portuguesa escrita e falada (PROPOR 2000), pp. 131–140. ICMC/USP, São Paulo (2000)

Proverb Variation:
Experiments on Automatic Detection
in Brazilian Portuguese Texts

Amanda Rassi[1,2], Jorge Baptista[2,3], and Oto Vale[1]

[1] Federal University of São Carlos - UFSCar,
Rodovia Washington Luís, km 235 - SP-310. São Carlos-SP, Brazil
{amandarassi85,otovale}@gmail.com
[2] University of Algarve - FCHS/CECL,
Campus de Gambelas, 8005-139 Faro, Portugal
{aprassi,jbaptis}@ualg.pt
[3] Institute of Systems and Computers Engineering - INESC-ID Lisboa/L2F,
Rua Alves Redol, n.° 9, 1000-029, Lisboa, Portugal

Abstract. This paper describes a methodology for automatically identifying proverbs and their variants in running texts. This methodology is based on existing compilations of proverbs, by exploring the regular syntactic structures that most proverbs present and intersecting syntactic structure with the lexical units of the proverbs. From the syntactic regularities we divided the data into 13 different classes. Finite-state automata is used to represent the regular patterns found in the classes. The results showed a precision rate of 74.68% tested in Brazilian Portuguese journalistic corpus.

Keywords: Proverbs, Variants, Automatic identification, Brazilian Portuguese.

1 Introduction

1.1 General Issues on Proverbs

This paper addresses proverbs, a term that covers a large variety of linguistic forms, also named parables, adages, aphorisms, maxims, anexins, and so on. Though there are conceptual differences among these terms, in practice, many authors ignore such distinctions. In this paper, we adopt the term *proverb* to designate a large number of linguistic expressions forming fixed word combinations, in spite of some lexical or structural variation.

Proverbs constitute a special type of multiword expressions, but they are distinct from frozen expressions (or idioms) for two reasons: (i) proverbs have the subject position necessarily filled by a fixed element [13, p.161], while the subject in fixed expressions usually varies and may be defined intensionally, by distributional constraints; and (ii) proverbs "always have an autonomous

J. Baptista et al. (Eds.): PROPOR 2014, LNAI 8775, pp. 137–148, 2014.

semantic value in communicative terms, unlike idioms that are only constituents of sentences and may never occur as a full sentence" [18].

The distinction is sometimes blurred by the fact that some expressions can be used both as proverbs or as idioms [26]; *e.g. Não adianta chorar pelo leite derramado* 'no use crying over spilled milk', used in an impersonal, proverbial mode, or *Ana chorou sobre o leite derramado* 'Ana cried over spilled milk', with an idiomatic format[1].

These multiword expressions have the same syntactic structure and the same words as ordinary, free sentences; however, they normally have a non compositional meaning, so they must be interpreted as a single unit of meaning. Neverthless, they often lack the presence of introductory expressions, that signal them as quotations; or are recast (and reshaped) in the ordinary stream of discourse, so it is necessary to recognize them in texts as multiword meaning units at a sentential/clausal level. Furthermore, proverbs are used in many contexts and can be found in many types of texts [12].

1.2 Theoretical Framework

Most studies and lexicographic work on proverbs consist basically of collecting and organizing them according to different criteria, but seldom was the issue of automatically spotting them in texts addressed in the literature [28]. For Portuguese proverbs, we only found [5] and some linguistic descriptions of proverbs and its classification [6,7]. Other works, adopting a systematic approach to the lexicon of proverbs are: in Spanish [1], French [9,8,10,11] and Italian [16].

In general, automatic processing for idiomatic expressions, fixed expressions, semi-fixed expressions, proverbs and other multiword expressions is still a hard task for Natural Language Processing [27]. Although there are many studies in NLP about the lexical aquisition of multiword expressions [14,15,17], it is still difficult to identify them automatically in natural language texts [3,4,21].

One of the NLP related issues raised by proverbs is the fact that they are very prone to certain types of formal variation, particularly the ellipsis of one of its clause-type components or their stylistic reformulation, in order to produce some perlocutionary effect. For example, a banking institution, in an advertisement of its products, recently "reinvented" the proverb *Tempo é dinheiro* 'Time is money' as *Tempo não é só dinheiro. É valor* 'Time is not just money. It is value'. This capacity of the proverbs to be reinterpreted and reformulated, which some linguists called "défigement" or "unfreezing", is an inherent part of the paremiologic dynamics in language.

1.3 Objectives

The goal of this paper is to develop a methodology for the automatic detection of proverbs and their variants in running texts. The task was tested on

---

[1] The translation, whenever possible, tried to present a similar proverb in English; otherwise it is just a literal translation; if necessary, some glosses are provided.

Brazilian Portuguese journalistic texts, and the paper shows the results of these experiments.

This approach lends itself to two main applications: (i) for lexicographic work, in order to build more complete dictionaries, and (ii) for Natural Language Processing, to improve linguistic resources, tools and applications, by allowing systems to signal these micro-texts as a special type of discursive elements.

The paper develops as follows: Section 2 describes the methodology, step by step; Section 3 discusses general issues of the task and some particular cases of each class; Section 4 presents the results of the experiments in Brazilian Portuguese texts; and Section 5 makes some final remarks and signals future work.

2   Methodology Description

In this section, we present the methodology used for identifying proverbs in texts automatically. The strategy can be explained as follows: based on existing compilations of proverbs, we explored the regular syntactic structures that most proverbs present; these regularities provided information to produce a formal (syntactical) classification; to retrieve proverbs and candidate variants from texts, a finite-state approach was adopted, taking advantage of the formal regularities underlying the proverbs' classification.

Then, this approach was tested in a Brazilian Portuguese corpus, namely PLN.Br Full corpus[2] [2], a journalistic corpus containing 103,080 texts, with 29,014,089 tokens, taken from *Folha de São Paulo*, a Brazilian daily newspaper, from 1994 to 2005. All these steps are described in the next subsections.

2.1   Collection of Proverbs

Number of collections proverbial sentences exist in many languages, and Portuguese is no exception. To produce a representative sample of proverbs in BP we used [24,29,31] and a dictionary of proverbs [20].

A substantial collection of over 3,502 proverbs (and their variants) has been gathered. The variants of each proverb were grouped together and one of them was selected to be considered the entry of our lexicon (or its *base-form*), based on its frequency among the sources consulted. Furthermore, we decided to use only Brazilian proverbs, so we discarded many entries that correspond to Portuguese or African proverbs written in Portuguese. After those steps, 614 base-forms were retained.

2.2   Classification of Proverbs

The list of proverbs (base-forms) was then classified into syntactic classes. This classification was based on the following criteria, applied in this order: (i) the

---

[2] PLN.Br Full corpus is available at http://www.linguateca.pt/acesso/corpus.php?corpus=SAOCARLOS.

number of propositions (one, two, or three clauses or clause-like units); (ii) coordination (in multiple-clause proverbs); (iii) order of the main *vs* the subordinate clauses (in multiple-clause proverbs); (iv) order of the constituents (in single-clause proverbs); (v) impersonal constructions; and (vi) obligatory negation.

Proverbs that did not fit in any of the categories above were added in a residual class. Table 1 shows the breakdown of the proverbs (base-forms) per class.

Table 1. Formal Classification of Brazilian Portuguese Proverbs

Class	Structure	Example (approximate translation)	Count
P1F1	$\emptyset$ V w (impersonal)	*Não há parto sem dor* 'There is no painless childbirth'	20
P1F2	N_0 $Vcop$ Adj/N w	*O silêncio é de ouro* 'Silence is golden'	53
P1F3	N_0 V w	*Uma mão lava a outra* 'One hand washes the other'	80
P1F4	N_0 Neg V w	*Cão que ladra não morde* 'A barking dog seldom bites'	53
P1F5	$Prep$ N_i N_0 V w	*Em terra de cego, quem tem um olho é rei* 'In the land of the blind, the one-eyed is king'	45
P2F1	F_1 $Conjs$-$comp$ F_2 (comparatives)	*Antes só que mal acompanhado* 'Better alone than in bad company'	39
P2F2	F_1 $Conjc$ F_2 (coordinated)	*Aqui se faz e aqui se paga* 'What goes around comes around'	71
P2F3	NP_1, NP_2	*Cada cabeça, uma sentença* 'Each head its sentence'	48
P2F4	Qu- F_1 F_2 (subordinated)	*Quem ri por último ri melhor* 'Who laughs last laughs best'	90
P2F5	F_1 $Conjs$ F_2 (subordinated)	*Pense duas vezes antes de agir* 'Look before you leap'	20
P2F6	$Conjs$ F_2, F_1 (fronted subord.)	*Quando o gato sai de casa, os ratos fazem festa* 'When the cat's away, the mice will play'	28
P3	F_1, F_2, F_3	*Mãos frias, coração quente, amor ardente* 'Cold hands, warm heart, burning love'	24
Residual	not specified	*Comer e coçar é só começar* 'To keep eating and scratching, just start'	43
Total			614

In this Table, the left column shows the conventional codes for designating each class; the structure of the proverbs' class is indicated as follows: *Adj* for adjective; *Conjc* for coordinative conjunctions; *Conjs* for subordinative conjunctions; *Conjs-comp* specifically for comparative conjunctions; F_1, F_2 and F_3 for the first, second and third clause (or clause-like units), respectively; N_0 for the subject; N_i for a noun in any syntactic position; NP_1 and NP_2 for nominal clauses; *Neg* for an obligatory negation item; *Prep* for preposition; *Qu*- for interrogative pronouns *quem* or *o que*, similar to *who* and *whom* in English; *V* for verb; *Vcop* for the copula *ser* 'to be'; *w* for any non-specified sequence of complements; and ∅ for an empty slot.

Proverbs with one single proposition are subclasses of P1 (see Table 1), proverbs with two propositions are subclasses of P2, and P3 includes proverbs with 3 propositions, but it was not subdivided.

Impersonal constructions (class P1F1) normally involve the verb *haver* 'there be' or the impersonal clitic pronoun -*se*, that imposes 3rd person-singular agreement to the verb, thus being indistinguishable from passive-like pronominal constructions. In Portuguese, only some few clear-cut cases of pronominal passives were found; e.g. *Entre mortos e feridos salvaram-se todos* 'Among dead and wounded all were saved'. Both syntactic strategies may be considered as a form of subject (agent) degenerescence, hence contributing to the generic referential effect of the proverbs. Specially in Brazilian Portuguese, the verb *ter* 'to have' is also used in impersonal constructions, such as in *Tem muito rei pra pouco súdito* 'There are too many kings and only few vassals'.

Sentences with copula *ser* 'to be' (class P1F2) usually present an adjectival or nominal predicate; these sometimes allow for mirror-permutation (*O amor é cego = Cego é o amor*[3] 'Love is blind = Blind is love').

Proverbs with obligatory negation (class P1F4) usually involve negation adverbs, *e.g. não* 'no/not', *nunca* 'never', *jamais* 'never', *nem* 'nor', *etc.*; negation has precedence over copula verbs, so that proverbs with negated copula were included in this class.

Nominal propositions (class P2F3), named NP_1 and NP_2, are treated as clausal propositions, even if they may contain no verbs and only have a 'clausal' or 'propositional content', such as in *Sorte no jogo, azar no amor* 'Lucky at cards, unlucky in love' (≈ *Quem tem sorte no jogo tem azar no amor* 'Who has luck at games has bad luck in love').

The proverbs were formalized into a tabular structure. The head noun (or pronoun) of noun phrases (NP) were identified (both the subject (N_0) and/or the complement (N_1), and eventual determiners (Det) or modifiers (Mod) are distributed across the corresponding columns. Eventual pre- or post-modifiers of verbs (*As más notícias chegam **depressa*** 'Bad news travel **fast**'), including obligatory auxiliary verbs (*Não deixe para amanhã o que você **pode** fazer hoje* 'Don't leave for tomorrow what you **can** do today'), and other elements, such as the impersonal pronouns (*Roupa suja **se** lava em casa* 'Don't wash your dirty laundry in public'), or obligatory negation (*Burro velho **não** aprende línguas* 'you can **not** teach an old dog new tricks') are also taken into consideration. Subordinative or coordinative elements are also provided with an adequate slot.

In this way, it is relatively simple to determine, and automatically extract, the *core* (or more representative) elements from each proverb, based on the classes' formal homogeneity.

---

[3] This is a romance title, available at: https://www.fanfiction.net/s/7279414/1/Cego-é-o-Amor [2014-03-08 13:11]

2.3   Determining Core Elements

In general, all lexically meaningful items were considered as core elements: nouns, verbs and adjectives. This includes the heads of all NPs in each proverb, which also involve all the variations grouped in those syntactic slots.

Sometimes, this head element is a pronoun (*e.g. Cada um por si* 'Every man for himself'); when the head is reduced (ellipsis) then the determiner or the modifier takes its place (*Um é pouco, dois é bom, três é demais* 'One is little, two is enough, three is too much' or *O ótimo é inimigo do bom* 'The perfect is the enemy of the good').

Depending on the formal class of the proverb, some other core elements may also be defined. For example, some proverbs do not present a main verb, so the determiners or the comparative conjunctions must be selected, along with the core nouns, such as in *Nem 8 nem 80* 'Neither 8 nor 80'; in this case, all those four words are selected as core elements.

2.4   Intersecting a Matrix of Core-Elements with Finite-State Transducers

The core elements associated to each proverb constitute a set of matrices of lexical items, one matrix per each class. Using UNITEX (version 3.1) linguistic development platform [22,23], a reference graph was built for each class.

Fig. 1 illustrates the reference graph for class P1F2, corresponding to single clause proverbs with the copula *ser* 'to be' as the main verb, *e.g. A vingança é um prato que se come frio* 'Revenge is a dish best served cold', which may allow for mirror-permutation (*e.g. O amor é cego = Cego é o amor* 'Love is blind = Blind is love).

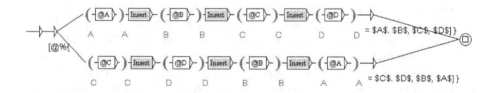

Fig. 1. Reference graph for class P1F2

This graph reads as follows: the system explores systematically each line in the table of the class core elements, replacing the variables @A, @B, *etc.*, by the corresponding content of columns A, B, *etc.* These input variables are then associated to output variables (the letters below the brackets), which are then to be reused in the output. In this case, in the output, the graph delimits the matched expression by brackets, and produces the content in a normalized form, introduced by the idiom number (the table's line number), represented by variable @%.

The shadowed boxes named *Insert* are subgraphs defining a window from 0 to 5 words and separators, allowed between the proverbs' core elements. This window insertion will be better explained in Section 3.

By intersecting the reference graph with the corresponding table, the system generates one subgraph for each line of the table, and a general resulting graph, containing all the subgraphs. The resulting graph can then be used to find patterns in texts. Next, we show a sample of a concordance of such matched strings from the PLN.Br corpus.

Table 2. Sample of a concordance of Class P1F2

1	[0002	carne é fraca= carne, é, fraca, , , ,]
2	[0003	esperança é a última que morre= esperança, é, última, morre, , ,]
3	[0009	Justiça é cega= Justiça, é, cega, , , ,]
4	[0009	Justiça não me olha porque é cega= Justiça, é, cega, , , ,]
5	[0014	pressa é inimiga da perfeição= pressa, é, inimiga, perfeição, , ,]
6	[0016	vida ser muito mais louca do que podemos imaginar= vida, ser, louca, , , ,]
7	[0017	vingança é prato que se deve comer frio= vingança, é, prato, frio, , ,]
8	[0017	Vingança é um prato que se serve frio= Vingança, é, prato, frio, , ,]
9	[0021	amar é cuidar= amar, é, cuidar, , , ,]
10	[0022	Amar É Sofrer= Amar, É, Sofrer, , , ,]
11	[0026	Desgraça pouca é bobagem= Desgraça, é, bobagem, , , ,]
12	[0027	Errar e horror é humano= Errar, é, humano, , , ,]
13	[0027	Errar é humano= Errar, é, humano, , , ,]
14	[0031	ninguém é profeta em sua terra= ninguém, é, profeta, terra, , ,]
15	[0032	amor é cego= amor, é, cego, , , ,]
16	[0033	buraco [...] é muito mais embaixo= buraco, é, embaixo, , , ,]
17	[0033	buraco é bem mais embaixo= buraco, é, embaixo, , , ,]
18	[0035	ótimo é inimigo do bom= ótimo, é, inimigo, bom, , ,]
19	[0036	pior cego é aquele que não quer ver= pior, cego, é, ver, , ,]
20	[0043	último será o primeiro= último, será, primeiro, , , ,]
21	[0044	pimenta nos olhos dos outros é refresco= pimenta, olhos, é, refresco, , ,]
22	[0046	querer é poder= querer, é, poder, , , ,]
23	[0046	querer não é poder= querer, é, poder, , , ,]
24	[0047	Recordar é viver= Recordar, é, viver, , , ,]
25	[0049	tempo é dinheiro= tempo, é, dinheiro, , , ,]

In this concordance, each line was numbered, followed by the number of the proverb type in the corresponding class, the actual words in the corpus and the core words that the transducer detected; empty variables are not represented (void commas). The table presents some interesting cases.

In line 4, for example, the proverb *A justiça é cega* 'Justice is blind' is the same as in line 3, but there is a wordplay in line 4, which is only possible because the proverb exists in the first place. Because of the window insertion (from 0 to 5 words between the core elements), the system captured the proverb in line 6, which is originally *A vida é louca* 'Life is crazy'.

The finite-state transducer also captured the expressions in lines 7 and 8, though they be variants of the same proverb *Vingança é um prato que se (come + serve) frio* 'Revenge is a dish best (eaten + served) cold'. Besides the core words *vingança* 'revenge', the verb *ser* 'to be' and *prato* 'dish', only the adjective *frio* 'cold' was considered as a core element, hence the verb variation can be captured.

Lines 12 and 13 present the same proverb *Errar é humano* 'To err is human', but, in line 12, the speaker introduced a new element for the proverb in a creative way *Errar e horror é humano* 'To err and horror is human', adapting the meaning of the proverb and the parophony of the words *errar* 'to err' and *horror* 'horror' to a different situation.

Lines 22 and 23 either captured the same proverb *Querer é poder* 'To want is to be able', normally as an affirmative sentence, but the speaker inserted a negative element in the original proverb, actually inverting the original/standard meaning in a creative way.

In this sense, it was possible to find other variants of proverbs than those we had previously retrieved from proverbs' collections and dictionaries.

3   Linguistic Remarks

This methodology allowed us to capture some variants of proverbs that were not in the previous lists, for example there are 12 variant forms of the same proverb *Antes tarde do que nunca* 'Better later than never':

(1). *(Antes + Melhor + é melhor) tarde (do + <E>) que (nunca + mais tarde)*[4]
    'It is better later than (never + even later)'

Most differences between variants of the same proverb consist in the variation of their grammatical elements, but they also vary on the lexical choices for some of their meaningful words.

(2). *A (ignorância + preguiça + ociosidade) é a mãe de todos os (erros + vícios + males)*
    '(Ignorance + sloth + idleness) is the mother of all (errors + vices + evils)'

Besides the variants of proverbs, we also found different proverbs than those we were expecting. For example, for the proverb *Quem sabe faz* 'Who knows makes', the system found both that form and another one *Quem sabe faz ao vivo* 'Who knows makes it live'. We considered them two different proverbs, and not only variants, because their overall meanings are different, so the method is also valid for searching new proverbs.

Another important decision for the task concerns the lemmatization of the meaningful words. Naturally, proper nouns were not lemmatized. Some verbs

---

[4] The items linked by "+" inside parentheses can comute in the given syntactic slot; the symbol <E> represents the empty string.

that only exist in the infinitive form, such as the proverbs *Querer é poder* 'To want is to be able', *Recordar é viver* 'To remember is to live' and *Amar é sofrer* 'To love is to suffer' were not lemmatized either.

Some proverbs that normally have 1 or 2 clauses may combine with others and generate a proverb with 3 or 4 clauses (or vice-versa, the longer form be splitted into their basic components). For example, the system found in the corpus two different proverbs *O silêncio é de ouro* 'Silence is golden' and *A palavra é de prata* 'The word is silver', but it found also the combination of those two single clauses as another proverb: *O silêncio é de ouro e a palavra é de prata* 'Silence is golden and the word is silver', so all those three forms were considered different entries in our lexicon.

Besides the core elements of each proverb, it is necessary to establish a window of possible insertions to allow for other elements (and punctuation marks) between the core elements. In a previous work [25], we defined a window containing 0 to 5 words between the core elements for all syntactic classes, and the precision rate was 60.15%. In this study, we tested the insertion of windows of different lengths for each class, by selecting the best performing window and precision rate raised from 60.15% to 74.68%. We used a window with 0-3 words for P1F1 and P2F3 classes, and another one with 0-5 words for the other classes.

4   Results

From the previous list of 614 proverbs, 557 matches were found in the PLN.Br corpus, from which 416 matches correspond to true positives. Since, to our knowledge, there is no available corpus annotated with proverbs and similar expressions, only precision was reported here (precision rate: 74.68%).

Table 3. Results of automatic identification of proverbs by class

Class	Proverbs	Matches	Types	True-Positives	False-Positives
P1F1	20	15	4	13	2
P1F2	53	91	21	75	16
P1F3	80	153	24	98	55
P1F4	53	61	15	61	0
P1F5	45	63	5	57	6
P2F1	39	40	7	39	1
P2F2	71	14	3	5	9
P2F3	48	40	8	15	25
P2F4	90	56	37	30	26
P2F5	20	3	1	3	0
P2F6	28	1	1	1	0
P3	24	0	0	0	0
Residual	43	20	8	19	1
Total	614	557	134	416	141

Table 3 shows the breakdown of these results by class. In spite of the number of matches, only 134 types (different proverbs) were found. The scarcity of the occurrence of proverbs in the corpus, as well as its reduced variety is most probably related to the journalist nature of the corpus. We decided to search these type of multi-word expression in journalistic corpus aiming at determining how frequently do they appear in this type of texts. It has been proved [30] that literary corpora contain a large number of proverbs, but the challenge consists in looking for them in non-literary texts as well.

5   Final Remarks and Future Works

One of the contributions of this work concerns the formal (syntactic) classification of proverbs in 13 classes; this classification may serve as a starting point for deeper analysis on each one of these proverbial structures, as it has been done for Spanish [1], French [10,11] and Italian [16].

We emphasize two advantages for the syntactic classification proposal: (i) the definition of an adequate extent of a window for insertions (words and punctuation), which vary depending on the formal class; and (ii) each class has specific properties, so we could apply specific transformations in each class, *e.g.* the mirror-permutation (P1F2 class), and the zeroing of negation elements (in P1F4 class) or of the main clause in double-clause classes (P2F1 and P2F2 classes).

Furthermore, we point out that (i) the methodology here presented to identify core elements' proverb is mostly language-independent and can be replicated for larger databases than the one used in this work; and (ii) finite-state automata can be applied to lexical matrices for automatic extraction of the proverbs and their variants in large corpus.

In future works, we intend to annotate a corpus or a sample of PLN.Br Full, aiming to compare the results between the automatic task and the human annotation, and then evaluate the work in terms of recall and F-measure. In the same way as we varied the window lenght by class, we want to experiment varying this lenght in different syntactic slots within the same structure and, eventually, determine which proverbs are more or less prone to formal variation.

For future works, we also intend to expand the database to include proverbs from European Portuguese [19], and then automatically build a proverbial database by using finite state automata and local grammars with discoursive markers that introduce proverbs and variants.

Acknowledgments. This work was supported by national funds through FCT - Fundação para a Ciência e a Tecnologia, under project PEst-OE/EEI/LA0021/ 2013 and by Capes/PDSE under Process BEX 12751/13-8. The authors would also like to thank the comments of the anonymous reviewers, which helped to improve this paper.

References

1. Brotons, M.L.N.: Las paremias y sus variantes: Análisis sintáctico, semántico y traductológico español/francés. Ph.D. thesis. Universidad de Alicante, Alicante, Spain (2008)
2. Bruckschein, M., Muniz, F., Souza, J.G.C., Fuchs, J.T., Infante, K., Muniz, M.: Gonçalez, P.N., Vieira, R., Aluisio, S.M.: Anotação linguística em xml do corpus PLN-BR. Série de relatórios do NILC, ICMC - USP (2008)
3. Bungum, L., Gambäck, B., Lynum, A., Marsi, E.: Improving word translation disambiguation by capturing multiword expressions with dictionaries. In: Proceedings of the 9th Workshop on Multiword Expression, Atlanta, Georgia, USA, pp. 21–30 (June 2013)
4. Caseli, H.M., Ramisch, C., Nunes, M.G.V., Villavicencio, A.: Alignment-based extraction of multiword expressions. Language Resources and Evaluation - Special Issue on Multiword Expression: Hard Going or Plain Sailing, 59–77 (2010)
5. Chacoto, L.: Estudo e formalização das propriedades léxico-sintácticas de expressões fixas proverbiais. Master's thesis. Faculdade de Letras da Universidade de Lisboa (1994)
6. Chacoto, L.: A sintaxe dos provérbios – As estruturas Quem/Quien en portugués e español. Cadernos de Fraseoloxía Galega 9, 31–53 (2007)
7. Chacoto, L.: Vale mais um gosto na vida que três vinténs na algibeira - Las estructuras comparativas en los proverbios portugueses. In: Aspectos Formales y Discursivos de Las Expresiones Fijas, pp. 87–103 (2008)
8. Conenna, M.: Sur un lexique-grammaire comparé de proverbes. Langages - Les expressions figées 90, 99–116 (1988)
9. Conenna, M.: Acerca del tratamiento informático de los proverbios. In: Léxico y Fraseología pp. 197–204 (1998)
10. Conenna, M.: Classement et traitement automatique des proverbes français et italiens. In: Lexique, Syntaxe et Sémantique: Mélanges offerts à Gaston Gross à l'occasion de son soixantième anniversaire. Special issue of Linguisticae Investigationes. BULAG, Numéro hors série, pp. 285–294 (2000)
11. Conenna, M.: Dictionnaire électronique de proverbes français et italiens. In: Englebert, A. (ed.) Actes du XXIIe Congrès International de Linguistique et de Philologie Romanes, pp. 137–145. Max Niemeyer Verlag, Bruxelles (2000)
12. Conenna, M.: Principes d'analyse automatique des proverbes. In: Leclère, C. (ed.) Syntax, Lexis & Lexicon-Grammar, Papers in honour of Maurice Gross, pp. 91–103. John Benjamins Publishing, Amsterdam (2004)
13. Gross, M.: Une classification des phrases figées du français. Révue Québécoise de Linguistique 11(2), 151–185 (1982)
14. Kordoni, V., Ramisch, C., Villavicencio, A. (eds.): Proceedings of the ACL Workshop on Multiword Expressions: From Parsing and Generation to the Real World (MWE 2011), Portland, OR, USA (June 2011)
15. Kordoni, V., Ramisch, C., Villavicencio, A. (eds.): Proceedings of the 9th Workshop on Multiword Expression, Atlanta, Georgia, USA (June 2013)
16. Lacavalla, C.B.: Lexique-grammaire des proverbes en Quand/Quando - Comparaison français-italien et représentation par grammaires locales. Ph.D. thesis. Università degli Studi di Bari, Bari, Itália (2007)
17. Laporte, É., Nakov, P., Ramisch, C., Villavicencio, A. (eds.): Proceedings of the COLING Workshop on Multiword Expressions: from Theory to Applications (MWE 2010), Beijing, China (August 2010)

18. Lopes, A.C.M.: Texto Proverbial Português - Elementos para u.ma análise semântica e pragmática. Ph.D. thesis. Universidade de Coimbra, Coimbra (1992)
19. Machado, J.P.: O grande livro dos provérbios. Coleção Estante Editorial. Editorial Notícias (1998)
20. Magalhães, Jr., R.M.: Dicionário brasileiro de provérbios, locuções e ditos curiosos: bem como de curiosidades verbais, frases feitas, ditos históricos e citações literárias, de curso corrente na língua falada e escrita. Documentário, 3rd edn., Rio de Janeiro (1974)
21. Palmer, M.: Complex predicates are multi-word expressions. In: Proceedings of the 9th Workshop on Multiword Expression, Atlanta, Georgia, USA, p. 31 (June 2013)
22. Paumier, S.: De la reconnaissance des formes linguistiques à l'analyse syntaxique. Ph.D. thesis. Université de Marne-la-Vallée (2003)
23. Paumier, S.P.: Unitex 3.1 - Manuel d'Utilisation (last version) edn. (2013)
24. Pinto, C.A.: Livro dos provérbios, ditados, ditos populares e anexins, 4th edn. Senac, São Paulo (2003)
25. Rassi, A.P., Baptista, J., Vale, O.: Automatic detection of proverbs and their variants. In: Proceedings of the III Symposium on Languages Technologies and Applications (SLATE 2014), Bragança, Portugal (June 2014)
26. Reis, S.M.M.: A correspondência entre provérbios e expressões fixas no Português Europeu. Master's thesis. Universidade do Algarve (2014)
27. Sag, I.A., Baldwin, T., Bond, F., Copestake, A., Flickinger, D.: Multiword expressions: A Pain in the Neck for NLP. In: Gelbukh, A. (ed.) CICLing 2002. LNCS, vol. 2276, pp. 1–15. Springer, Heidelberg (2002)
28. Sidhu, B.K., Singh, A., Goyal, V.: Identification of proverbs in Hindi text corpus and their translation into Punjabi. Journal of Computer Science and Engineering 2(1), 32–37 (2010)
29. Steinberg, M.: 1001 provérbios em contraste: Provérbios ingleses e brasileiros. Ática, São Paulo (1985)
30. Teixeira, J.: Mecanismos metafóricos e mecanismos cognitivos: Provérbios e publicidade. In: Actas del VI Congreso de Lingüística General, Madrid, pp. 2271–2280 (2007)
31. Teixeira, N.C.: O grande livro de provérbios. Leitura, Belo Horizonte (1942)

Using Cross-Linguistic Knowledge to Build VerbNet-Style Lexicons: Results for a (Brazilian) Portuguese VerbNet

Carolina Scarton[1,2], Magali Sanches Duran[2], and Sandra Maria Aluísio[2]

[1] Department of Computer Science, University of Sheffield,
Regent Court, 211 Portobello, Sheffield, S1 4DP, UK
c.scarton@sheffield.ac.uk
[2] NILC, ICMC, University of São Paulo,
Avenida Trabalhador são-carlense, 400 - Centro, 13566-590, São Carlos - SP
magali.duran@uol.com.br, sandra@icmc.usp.br

Abstract. In this paper, we present a new language-independent method to build VerbNet-based lexical resources. As a proof of concept, we show the use of this method to build a VerbNet-style lexicon for Brazilian Portuguese. The resulting resource was built semi-automatically by using existing lexical resources for English and Portuguese and knowledge extracted from corpora. The results achieved around 60% of f-measure when compared with a gold standard for Brazilian Portuguese, which is also described in this paper. The method proposed here also outperformed state-of-art machine learning method (verb clustering) by around 20% of f-measure.

Keywords: VerbNet.Br, VerbNet, lexico-semantic information, Levin's verb classes, verb clustering.

1 Introduction

The task of building and making available computational lexical resources (CLRs) is relevant to many fields of Natural Language Processing (NLP). A CLR with information about verbs, as addressed herein, is useful for information retrieval [1], semantic role labeling [2,3,4], word sense disambiguation [5,6,7,8] to cite just a few examples, since verbs carry a lot of syntactic and semantic information, being the verb meaning the main responsible for the sentence meaning[1].

English language has a tradition in CLRs and some of those which aggregate information about verbs are: WordNet [9], PropBank [10], FrameNet [11] and VerbNet [12]. Brazilian Portuguese (BP) has some initiatives in CLRs like WordNet.Br [13], FrameNet Brasil [14], FrameCorp [15] and PropBank.Br [16], all under development. Wordnet and Propbank inspired similar European initiatives ([17] and [18], respectively).The information these kind of resources aggregate follows different theories and each of them groups the verbs together in a different way.

---

[1] http://www.stanford.edu/~bclevin/lsa09intro-sr.pdf

J. Baptista et al. (Eds.): PROPOR 2014, LNAI 8775, pp. 149–160, 2014.
© Springer International Publishing Switzerland 2014

WordNet (probably, the most used CLR) is a lexical resource that organizes the lexicon following semantic relations, like synonymy, antonymy and hypernymy. The VerbNet, on its turn, is a lexical resource that organizes the verbal lexicon into classes that share similar semantic components at the same time that share the same syntactic alternations, by following Levin's theory [19]. One main difference of WordNet and VerbNet is the criteria for clustering verbs, which in WordNet do not include any syntactic restriction, whereas this is the very contribution of VerbNet, that is, to ensure that verbs of a same class are interchangeable in the same syntactic constructions

The manual building of CLRs is very time consuming and represents a heavy workload. In order to overcome these shortcomings, there are some initiatives to build these resources automatically. Two main recognized approaches are used to do this. The first one uses unsupervised machine learning (clustering) to build CLRs directly from corpora. The main advantages of such approach is its quickness and language independence (clustering methods are generic). The second approach uses resources available for a given language (usually English) to support the building of a new resource for another language. Taking advantage of English CLRs and their theories is relevant not only under the aspect of knowledge reuse, but also under the aspect of building aligned resources (see [20,13]). There are basically two ways to do this. One of them is to translate the English resource into another language. However, translation presents a disadvantage: events that are lexicalized in a way different from English remain undiscovered in the target language. The other way is to reuse the data structure, the theoretical principles and other related issues to extract the lexical items directly from the target language.

The research we report in this paper falls within the second way and presents a novel and language-independent method to semi-automatically build VerbNet-based CLRs, directly aligned to VerbNet, taking advantage of the alignments between existing CLRs in English and in the target language and relying on the hypothesis that verb classes present cross-linguistic features. The cross-linguistic potential of VerbNet classes was stated before for Italian [21] and for French [22]. Besides that, Kipper [12] also identified that this cross-linguistic potential appears for Portuguese, by experimenting with the class *carry-11.4* of VerbNet and Scarton et al. [23] applied algorithms and features used previously for verb clusetering for English into verb clustering for Portuguse. Moreover, as the method builds the new resource aligned to VerbNet, the tasks of validation and evaluation of the resource are easier than in cluster-based methods, since the information about the classes is explicit.

The proof of concept of our method takes Portuguese as language to study. Hereafter, we will refer to Brazilian Portuguese (BP) variant since we only used BP data. Portuguese speakers know that European Portuguese (EP) and BP have lexical differences and a CLR that only considers data from one variant (BP in this case) will only include word meanings of this variation. However, it is worth mentioning that the resource can be easily extended to include EP lexical items, by using data from EP corpora and resources. A VerbNet-style

CLR in Portuguese is relevant, since it contains information about verbs not provided by the CLRs above mentioned. Besides that, VerbNet is a key resource of SemLink Project [24], which links several CLRs like the cited above (VerbNet, WordNet, FrameNet and PropBank), and a Verbnet for Portuguese will enable the same model of linkage. The linkage of CLRs is useful to improve tasks in NLP, since each CLR presents different and complementary information.

In order to evaluate the results of the semiautomatic method, the verb clustering results for Portuguese of Scarton et al.[23] were used. As we show in Section 4, our semiautomatic method (with more deep linguistic knowledge) outperforms the clustering results.

We organized the remaining of this paper as follows. In Section 2 we present some related work. Section 3 presents the method for building the Brazilian Portuguese VerbNet (called VerbNet.Br[2] [25,26]). Section 4 contains the resource and its evaluation. Finally, in Section 5, we present some conclusions and future work.

2   Related Work

VerbNet deals with the syntactic-semantic interface of verbs to group them into classes. It is based on Levin's verb classes [19] and follows the hypothesis that verbs that share syntactic behavior should present argument structures with the same semantic roles. For example, the verbs "to spray" and "to load" in sentences (1) and (2) illustrate the locative alternation (extracted from [19], p. 2)).

1. (a) Sharon sprayed water on the plants.
   (b) Sharon sprayed the plants with water.
2. (a) The farmer loaded apples into the cart.
   (b) The farmer loaded the cart with apples.

In (1) and (2), the syntactic patterns (or subcategorization frames - SCFs) of (1a) and (2a) are similar: NP V NP PP, in which NP is a noun phrase, V is the verb and PP is a prepositional phrase with preposition "on" in (1a) and "into" in (2a) (despite the fact that the prepositions are different, they both introduced a place). The same happens for (1b) and (2b) in which the common SCF is NP V NP PP (the preposition "with" appeared in both cases).

Each class in VerbNet is described by its members (verbs that belongs to the class); by the thematic roles (roles that the verb arguments can assume, e.g. "Agent", "Theme"); by the selectional restrictions (restrictions for the thematic roles, e.g. "+animate", "-region"); by the syntactic frames (SCFs), and by semantic predicates, that are related to the SCFs, and describe the event that the verb members perform. VerbNet also has mappings to other CLRS like WordNet, FrameNet and PropBank [24].

There are also some studies aiming to build VerbNet-style lexicons automatically, using machine learning techniques. For English language, Kipper [12]

---

[2] http://www.nilc.icmc.usp.br/portlex/index.php/en/

explores clustering techniques to extend VerbNet verbs coverage, by using Prop-Bank corpus and the argument annotation as features for the clustering.

Sun et al. [27] and Sun and Korhonen [28] use classification and clustering to group English verbs into classes following VerbNet-style. The first explore four classification methods and one clustering method to cluster the verbs using SCFs as features and 17 classes as gold standard (GS). The best result is achieved by the Gaussian classification method (62.5% of f-measure). The second explore two methods of clustering considering the same GS as the previous one. They use features with information about lexical preferences, SCFs and selectional preferences. The best results are achieved with a spectral cluster method and selectional-preferences-based features (80.35% of f-measure).

For languages other than English, there are initiatives that explored the cross-linguistic potential of Levin classes. Merlo et al. [21] classify Italian verbs in three main classes of Levin (change of state, object drop and psych verbs), achieving the best result of 86.4% of accuracy. Ferrer [29] uses SCFs as features to cluster Spanish verbs following Levin hypothesis. The best result was 0.37 of average Silhouette. Sun et al. [22] cluster French verbs in VerbNet-style classes, following [28], achieving the best results with a spectral cluster method and features based on collocations (55.1% of f-measure). Falk et al. [30] propose a new method for clustering verbs, following the VerbNet-style, by using syntactic features from manually validated CLRs and semantic information acquired by translating the English verbs of VerbNet classes into French. The best result is 70% of f-measure. Scarton et al. [23] used the algorithm and features of Sun et al. [22] and a new cluster algorithm (called DCD) to cluster verbs in Portuguese. The best result was 42.77% of f-measure with spectral cluster algorithm and selectional-preferences-based features.

3   The Generic Method for Building VerbNet-Style Lexicons

To build the VerbNet.Br, three existing CLRs are used: VerbNet, WordNet and WordNet.Br. From VerbNet, the classes and theory were extended to Verb-Net.Br. From WordNet.Br, the BP verbs were selected as candidate members of VerbNet.Br. WordNet was only used to link VerbNet and WordNet.Br.

Our method is divided in four stages (Figure 1). Stage 1 uses manual work to define SCFs for VerbNet.Br classes (Section 3.1). Stage 2 searches in corpora for SCFs in BP (Section 3.2). Stage 3 uses the alignments among VerbNet, WordNet and WordNet.Br to define candidate members for VerbNet.Br (Section 3.3). Finally, Stage 4 selects the members for VerbNet.Br (by combining the other stages) (Section 3.4).

3.1   Stage 1: Manual definition of SCFs in VerbNet.Br classes

The first stage is manual and consists in reusing SCFs from VerbNet classes into VerbNet.Br. In other words, we "translate" the SCFs from English into Por-tuguese. This stage subsumes that English and Portuguese share cross-linguistic

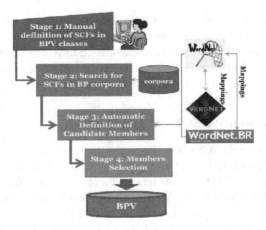

Fig. 1. VerbNet.Br method architecture

features of verbal classes. Only the SCFs that are productive in BP are "translated". It is worth emphasizing that only the SCFs are "translated", not the verbs (the verbs acquisition for VerbNet.Br is explained in Section 3.3). For example, the class *put-9.1* has two syntactic frames, as the examples for "to put" (*colocar* in Portuguese) ([19], p. 111):

3. I put the book on/under/near the table. / *Eu coloquei o livro em/sobre/sob/ entre a mesa.*
4. I put the book here/there. / *Eu coloquei o livro aqui/lá.*

In this case, both sentences present the same syntactic behavior in English and in Portuguese. For sentences in (3), the SCF is V NP PP and for sentences in (4), it is V NP ADVP, in which ADVP is an adverbial phrase. Then, we only translate the preposition and the SCFs V NP PP[*em/sobre/sob/entre*] and V NP ADVP became SCFs of class *put-9.1* in VerbNet.Br[3]. On the other hand, the class *send-11.1* of VerbNet presents a SCF that does not appear in Portuguese, as the example for the verb "to send" (*enviar* in Portuguese) ([19], p.132):

5. Nora sent the book to Peter / *Nora enviou o livro para Peter.*
6. Nora sent Peter the book. / *Nora enviou Peter o livro.*

Portuguese do not present the SCF of (6) for verbs like *enviar*. Therefore, only the first sentence has the SCF translation: V NP PP[to] into V NP PP[*para*] and the class *send-11.1* in VerbNet.Br does not include the SCF of (6).

Classes for which no SCF could be "translated" are marked as 'all', meaning that, in the last stage (Section 3.4) all candidate members should be considered

---

[3] SCFs in the whole method can be parameterized or non-parameterized by prepositions and do not consider subjects, since Portuguese admits the omission of the subject.

as members, despite the SCF they expressed. Such classes that lack SCF are more susceptible to low precision, as SCF is a feature used to filter candidates.

VerbNet contains 274 first level classes, but we have only considered the 213 main classes of VerbNet, compatible to the Levin classes (considering only the first level extensions proposed by [12] until class number 57).

3.2 Stage 2: Search for SCFs in BP Corpora

In this stage we search SCFs for BP verbs in corpora, in order to compare them with the "translated" SCFs (Stage 1). To do that, we use the system reported in Zilio et al. [31] that extracts SCFs for BP verbs from corpora annotated by PALAVRAS parser [32]. We use three corpora to ensure a considerable accuracy: Lácio-Ref [33] (approximately 9 million words), PLN-BR-FULL [34] (approximately 29 million words) and Revista Pesquisa FAPESP corpus[4] [35] (approximately 6 million words). As an example, consider the verb *colocar* (to put) of class *put-9.1*. Two examples, extracted from corpus, matching the SCFs of the respective class are: V NP PP[em] - *Durante o protesto, os 12 ambientalistas colocaram faixas e máscaras em algumas estátuas.* (During the protest, the 12 environmentalists put banners and masks on some statues.); and, V NP PP[sobre] - *Os militantes negros colocaram uma bandeira sobre a mesa.* (The black militants put a flag on the table.)

This stage was idealized as an attempt to overcome the differences among VerbNet and WordNet theories, which could cause the selection of wrong members in Stage 3 (Section 3.3).

3.3 Stage 3: Automatic Definition of Candidate Members

This stage explores the alignments among WordNet, VerbNet and WordNet.Br.

WordNet presents deep semantic relations between the verbs, being synonym the main relation, since the synsets are built based on this relation. For our research, we uses WordNet 2.0 version with 152,059 lemmas from which 11,306 lemmas and 13,508 synsets relate to verbs), since WordNet.Br is aligned to this version. VerbNet (presented in Section 2) contains 274 first level classes, 5,257 verb senses and 3,769 verb lemmas. It is also worth mentioning that we used the VerbNet 3.1 version with the mappings to WordNet developed within SemLink project.

The WordNet.Br project first built linguistically motivated synsets for BP and then identified correspondent synsets in WordNet (version 2.0). It is worth emphasizing that the BP synsets are not a translation from English synsets: they were originally built in BP by linguists using BP lexicographical material. After that, all the synsets of WordNet.Br were aligned to one or more synsets of WordNet. Since it was not possible to define only perfect alignments, they can be of four kinds (following [36]) and we only considered the ones related to synonym: EQ_SYNONYM: representing a perfect synonym between the synsets across the

---

[4] http://revistapesquisa.fapesp.br/en/

wordnets; and EQ_NEAR_SYNONYM: many-to-many alignment representing an imperfect synonym, but still a synonym.

Since each verb member in VerbNet can be mapped to WordNet synsets and WordNet.Br has its synsets aligned to one or more synsets of WordNet, it is possible to navigate on these links and define candidate members for VerbNet.Br. For example, the verb "to put" (*colocar*) of class *put-9.1* is mapped into a synset in WordNet that has the gloss "put into a certain place or abstract location". This synset is stored with the inter-lingual index (ILI) number 01452069. It is aligned to synset number 3314 in WordNet.Br, with the gloss "colocar em um certo lugar ou localização abstrata". This synset contains three verbs: *colocar* (to put, in this sense), *localizar* (to place, in this sense) and *situar* (to situate, in this sense). Therefore, these three verbs are considered as candidate members of *put-9.1* class in VerbNet.Br.

3.4 Stage 4: Selection of Classes Members

The Stage 4 combines the other three stages in order to select the classes members for the final resource. Since WordNet.Br has not been fully validated yet and there are differences between VerbNet and WordNet theories, a stage that filters the verbs is useful to eliminate spurious candidate members. We execute four experiments: two experiments consider the manual stage and the other two do not consider it. As baseline we consider all candidate members with frequency higher than 10, referred herein as EXP0.

For all experiments introduced in next sections, in order to avoid errors of parsing and/or extraction of SCFs, we only consider as candidate members verbs presenting frequency higher than 10 and SCFs with frequency higher than 5 in the corpora presented in Section 3.2. The results for all experiments are presented in Section 4.

Experiments that considered Stage 1: Experiments that considered Stage 1 use information defined manually to select the members of VerbNet.Br. We evaluate with two scenarios: one using SCFs parameterized by prepositions (EXP1) and another using SCFs non-parameterized by prepositions (EXP2). Parametrization is a more restrict way to select members (only verbs that presents a given preposition are selected) and it is expected that this methods will select less verbs than the non-parameterized one. However, it is also expected that the parameterized method leads to higher precision.

These experiments are implemented as follows: for each candidate member (defined in Stage 3), all the SCFs extracted from corpus (Stage 2 parameterized or not) are recovered and compared with the SCFs translated for the class (Stage 1 parameterized or not). If there is a match (with 10% of threshold empirically defined) the verb is selected. For example, consider the class *put-9.1*. Only the SCF presented in (1) could be used for this class, since the tool used to extract SCFs in Stage 2 did not identify ADVPs. Therefore, all candidate members of *put-9.1* should present the SCF of (1) to be selected.

Two candidate members of *put-9.1* are: *colocar* (to put) and *sepultar* (to bury). For both experiments, EXP1 and EXP2, the verb *colocar* (to put) presents the required SCF, respectively V NP PP[*em*] and V NP PP, therefore it is selected in both cases. The verb *sepultar*, on the other hand, do not present any SCF related to this class and, for this reason, it is not selected.

Experiments without Stage 1: Two experiments without the manual stage (Stage 1) are also conducted. In this approach, we select for each class the most frequent SCF. This means that, to represent each class, we select the SCF presented by the majority of candidate members within the given class. Then, the verbs presenting such SCF are selected. We also experiment with SCFs parameterized (EXP3) and non-parameterized (EXP4) by prepositions.

For example, for the class *put-9.1* the SCF selected for the parameterized version is V NP (since the most frequent SCF presented no preposition). On the other hand, the SCF selected for the non-parameterized version is the same: V NP. The verb *colocar* (to put) is selected again; although, the verb *sepultar* (to bury) is wrongly selected in this case.

4   Results and Evaluation

4.1   Gold Standard for Brazilian Portuguese

We use the second version of the GS described in Scarton et al. [23].

The first version of this GS was based on the English GS [27] which contains 17 classes of VerbNet with 12 verbs per class. Following the method described in Sun et al. [22] for building the French GS, a native speaker of Portuguese and expert in Levin's theory translated the verbs and defined some Portuguese syntactic restrictions for each class. 203 verbs in 16 classes (12.69 verbs per class) were considered, one class less than English GS, because the class *peer-30.3* was incompatible with Portuguese. The second version of this GS includes candidate members of VerbNet.Br (acquired in Stage 3) that did not appear in the first version. The candidate members in the GS were validated by a linguistic expertise. The new GS total 540 lemmas of verbs grouped into 16 classes (33.75 verbs per class).

4.2   Evaluation

Since we work with verb selection for classes, we use supervised ML measures considering in the positive class the verbs of the class and in the negative class the verbs that do not belong to the class. Therefore, for each class, we define the true positives (TP), false positives (FP), true negatives (TN) and false negatives (FN). We apply the evaluation measures in the baseline (EXP0, i.e., all the candidates obtained from mappings) and in all the experiments for every class of the GS.

To evaluate the significance of the results of the four experiments against EXP0, we apply the two-tailored pairwise t-test with p>0.05. To evaluate the significance of the results among the four experiments, we use the ANOVA test, also with p>0.05. Both tests are applied for precision, recall and f-measure. We also considered the results for verb clustering with the MNCut Spectral Cluster algorithm and SCFs non-parameterized (SPEC-1) and parameterized (SPEC-2) as features, reported by Scarton et al. [23]. Table 1 shows the results for global precision, recall and f-measure for all experiments and the baseline (EXP0). It also shows the results for verb clustering considering both SCFs feature-sets.

Table 1. Evaluation of VerbNet.Br

	EXP0	EXP1	EXP2	EXP3	EXP4	SPEC-1	SPEC-2
Precision	41.51	**44.99**	43.36	43.30	43.26	32.17	37.59
Recall	**100**	92.87	96.17	92.40	92.40	32.58	39.14
F-measure	58.67	**60.62**	59.77	58.97	58.93	33.62	42.27

The best result for precision is achieved by EXP1 (44.99%), although, there is no significant difference between EXP1-EXP4 and the baseline. In addition, the results for EXP1-EXP4 do not lead to a conclusion either way (ANOVA).

As we expected, for recall, EXP0 achieves the best result (since all the candidate members are selected as members, there is no false negative in this case) and there is significant difference between EXP0 and the others (t-test). For the experiments, the best result is achieved by EXP2, and it is also expected that EXP2 presented a better recall than EXP1, since the non-parameterized version tends to select more verbs, because it is less restrictive. However, there is no significant difference between the experiments (ANOVA).

EXP1 presents the best global result (60.62%) for f-measure, although, no significant difference is found between EXP1-EXP4 and EXP0. There is also no significant difference between EXP1-EXP4 (ANOVA).

Comparing the results of VerbNet.Br and clustering, even the baseline outperforms the clustering results (for f-measure, clustering show results lower than 45% whereas EXP0-EXP4 show results higher than 58%). It is expected, since the method presented in Section 3 uses more linguistic knowledge.

Since there is no significant difference between each experiment and the baseline EXP0, it would have been enough to use only the alignments among VerbNet, WordNet and WordNet.Br to achieve good results. However, there are some differences between EXP0, EXP1-EXP2 and EXP3-EXP4 that must be discussed. EXP3 and EXP4 are completely automatic and, therefore, are less time consuming than the others. On the other hand, EXP1 and EXP2 contain information that enable inheriting more data from VerbNet. For example, the translated SCFs in Stage 1 could inherit all the semantic predicates of their correspondent patterns in VerbNet.

5   Conclusion and Future Work

The first results for a BP VerbNet, built through a semi-automatic method, based on existing resources is presented in this paper. The experiments showed that the pair of languages Portuguese and English share most of the verb classes of the GS, which reinforces the hypothesis that Levin classes may have a large cross-linguistic potential.

As discussed in Section 4.2, the advantage of using the manual stage (Stage 1) is to inherit the information of semantic predicates. On the other hand, EXP3 and EXP4 carry interesting information about Portuguese verb classes: the SCFs that the major part of the candidate members presents. An interesting test would be to combine both kinds of experiments and evaluate their results. This is left for future work. Another interesting evaluation would be to align VerbNet.Br with PropBank.Br, by using the alignments among VerbNet, PropBank and PropBank.Br and this is also a future work.

A relevant advantage of this method is being language-independent. It can be used to build a VerbNet-like resource for other languages, provided that such languages have a wordnet aligned to WordNet and a set of SCFs per verb (or, at least, a tool to extract them from corpora).The best result (60.62% of f-measure), that is better than the results for verb clustering for Portuguese (SPEC-1 - 33.62%, SPEC-2 - 42.27% and also the best result of 42.77%), is also higher than the best result achieved for French verb clustering in Sun et al. (2010) (55.1%). However, it is lower than the one acquired for English in Sun and Korhonen (2009) (80.4%) and for French in Falk et al. (2012) (70%). Nevertheless, it is worth mentioning that these results for other languages were built by methods based on clustering. These methods carry a heavy workload on the evaluation of the acquired clusters, since they are grouped without any clue to the classes related to them. Our method is better in this sense, since we can directly address the classes, based on VerbNet.

The work reported herein is a computational effort to design a new resource reusing preexisting lexical resources. Clearly, the results do not constitute a final version of the new resource, but they lessen the linguistic effort that would be necessary to build a Verbnet-like lexicon from scratch. Results obtained for other languages using other approaches have benefited from the existence of lexical resources that we did not have for Portuguese when we made the experiments. This is the case, for example, of the dictionary of SCFs available for French verbs. Finally, it is important to stress that initiatives like our VerbNet.Br, PropBank.Br and WordNet.Br, inspired in English lexical resources, do not ignore the point of view of Portuguese language, as a superficial analysis might have suggested. Such resources share the theoretical motivation, the set of semantic roles and the database structure of their English counterparts. However, the lexical units that integrate them came from lexicographic material originally elaborated in Portuguese (in the case of WordNet.Br), extracted from corpus in Portuguese (in the case of PropBank.Br) and from WordNet.Br (in the case of VerbNet.Br).

Acknowledgements. This work was supported by FAPESP/Brazil (No. 2010/03785-0 and No. 2011/22882-0) and EXPERT (EU Marie Curie ITN No. 317471) project.

References

1. Crouch, D., King, T.H.: Unifying Lexical Resources. In: Interdisciplinary Workshop on the Identication and Representation of Verb Features and Verb, Saarbruecken, Germany, pp. 32–37 (2005)
2. Yi, S., Palmer, M.: Pushing the boundaries of Semantic Role Labeling with SVM. In: ICON 2004, Hyderabad, India (2004)
3. Swier, R., Stevenson, S.: Unsupervised Semantic Role Labelling. In: EMNLP 2004, Barcelona, Spain, pp. 95–102 (2004)
4. Yi, S., Lopper, E., Palmer, M.: Can Semantic Roles Generalize Across Genres? In: NAACL HLT 2007, Rochester, NY, USA, pp. 548–555 (2007)
5. Girju, R., Roth, D., Sammons, M.: Token-level Disambiguation of Verbnet Classes. In: Interdisciplinary Workshop on the Identification and Representation of Verb Features and Verb Classes, Saarbruecken, Germany (2005)
6. Abend, O., Reichart, R., Rappoport, A.: A Supervised Algorithm for Verb Disambiguation into Verbnet Classes. In: LREC 2008, Manchester, UK, pp. 9–16 (2008)
7. Chen, L., Eugenio, B.D.: A Maximum Entropy Approach to Disambiguating Verbnet Classes. In: 2nd Interdisciplinary Workshop on Verbs, The Identification and Representation of Verb Features, Pisa, Italy (2010)
8. Brown, S.W., Dligach, D., Palmer, M.: Verbnet Class Assignment as a WSD Task. In: IWCS 2011, Oxford, UK, pp. 85–94 (2011)
9. Fellbaum, C.: WordNet: An electronic lexical database. MIT Press, Cambridge (1998)
10. Palmer, M., Gildea, D., Kingsbury, P.: The Proposition Bank: A Corpus Annotated with Semantic Roles. Computational Linguistics 31(1), 71–106 (2005)
11. Baker, C.F., Fillmore, C.J., Lowe, J.F.: The Berkeley Framenet Project. In: 36th Annual Meeting of the Association for Computational Linguistics and 17th International Conference on Computational Linguistics, pp. 86–90. University of Montréal, Canadá (1998)
12. Kipper-Schuler, K.: Verbnet: A broad coverage, comprehensive verb lexicon. Doctor of philosophy. University of Pennsylvania (2005)
13. Dias da Silva, B.C., Felippo, A.D., Nunes, M.G.V.: The Automatic Mapping of Princeton Wordnet lexical-conceptual relations onto the Brazilian Portuguese Wordnet database. In: Proc. LREC 2008, Marrakech, Morocco, pp. 1535–1541 (2008)
14. Salomao, M.M.: Framenet Brasil: Um trabalho em progresso. Revista Calidoscópio 7(3), 171–182 (2009)
15. Bertoldi, A., Chishman, R.: Frame semantics and legal corpora annotation: Theoretical and applied challenges. Linguistic Issues in Language Technology 7(9) (2012)
16. Duran, M.S., Aluisio, S.M.: Propbank-br: A brazilian treebank annotated with semantic role labels. In: LREC 2012, Istanbul, Turkey (2012)
17. Marrafa, P.: Portuguese wordnet: general architecture and internal semantic relations. DELTA 18, 131–146 (2002)
18. Branco, A., Carvalheiro, C., Pereira, S., Avels, M., Pinto, C., Silveira, S., Costa, F., Silva, J., Castro, S.: A propbank for portuguese: The cintil-propbank. In: Proc. LREC 2012, Istanbul, Turkey, pp. 1516–1521 (2012)

19. Levin, B.: English Verb Classes and Alternation, A Preliminary Investigation. The University of Chicago Press, Chicago (1933)
20. Palmer, M.: Semlink: Linking propbank, verbnet and framenet. In: Generative Lexicon Conference, Pisa, Italy (2009)
21. Merlo, P., Stevenson, S., Tsang, V., Allaria, G.: A multilingual paradigm for automatic verb classification. In: ACL 2002, Philadelphia, PA, pp. 207–214 (2002)
22. Sun, L., Korhonen, A., Poibeau, T., Messiant, C.: Investigating the cross-linguistic potential of Verbnet-style classification. In: COLING 2010, Beijing, China, pp. 1056–1064 (2010)
23. Scarton, C., Sun, L., Kipper-Schuler, K., Duran, M.S., Palmer, M., Korhonen, A.: Verb Clustering for Brazilian Portuguese. In: Gelbukh, A. (ed.) CICLing 2014, Part I. LNCS, vol. 8403, pp. 25–39. Springer, Heidelberg (2014)
24. Loper, E., Yi, S., Palmer, M.: Combining lexical resources: Mapping between propbank and verbnet. In: 7th International Workshop on Computational Linguistics, Tilburg, Netherlands (2007)
25. Scarton, C.: Verbnet.br: Construção semiautomática de um léxico computacional de verbos para o Português do Brasil. In: STIL 2011, Cuiabá, MT, Brazil (2011)
26. Scarton, C., Aluísio, S.M.: Towards a cross-linguistic Verbnet-style lexicon to Brazilian Portuguese. In: CREDISLAS 2012, in Conjunction with LREC 2012, Istanbul, Turkey (2012)
27. Sun, L., Korhonen, A., Krymolowski, Y.: Verb class discovery from rich syntactic data. In: The 9th International Conference on Computational linguistics and Intelligent Text Processing, Haifa, Israel, pp. 16–27 (2008)
28. Sun, L., Korhonen, A.: Improving verb clustering with automatically acquired selectional preferences. In: EMNLP 2009, Singapore, pp. 638–647 (2009)
29. Ferrer, E.E.: Towards a semantic classification of spanish verbs based on subcategorisation information. In: The Workshop on Student Research, in Conjunction with ACL 2004, Barcelona, Spain, pp. 163–170 (2004)
30. Falk, I., Gardent, C., Lamirel, J.C.: Classifying french verbs using french and english lexical resources. In: ACL 2012, Jeju, Republic of Korea, pp. 854–863 (2012)
31. Zilio, L., Zanette, A., Scarton, C.: Automatic extraction of subcategorization frames from portuguese corpora. In: Aluisio, S.M., Tagnin, S.E.O. (eds.) New Languages Technologies and Linguistic Research: A Two-Way Road, pp. 78–96. Cambridge Scholars Publishing (2014)
32. Bick, E.: The Parsing System Palavras: Automatic Grammatical Analysis of Portuguese in a Constraint Grammar Framework. Doctor of philosophy. University of Aarhus (2005)
33. Aluísio, S.M., Pinheiro, G.M., Manfrim, A.M.P., Genovês Jr., L.H.M.G., Tagnin, S.E.O.: The Lácio-web: Corpora and Tools to Advance Brazilian Portuguese Language Investigations and Computational Linguistic Tools. In: LREC 2004, Lisbon, Portugal, pp. 1779–1782 (2004)
34. Muniz, M., Paulovich, F.V., Minghim, R., Infante, K., Muniz, F., Vieira, R., Aluísio, S.: Taming the tiger topic: An xces compliant corpus portal to generate subcorpus based on automatic text topic identification. In: CL 2007, Birmingham, UK (2007)
35. Aziz, W., Specia, L.: Fully automatic compilation of a Portuguese-English parallel corpus for statistical machine translation. In: STIL 2011, Cuiabá, MT (October 2011)
36. Vossen, P.: Eurowordnet: A multilingual database of autonomous and language specific wordnets connected via an interlingual-index. International Journal of Linguistics 17 (2004)

The Creation of Onto.PT:
A Wordnet-Like Lexical Ontology for Portuguese

Hugo Gonçalo Oliveira

CISUC, Dept. of Informatics Engineering, University of Coimbra, Portugal
hroliv@dei.uc.pt

Abstract. A wordnet is an important tool for developing natural language processing applications for a language, but the manual creation of such a resource limits its development. This dissertation studied the automatic construction of Onto.PT, a large Portuguese wordnet, aiming to minimise the main limitations of existing Portuguese wordnets. On this context, we propose ECO, an approach for creating wordnets automatically from text – relation instances are extracted, synonymy clusters (synsets) are discovered, and the remaining relations are then attached to suitable synsets. This document also reports on the contents of Onto.PT, its comparison to other wordnets, and its evaluation.

1 Introduction

The existence of a broad-coverage lexical-semantic knowledge base has a positive impact on the computational processing of its target language. This is the case of Princeton WordNet (hereafter, WN.Pr), for English, which has successfully been used in a wide range of natural language processing (NLP) tasks. WN.Pr is, however, created manually by experts so, despite ensuring highly reliable contents, its creation is expensive, time-consuming and has negative consequences on its coverage and growth. For Portuguese, there are several wordnets, but none is as successful as WN.Pr is for English, as they all have limitations, that go from coverage to availability issues.

We set the final goal of this dissertation, described in detail in the author's PhD thesis [1], to the automatic construction of Onto.PT, a lexical ontology for Portuguese, structured in a similar fashion to WordNet. Onto.PT contains synsets – groups of synonymous words which are lexicalisations of a natural language concept – and semantic relations, held between synsets. For this purpose, we developed ECO, a model focused on the development of computational tools for the acquisition and organisation of lexical-semantic knowledge from text.

ECO starts by exploring textual sources for the extraction of relations, connecting lexical items according to their possible senses. As natural language is ambiguous, a lexical item, identified by its orthographical form, is sometimes not enough to denote a concept. Therefore, two additional steps are performed towards a sense-aware resource. First, the synsets of a synset-based thesaurus are augmented with new synonyms acquired in the first step, and new synsets

J. Baptista et al. (Eds.): PROPOR 2014, LNAI 8775, pp. 161–169, 2014.

are discovered from the remaining synonymys, after the identification of word clusters. Second, the full set of extracted relations is exploited for attaching the arguments of the non-synonymy relations to the most suitable synsets.

After enumerating some related work, this document addresses ECO and the creation of Onto.PT. An overview of Onto.PT v0.35 is then provided, along with its comparison to other wordnets, figures on its global coverage, and on its manual evaluation. Onto.PT, which can be further augmented, is freely available for download and can be used in a wide range of NLP tasks for Portuguese, as WN.Pr is for English. Despite the current limitations of an automatic creation approach, we are positive that it will contribute for advancing the state-of-the-art of the computational processing of Portuguese.

2   Related Work

This section enumerates the limitations of other Portuguese wordnets, and then describes work on lexical-semantic information extraction.

2.1   Portuguese Wordnets

There are four known Portuguese wordnets – Wordnet.PT (WN.PT) [2], Wordnet.Br (WN.Br) [3], MultiWordNet.PT (MWN.PT)[1] and OpenWordnet.PT (OWN.PT) [4]. All of them have strong limitations, listed below:

- **Size:** they are substantially smaller than WN.Pr.
- **Coverage:** they cover a small relation set, MWN.PT covers only nouns.
- **Creation:** they are handcrafted, which limits their growth/maintenance; all but WN.PT rely on the translation of WN.Pr (MWN.PT, OWN.PT), or its relations (WN.Br). In MWN.PT, this results in explicit lexical gaps;
- **Availability:** only OWN.PT is freely available. MWN.PT is available under a paid license, and WN.PT and WN.Br are not available for download.

With the development of Onto.PT, we wanted to tackle most of those limitations. So, Onto.PT would: (i) be larger than other Portuguese wordnets; (ii) cover a wide range of semantic relations; (iii) be created automatically from Portuguese resources; (iv) be freely available.

2.2   Lexical-Semantic Information Extraction

Research on the automatic acquisition of lexical-semantic knowledge from text dates back to the 1970s. Electronic dictionaries were the primary sources exploited, because they cover words and meanings extensively, they are created by experts, and they use systematic definitions. Continued work on information extraction (IE) from dictionaries led to to the automatic creation of MindNet [5], an independent LKB. Since then, dictionaries have been used in the acquisition

---

[1] See http://mwnpt.di.fc.ul.pt/

of ontologies (e.g. [6]) and other IE tasks, with some recent attention given to the collaboratively created Wiktionary (e.g. [7]).

Despite several automatic attempts to the creation of a broad-coverage LKB, for English, WN.Pr, a manual effort, ended up the leading resource of this kind. Though, given the limitations of handcrafting knowledge bases, research on lexical-semantic IE proceeded, whether for the creation of new LKBs, or for their enrichment, towards minimising information sparsity issues.

With the availability of huge amounts of text on most domains, researchers turned on to IE from less structured sources. While targeting textual corpora, unsupervised procedures were developed for clustering words according to their distributional similarity [8], useful for identifying synonyms and creating thesauri. Most research on relation extraction from text is inspired by Hearst [9], who discovered discriminating patterns from seed relation instances, then used to extract new instances. The previous method was combined with distributional measures to improve extraction [10], and generalised to fully [11] and weakly-supervised approaches [12] for relation extraction. The latter were also applied in relation extraction from the Web, where great amounts of facts were learned and integrated in large knowledge bases [13].

To increase their coverage, wordnets can also be linked to other resources, such as FrameNet and VerbNet [14], Wikipedia [15], or to all the previous and Wiktionary [16]. Due to its relevance to this work, the task of ontologising [17], which maps terms to synsets/concepts, should also be mentioned.

Most of the previous works either extract knowledge to enrich an existing resource, or merge resources. The other few are not concerned with the creation of a wordnet from scratch – some are not restricted to lexical-semantic knowledge, and most do not structure knowledge as a wordnet. Alternatively, we propose an approach that may exploit existing resources, but can be used to create a wordnet automatically, completely from scratch, and in an unsupervised way.

3   ECO: The Construction of Onto.PT

ECO [18] is the model we propose for creating a wordnet automatically from textual sources in a target language. It combines IE techniques to acquire, organise and integrate lexical-semantic knowledge, along three main steps, as shown in figure 1. This is a flexible way of coping with information sparsity, since it allows for the extraction of knowledge from heterogeneous sources (e.g. dictionaries, thesauri, corpora), and provides a way to harmoniously integrate all the acquired information in a common knowledge base.

Our implementation towards the construction of Onto.PT is illustrated in figure 2. From a dictionary definition, two relation instances (hereafter, tb-triples) are extracted. Available synsets are then augmented with the synonymy instance (hereafter, synpair). In the last step, the arguments of the non-synonymy tb-triples are attached to synsets, resulting in a synset relation instance (hereafter, sb-triple). Details on these steps are described in the following subsections.

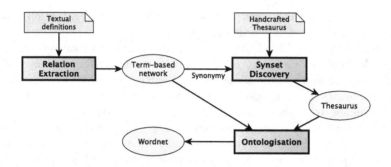

Fig. 1. Diagram of the ECO approach

Extraction		
gado	s.m.	conjunto de animais criados para diversos fins; rebanho
	$tb - triple\_1$	= rebanho SINONIMO_DE gado
	$tb - triple\_2$	= animal MEMBRO_DE gado
Clustering		
	$synset_1$	= {manada, rebanho, mancheia, boiada}
$synset_1 + tb - triple_1$		= {manada, rebanho, mancheia, boiada, gado}
Ontologising		
	$synset_2$	= {bicho, animal, alimal, béstia, minante}
	$sb - triple_1$	= $synset_2$ MEMBRO_DE $synset_1$

Fig. 2. Example of the three ECO steps

Extraction. Tb-triples are acquired from dictionaries, using handcrafted grammars[2], originally created in the scope of PAPEL [19]. After confirming that most regularities were preserved in the definitions of Dicionário Aberto [20] and Wiktionary.PT[3], the same grammars were used for extracting tb-triples from these dictionaries and merging them with PAPEL's lexical network [21].

Clustering. Clusters of synonymous lexical items, discovered among the extracted synpairs, are used as synsets. This could be done in the full lexical network [22], but TeP [23], a handcrafted thesaurus, was our starting point. OpenThesaurus.PT[4], four times smaller and not created by experts, was converted to synpairs and used together with those extracted, in two substeps [24]:

1. **Synset enrichment:** synpairs are assigned to the most similar TeP synset with one of the synpair items. The higher the similarity between synpair and synset, computed with the average cosine of the adjacencies of each involved word, the higher the confidence on their assignment.
2. **Clustering:** remaining synpairs are either not similar enough or have both items not covered by TeP. They establish a smaller network, then exploited for discovering new synsets, using a graph-clustering procedure.

---

[2] PAPEL grammars available from http://www.linguateca.pt/PAPEL/

[3] Wiktionary.PT is available from http://pt.wiktionary.org/

[4] OpenThesaurus.PT is available from http://openthesaurus.caixamagica.pt/

Ontologising. Algorithms were developed for exploiting the lexical network acquired during the extraction step and mapping the arguments of the non-synonymy tb-triples to suitable synsets [25]. After a performance comparison, hypernymy was ontologised with the average cosine (AC) between the adjacencies of the words involved. For the remaining relations, each argument was fixed for mapping the other to the synset with most words related the same way to the fixed argument. When confidence was low, AC was applied.

4 Onto.PT and Its Evaluation

From its webpage – `http://ontopt.dei.uc.pt/` – Onto.PT is freely available as a RDF/OWL model, and can be queried online through OntoBusca, a web interface. This section is about its contents and evaluation.

4.1 Quantitative Data

Onto.PT v0.35 contained about 109,000 synsets, of which about 105,000 were involved in at least one sb-triple. As for relations, it contained about 173,000 (+inverse) sb-triples with the same types covered by PAPEL. Almost half of them are hypernymy ($\approx$80,300) and property-of is the second relation with most sb-triples ($\approx$25,100 between adjectives and verbs, and $\approx$9,700 between adjectives and nouns). The remaining relations are purpose-of ($\approx$15,300), causation ($\approx$9,800), part-of ($\approx$8,500), member-of ($\approx$7,000), antonymy ($\approx$6,000, some from TeP), manner-of ($\approx$3,400), producer-of ($\approx$2,100), quality-of ($\approx$2,000), contained-in ($\approx$600), state-of ($\approx$500), place-of ($\approx$300), manner-without ($\approx$200).

4.2 Comparison with Portuguese Wordnets

Table 1 puts Onto.PT side-by-side the other Portuguese wordnets with some figures on their size, in terms of covered lexical items, included synsets and semantic relations. These numbers may be compared to those of WN.Pr, to show that they are substantially smaller, except for Onto.PT. Despite being the second youngest Portuguese wordnet (OWN.PT is the youngest), its size is comparable to WN.Pr, and it covers a richer set of relations. We should recall that, among others, Onto.PT v0.35 integrates the synsets of WN.Br (TeP), so it was expected to be larger than those two.

Table 1. Portuguese wordnets vs. Princeton WordNet

Resource	Lexical items	Synsets	Relations
WN.PT	11k	13k	40k
WN.Br	44k	20k	N/A
MWN.PT	16k	17k	69k
OWN.PT	48k	39k	83k
Onto.PT	162k	109k	340k
WN.Pr 3.0	155k	118k	285k

Although size is not the most important property of a wordnet, these numbers show the benefits of automatic creation. Besides typically larger resources, automatic approaches provide a faster creation, an easier maintenance, and a higher growth potential, in a trade-off on the virtual 100% reliability. For Portuguese, selecting the most adequate(s) wordnet(s) to use in a project should consider, among others, the language coverage needs, the tolerance to errors and the available budget.

4.3   Global Coverage

We assessed Onto.PT's coverage of the 164 base concepts provided by the Global WordNet Association[5] and referred as the "most important" in the wordnets of English, Spanish, Dutch and Italian. They are divided into 98 abstract and 66 concrete concepts, and are represented as WN.Pr 1.5 synsets. Based on rough manual matches with the Onto.PT v0.35 synsets, we concluded that this resource covers 92 abstract and 61 concrete synsets (93%).

4.4   Manual Evaluation

Onto.PT v0.35 was the target of an extensive manual evaluation, where the synsets were, first, considered alone and, then, in the sb-triples. Two random samples of 300 sb-triples, one with hypernymy relations and another with all the other were selected. Each synset with more than one word was detached from its sb-triple, reduced to its top-3 most frequent words in the AC/DC corpora [26], and presented, independently, to two human judges. From the 774 distinct synsets, 572 (73.9%) and 58 (7.5%) were respectively classified as correct and incorrect by both judges. They did not agree in the remaining 144 (18.6%).

Sb-triples connecting one synset classified twice as incorrect were removed from the original samples. The remaining sb-triples (247 hypernymy, 267 other) were presented, independently, to two human judges. The proportion of correct hypernymy tb-triples was 78%-79%, depending on the judge, with $\kappa = 0.47$. For the other relations, these proportions were 88%-92%, with $\kappa = 0.48$. If the sb-triples connecting incorrect synsets are considered, these numbers drop to ≈65% (hypernymy) and 78-82% (other). The analysis of these results led to minor corrections in the most recent version of Onto.PT [27].

5   Concluding Remarks

We have described the construction of Onto.PT, a new wordnet for Portuguese which is now a valuable public alternative in the paradigm of Portuguese lexical-knowledge bases. Onto.PT is created with ECO, an alternative to the time-consuming manual construction of wordnets, which intends to minimise information sparsity issues by exploiting several heterogeneous textual resources.

---

[5] Base concepts available from http://www.globalwordnet.org/gwa/ewn_to_bc/corebcs.html

Of course, this is performed in a trade-off of lower reliability than that of hand-crafted resources. So far, Onto.PT has been used for synonymy expansion [28], and cloze question answering [29], where it played the role of a word sense inventory. Nevertheless, the evaluation results show that there is still a long way to go until we have a highly-reliable resource. Other important contributions of this work are the resources obtained from each ECO step, namely: a large lexical network [21], a fuzzy [22] thesaurus, and a simple thesaurus [24].

We should recall that any resource created by ECO cannot be seen as static. So, in the future, new instances of Onto.PT can be created, not only by improving issues identified along the performed evaluations, but also by exploiting other resources. In fact, after Hugo's dissertation, there were recent developments on Onto.PT [27], including the assignment of dictionary definitions as synset glosses and the integration of OWN.PT synsets. Other cues for further work include studying the integration of more resources (e.g. Wikipedia), the assignment of confidence values to its contents, and the representation in other data formats.

Acknowledgements. Hugo's PhD was developed at CISUC, supervised by Paulo Gomes, and supported by FCT scholarship grant SFRH/BD/44955/2008. The development of Onto.PT is currently supported by the iCIS project (CENTRO-07-ST24-FEDER-002003), co-financed by QREN, in the scope of the Mais Centro Program and European Union's FEDER.

References

1. Gonçalo Oliveira, H.: Onto.PT: Towards the Automatic Construction of a Lexical Ontology for Portuguese. PhD thesis. University of Coimbra (2013), http://eden.dei.uc.pt/~hroliv/pubs/GoncaloOliveira_PhdThesis2012.pdf
2. Marrafa, P.: Portuguese WordNet: General architecture and internal semantic relations. DELTA 18, 131–146 (2002)
3. Dias-da-Silva, B.C.: Wordnet.Br: An exercise of human language technology research. In: Procs of 3rd International WordNet Conference, GWC 2006, South Jeju Island, Korea, pp. 301–303 (January 2006)
4. de Paiva, V., Rademaker, A., de Melo, G.: OpenWordNet-PT: An open brazilian wordnet for reasoning. In: Procs of 24th International Conferene on Computational Linguistics, COLING (Demo Paper) (2012)
5. Richardson, S.D., Dolan, W.B., Vanderwende, L.: MindNet: Acquiring and structuring semantic information from text. In: Procs of 17th International Conference on Computational Linguistics, COLING 1998, pp. 1098–1102 (1998)
6. Nichols, E., Bond, F., Flickinger, D.: Robust ontology acquisition from machine-readable dictionaries. In: Procs of 19th International Joint Conference on Artificial Intelligence, IJCAI 2005, pp. 1111–1116. Professional Book Center (2005)
7. Zesch, T., Müller, C., Gurevych, I.: Extracting lexical semantic knowledge from Wikipedia and Wiktionary. In: Procs of 6th International Conference on Language Resources and Evaluation, LREC 2008, Marrakech, Morocco (2008)
8. Lin, D.: Automatic retrieval and clustering of similar words. In: Procs of 17th International Conference on Computational linguistics, COLING 1998, pp. 768–774. ACL Press, Montreal (1998)

9. Hearst, M.A.: Automatic acquisition of hyponyms from large text corpora. In: Procs of 14th Conference on Computational Linguistics, COLING 1992, pp. 539–545. ACL Press (1992)
10. Caraballo, S.A.: Automatic construction of a hypernym-labeled noun hierarchy from text. In: Procs of 37th Annual Meeting of the Association for Computational Linguistics, pp. 120–126. ACL Press (1999)
11. Snow, R., Jurafsky, D., Ng, A.Y.: Learning syntactic patterns for automatic hypernym discovery. In: Advances in Neural Information Processing Systems, pp. 1297–1304. MIT Press, Cambridge (2005)
12. Pantel, P., Pennacchiotti, M.: Espresso: Leveraging generic patterns for automatically harvesting semantic relations. In: Procs of 21st International Conference on Computational Linguistics and 44th Annual Meeting of the Association for Computational Linguistics, pp. 113–120. ACL Press, Sydney (2006)
13. Etzioni, O., Fader, A., Christensen, J., Soderland, S.: Mausam: Open information extraction: The second generation. In: Procs of 22nd International Joint Conference on Artificial Intelligence, IJCAI 2011, pp. 3–10. IJCAI/AAAI, Barcelona (2011)
14. Shi, L., Mihalcea, R.: Putting pieces together: Combining FrameNet, VerbNet and WordNet for robust semantic parsing. In: Gelbukh, A. (ed.) CICLing 2005. LNCS, vol. 3406, pp. 100–111. Springer, Heidelberg (2005)
15. Navigli, R., Ponzetto, S.P.: BabelNet: The automatic construction, evaluation and application of a wide-coverage multilingual semantic network. Artificial Intelligence 193, 217–250 (2012)
16. Gurevych, I., Eckle-Kohler, J., Hartmann, S., Matuschek, M., Meyer, C.M., Wirth, C.: UBY - a large-scale unified lexical-semantic resource. In: Procs of 13th Conference of the European Chapter of the Association for Computational Linguistics, EACL 2012, pp. 580–590. ACL Press, Avignon (2012)
17. Pennacchiotti, M., Pantel, P.: Ontologizing semantic relations. In: Procs of 21st International Conference on Computational Linguistics and 44th Annual Meeting of the Association for Computational Linguistics COLING/ACL, pp. 793–800. ACL Press (2006)
18. Gonçalo Oliveira, H., Gomes, P.: ECO and Onto.PT: A flexible approach for creating a Portuguese wordnet automatically. Language Resources and Evaluation 48(2), 373–393 (2014)
19. Gonçalo Oliveira, H., Santos, D., Gomes, P., Seco, N.: PAPEL: A dictionary-based lexical ontology for Portuguese. In: Teixeira, A., de Lima, V.L.S., de Oliveira, L.C., Quaresma, P. (eds.) PROPOR 2008. LNCS (LNAI), vol. 5190, pp. 31–40. Springer, Heidelberg (2008)
20. Simões, A., Sanromán, Á.I., Almeida, J.J.: Dicionário-Aberto: A source of resources for the Portuguese language processing. In: Caseli, H., Villavicencio, A., Teixeira, A., Perdigão, F. (eds.) PROPOR 2012. LNCS (LNAI), vol. 7243, pp. 121–127. Springer, Heidelberg (2012)
21. Gonçalo Oliveira, H., Antón Pérez, L., Costa, H., Gomes, P.: Uma rede léxico-semântica de grandes dimensões para o português, extraída a partir de dicionários electrónicos. Linguamática 3(2), 23–38 (2011)
22. Gonçalo Oliveira, H., Gomes, P.: Automatic Discovery of Fuzzy Synsets from Dictionary Definitions. In: Procs of 22nd International Joint Conference on Artificial Intelligence, IJCAI 2011, pp. 1801–1806. IJCAI/AAAI, Barcelona (2011)
23. Maziero, E.G., Pardo, T.A.S., Felippo, A.D., Dias-da-Silva, B.C.: A Base de Dados Lexical e a Interface Web do TeP 2.0 - Thesaurus Eletrônico para o Português do Brasil. In: VI Workshop em Tecnologia da Informação e da Linguagem Humana, TIL, pp. 390–392 (2008)

24. Gonçalo Oliveira, H., Gomes, P.: Towards the automatic enrichment of a thesaurus with information in dictionaries. Expert Systems: The Journal of Knowledge Engineering 30(4), 320–332 (2013)
25. Gonçalo Oliveira, H., Gomes, P.: Ontologising semantic relations into a relationless thesaurus. In: Procs of 20th European Conference on Artificial Intelligence (ECAI 2012), pp. 915–916. IOS Press, Montpellier (2012)
26. Santos, D., Bick, E.: Providing Internet access to Portuguese corpora: the AC/DC project. In: Proc 2nd Language Resources and Evaluation, LREC 2000, pp. 205–210. ELRA, Athens (2000)
27. Gonçalo Oliveira, H., Gomes, P.: Onto.PT: Recent developments of a large public domain portuguese wordnet. In: Procs of the 7th Global WordNet Conference, GWC 2014, Tartu, Estonia, pp. 16–22 (2014)
28. Rodrigues, R., Gonçalo Oliveira, H., Gomes, P.: Uma abordagem ao Págico baseada no processamento e análise de sintagmas dos tópicos. Linguamática 4(1), 31–39 (2012)
29. Gonçalo Oliveira, H., Coelho, I., Gomes, P.: Exploiting Portuguese lexical knowledge bases for answering open domain cloze questions automatically. In: Proc 9th Language Resources and Evaluation Conference, ELRA, Reykjavik (2014)

Comparing Semantic Relatedness between Word Pairs in Portuguese Using Wikipedia

Roger Granada[1], Cassia Trojahn[2], and Renata Vieira[3]

[1] PUCRS & IRIT - Toulouse, France
roger.granada@acad.pucrs.br
[2] UTM & IRIT - Toulouse, France
cassia.trojahn@irit.fr
[3] PUCRS - Porto Alegre, Brazil
renata.vieira@pucrs.br

Abstract. The growth of available data in digital format has been facilitating the development of new models to automatically infer the semantic similarity between word pairs. However, there are still many natural languages without sufficient resources to evaluate measures of semantic relatedness. In this paper we translated word pairs from a well-known baseline for evaluating semantic relatedness measures into Portuguese and performed a manual evaluation of each pair. We compared the correlation with similar datasets in other languages and generated LSA models from Wikipedia articles in order to verify the pertinence of each dataset and how semantic similarity conveys across languages.

Keywords: Semantic relatedness, semantic similarity, similarity dataset.

1 Introduction

Discovering similar words in a document collection is still an open problem. The idea of semantic similarity was expressed by Zellig Harris [6] when he formulated the distributional hypothesis. This hypothesis is based on the idea that words that occur in the same contexts tend to have similar meanings. Models built on this assumption are called Distributional Similarity Models (DSMs) and take into account the co-occurrence distributions of the words in order to cluster them together. Several implementations of DSMs have been proposed in the last decades [3,5,8,10,15] and have being used in tasks such as query expansion [1], building bilingual comparable corpora [16], clustering [2], discovering of meaning of noun compounds [14] *etc.*

Although there are many proposals on DSMs, their practical applicability depends on their evaluation. However, evaluation is still an open issue since manual evaluation is a time consuming task and automatic evaluation requires a gold-standard. An approach to overcome this problem is to manually generate a gold-standard containing pairs of terms and a score associated to each pair [4,11,13].

An important resource for English has been defined by Rubenstein and Goodenough [13]. This dataset (from now on called as RG65) was developed to evaluate

J. Baptista et al. (Eds.): PROPOR 2014, LNAI 8775, pp. 170–175, 2014.

semantic similarity measures and contains judgements from 51 human subjects for 65 word pairs. Judgements are scaled from 0 to 4 according to their similarity of meaning, where the greater the similarity between the words, the higher the score. Thus, 0 representing no similarity between words and 4 perfect similarity. The average correlation over the subjects was quite high, achieving $r = .85$.

Miller and Charles [11] repeated the experiments using a subset of RG65 dataset containing 30 word pairs. These pairs were selected according with their score in the original RG65 dataset: 10 pairs have high level of similarity scores (scores between 3 and 4), 10 pairs have intermediate level (scores between 1 and 3) and 10 pairs have low level (scores between 0 and 1). This new dataset (MC30) was evaluated by 38 human subjects who were asked to evaluate specifically the similarity of meaning and to ignore any other semantic relations. Comparing the results obtained using the MC30 dataset with the results obtained by Rubenstein and Goodenough using RG65 dataset the correlation achieved was $r = .97$.

Finkelstein *et al.* [4] expanded the initial MC30 dataset, increasing significantly the number of word pairs. WordSimilarity-353 or just WordSim-353[1] contains 353 pairs of words divided in two sets. The first set contains 153 word pairs along with their similarity scores assigned by 13 subjects. The second set contains 200 word pairs, with their similarity scores assessed by 16 subjects. The subjects were instructed to evaluate the word pairs on a scale ranging from 0 to 10 according to their relatedness, being 0 totally unrelated words and 10 very closely related or identical words. The correlation between MC30 and WorsSim-353 datasets is also quite high, having a Pearson correlation of $r = .95$.

In order to evaluate similarity measures in other natural languages, a translation of some datasets has been made. Joubarne and Inkpen in [9] translated the RG65 dataset into French in order to measure the semantic similarity using second-order co-occurrence measures. After translating all word pairs, 18 human subjects who are French native speakers evaluated the similarity between the word pairs. As the work by Rubenstein and Goodenough [13], evaluators judge the word pairs in a scale ranging from 0 to 4. According to the authors, there was a good agreement amongst the evaluators for 71% of the word pairs and a high disagreement for 23% of the cases. The correlation between RG65 original dataset and the French dataset (JI65) achieved $r = .91$.

Following the work by Joubarne and Inkpen [9], this work attempts to translate into Portuguese all pairs from RG65 and evaluate them using 50 human subjects. Human scores are compared with previous works and an automatic evaluation is performed by comparing LSA generated models from Wikipedia articles with each dataset. These experiments verify the pertinence of each dataset and how the semantic similarity conveys across languages.

2   Data and Methods

In order to generate a dataset for evaluating similarity measures using Portuguese, all word pairs from RG65 were translated into Portuguese by two native

---

[1] http://www.cs.technion.ac.il/~gabr/resources/data/wordsim353/

speakers with proficiency in English. Each pair of words was translated separately and their relatedness score in RG65 was used as a hint to disambiguate words when multiple translations were possible. In some cases, the direct translation of each word from the pair resulted in one word, *e.g.*, the pair *cock* and *rooster* from RG65 has the same word *galo* as translation into Portuguese. As performed by Joubarne and Inkpen [9], in these cases the same word was kept as the translation of the word pair.

The evaluation process was performed by 50 undergraduate and graduate students who were asked to evaluate each pair according with their semantic relatedness. Following Rubenstein and Goodenough [13] our scores also range from 0 to 4. Results were averaged over all 50 subjects and the whole dataset (hereafter named as PT65) is freely available[2]. The average agreement among subjects was $r = .71$ having a standard deviation $\sigma = .13$. We use Pearson (r) and Spearman (ρ) correlation coefficients to measure the relation between scores, since Pearson correlation is highly dependent on the linear relationship between the distributions in question and Spearman mainly emphasizes the ability of the distributions to maintain their relative ranking. Table 1 presents the correlation scores between datasets for 65 word pairs datasets and for 30 word pairs datasets.

Table 1. Correlation between datasets evaluations

	r		ρ			r			ρ		
	JI65	PT65	JI65	PT65		RG30	JI30	PT30	RG30	JI30	PT30
RG65	.91	.90	.91	.83	**MC30**	.97	.92	.87	.95	.89	.87
JI65	-	.89	-	.85	**RG30**		.91	.89	-	.90	.90
					JI30			.86	-	-	.87

As reported by Miller and Charles [11] the correlation between MC30 and RG30 dataset was $r = .97$. RG65 and PT65 datasets achieved $\rho = .83$, the lowest correlation among datasets. On the other hand, their correlation using Pearson achieved $r = .90$, which is almost the same of the French dataset. Although the correlations of the Portuguese dataset have the lowest scores, their values are still relatively high (greater than $r = .80$).

In order to evaluate the pertinence with respect to the representativity of the word pairs in the languages, experiments using these datasets and Wikipedia dumps dating from February 2013 were performed. Each Wikipedia dump was pre-processed by WikiExtractor[3] (version 2.6) in order to extract and clean its content.

Each Wikipedia article was tokenized and a bag-of-words model was generated. Thus, each article is represented as an attribute vector of words that occur in the corresponding article. In order to remove noisy words, a threshold was applied removing words that appear less than 10 times in the whole Wikipedia.

---

[2] http://www.inf.pucrs.br/linatural/wikimodels/similarity.html
[3] http://medialab.di.unipi.it/wiki/Wikipedia_Extractor

The resulting vectors are weighted using Term Frequency - Inverse Domain Frequency (TFIDF) scheme and transformed into LSA models using the Gensim [12] tool.

LSA model uses Singular Value Decomposition (SVD) on a word-document matrix to extract its reduced representation by truncating the matrix to a certain size (also called the semantic dimension of the model). It is justified because it often improves the quality of the semantic space [10]. In this model, two words end up with similar vectors if they co-occur multiple times in similar documents. Thus, a similarity measure can be used between word vectors in order to measure the similarity between word pairs.

3   Experiments and Results

In our experiments, the LSA model was generated reducing the original matrix to a matrix containing 250 dimensions, $i.e.$, rank $k = 250$, and the cosine of the angle between word vectors was used to measure their similarity. Scores from Wikipedia were compared in terms of Pearson and Spearman correlations as presented on Table 2. The correlation between scores in different languages allows to see whether it is possible to transfer semantic similarity across languages.

Table 2. Pearson and Spearman correlation between datasets and Wikipedia data

		RG65	JI65	PT65	MC30	RG30	JI30	PT30
	Wikipedia (EN)	.65	.59	.63	.69	.72	.62	.70
Pearson (r)	Wikipedia (FR)	.57	.55	.53	.55	.53	.48	.50
	Wikipedia (PT)	.47	.56	.57	.67	.63	.72	.77
	Wikipedia (EN)	.69	.61	.61	.71	.77	.66	.67
Spearman (ρ)	Wikipedia (FR)	.52	.50	.38	.52	.51	.46	.39
	Wikipedia (PT)	.43	.42	.53	.66	.66	.69	.79

Joubarne and Inkpen in [9] suggest that it might be possible to transfer semantic similarity across languages. As Joubarne and Inkpen, the correlation found in our experiments using data in French suggests that it would be possible to transfer semantic similarity across languages. For example, looking at Table 2 it would be possible to use RG65 scores to find similar terms in French Wikipedia, since it achieved almost the same correlation when compared with JI65 dataset. On the other hand, scores found using Wikipedia in Portuguese achieved the highest correlation using Portuguese datasets, which is an evidence that using translated words evaluated by native speakers would get better results when compared with approaches that transfer the human scores across languages.

Observing the distributional similarity between the evaluations, the correlation using English and Portuguese Wikipedias has a similar behavior. Both languages presented an increase in correlation scores when the number of terms decreased, *i.e.*, when changing datasets from 65 to 30 word pairs. On the other hand, French Wikipedia had a decrease in correlation scores when the number of terms decreased (except for Spearman score using Portuguese dataset which increased $\rho = .01$). This decrease might be due to the fact that the MC30 dataset contains terms that are less related in the French Wikipedia.

Our correlation scores are close to the scores achieved by Hassan and Milhalcea [7] when using the MC30 dataset to evaluate a method based on Explicit Semantic Analysis (ESA). In that work, the authors achieved a Pearson correlation of $r = .58$ and a Spearman correlation of $\rho = .75$ for the English Wikipedia. In our experiments the correlation between the MC30 dataset and the English Wikipedia achieved a Pearson correlation of $r = .69$ and a Spearman correlation of $\rho = .71$. A comparison using other languages is not applicable since Hassan and Milhalcea used Arabic, Romanian and Spanish Wikipedias while our work used French and Portuguese Wikipedias. Joubarne and Inkpen in [9] used French to evaluate an automatic similarity measure, but unfortunately a comparison is not possible since they used Google n-grams as corpus.

4 Conclusions

In this paper we have proposed a resource that can be used as gold-standard for evaluating semantic similarity and relatedness between words, which results from the manual translation into Portuguese of a well-known baseline in English. The evaluation scores were compared with similar proposals in the literature which aimed at translating the English baseline in other languages, such as French.

Automatic evaluation was also performed by comparing LSA models based on Wikipedia articles with each proposed dataset. In this experiment we observed that it might be possible to transfer semantic similarity across languages, but for Portuguese, a manual evaluation of the translated word pairs has better results. We believe that this resource in Portuguese is specially useful as gold-standard for evaluating Distributional Similarity Models, supporting the automatic evaluation of such approaches.

Similarly to Hassan and Milhalcea [7], an approach to measure semantic similarity across languages would be to use the generated datasets to tests cross-lingual similarity using Wikipedia. Unlike Hassan and Milhalcea, instead of using only the English dataset (RG65), one could use both datasets (*e.g.*, RG65 for English and PT65 for Portuguese) and the evaluation score would be the mean of both evaluation scores.

Acknowledgments. This work is partially supported by the CAPES-COFECUB Cameleon project number 707/11.

References

1. Chen, L., Chen, S.: A New Approach for Automatic Thesaurus Construction and Query Expansion for Document Retrieval. International Journal of Information and Management Sciences 18(4), 299–315 (2007)
2. Di Marco, A., Navigli, R.: Clustering and Diversifying Web Search Results with Graph-Based Word Sense Induction. Computational Linguistics 39(3), 709–754 (2013)
3. Erk, K.: Vector Space Models of Word Meaning and Phrase Meaning: A Survey. Language and Linguistics Compass 6(10), 635–653 (2012)
4. Finkelstein, L., Gabrilovich, E., Matias, Y., Rivlin, E., Solan, Z., Wolfman, G., Ruppin, E.: Placing Search in Context: The Concept Revisited. ACM Transactions on Information Systems 20(1), 116–131 (2002)
5. Granada, R.L., Vieira, R., Strube de Lima, V.L.: Evaluating co-occurrence order for automatic thesaurus construction. In: IEEE 13th International Conference on Information Reuse and Integration (IRI), pp. 474–481 (2012)
6. Harris, Z.S.: Distributional structure. Words 10(23), 146–162 (1954)
7. Hassan, S., Mihalcea, R.: Cross-lingual Semantic Relatedness Using Encyclopedic Knowledge. In: EMNLP 2009, pp. 1192–1201. Association for Computational Linguistics, Stroudsburg (2009)
8. Iosif, E., Potamianos, A.: Similarity computation using semantic networks created from web-harvested data. Natural Language Engineering, 1–31 (2014)
9. Joubarne, C., Inkpen, D.: Comparison of Semantic Similarity for Different Languages Using the Google N-gram Corpus and Second-Order Co-occurrence Measures. In: Butz, C., Lingras, P. (eds.) Canadian AI 2011. LNCS (LNAI), vol. 6657, pp. 216–221. Springer, Heidelberg (2011)
10. Landauer, T.K., Dumais, S.T.: A solution to Plato's problem: The latent semantic analysis theory of the acquisition, induction, and representation of knowledge. Psychological Review 104(2), 211–240 (1997)
11. Miller, G.A., Charles, W.G.: Contextual correlates of semantic similarity. Language & Cognitive Processes 6(1), 1–28 (1991)
12. Rehurek, R., Sojka, P.: Software Framework for Topic Modelling with Large Corpora. In: Proceedings of the LREC 2010 Workshop on New Challenges for NLP Frameworks, pp. 45–50. ELRA, Valletta (2010)
13. Rubenstein, H., Goodenough, J.B.: Contextual correlates of synonymy. Communications of the ACM 8(10), 627–633 (1965)
14. Utsumi, A.: A semantic space approach to the computational semantics of noun compounds. Natural Language Engineering 20(2), 185–234 (2014)
15. Yang, D., Powers, D.M.W.: Automatic thesaurus construction. In: 31st Australasian conference on Computer science – ACSC 2008, pp. 147–156. Australian Computer Society, Inc., Darlinghurst (2008)
16. Zhu, Z., Li, M., Chen, L., Yang, Z.: Building Comparable Corpora Based on Bilingual LDA Model. In: 51st Annual Meeting of the Association for Computational Linguistics, pp. 278–282. Association for Computational Linguistics, Sofia (2013)

On the Utility of Portuguese Term-Based Lexical-Semantic Networks

Hugo Gonçalo Oliveira

CISUC, Dept. of Informatics Engineering, University of Coimbra, Portugal
hroliv@dei.uc.pt

Abstract. This paper discusses the utility of term-based lexical networks, with focus on Portuguese. Although less complex and often undervalued towards wordnets, for Portuguese, these resources have been used in varied natural language processing tasks. We enumerate those, and then introduce a larger resource of this kind together with two additional tasks where it was useful.

Keywords: Lexical knowledge bases, words, semantic relations, cloze question answering, poetry generation.

1 Introduction

When we think of machines understanding the meaning of natural language, semantic relations are among the first structures that come to mind. They connect words according to their meanings and may be integrated in lexical-semantic knowledge bases, where words are organised according to their senses. WordNet [1] is the paradigmatic resource of this kind, for English. Its model is almost a standard and has, in fact, been adapted for several non-English languages [2]. Wordnets are structured on synsets, which group word senses and may be seen as the possible lexicalisations of a natural language concept. According to the meaning they transmit, synsets may be connected to other synsets by semantic relations, such as hypernymy and part-of.

For a long time, Portuguese suffered from not having a freely available comprehensive wordnet. Researchers had to take shortcuts or to be creative, in order to process Portuguese at the semantic level. This is not the current situation because, in the last two years, projects as OpenWordNet.PT [3] and Onto.PT [4] have released freely available Portuguese wordnets. Before them, we saw the emergence of PAPEL [5], which may be seen as a middle-way towards a wordnet. Though extracted automatically from a proprietary Portuguese dictionary, PAPEL is free, structured on words and relations, established according to their meanings, but it does not handle word senses. While this may be seen as an important limitation in tasks where it is critical to distinguish between different senses of the same word, several works ended up using PAPEL for a varied range of purposes. Its wide application from question-answering [6] to the creation of sentiment lexicons [7] confirms that, besides a step towards a

J. Baptista et al. (Eds.): PROPOR 2014, LNAI 8775, pp. 176–182, 2014.

wordnet-like structure, alone, term-based lexical knowledge-bases (TBLNs) are still very useful resources.

This paper highlights the utility of TBLNs, with focus on Portuguese. It starts by enumerating several works where PAPEL was used. Then, it briefly introduces CARTÃO, an enlarged PAPEL. Before concluding, two tasks where CARTÃO has recently revealed to be very useful are described.

2 Using Term-Based Lexical-Semantic Networks

Since the beginning of the project PAPEL [5], our goal was to build a lexical resource that would share some similarities with a wordnet (words, semantic relations, free), but that would be created automatically, be less complex, and where lexical items would not be divided into word senses. One of the main reasons for the latter choice relied on the fact that, from a linguistic point of view, word senses are not discrete and cannot be separated with clear boundaries [8]. Sense division in dictionaries and lexical ontologies is most of the times artificial and sense granularity is often different from lexicographer to lexicographer. Moreover, the study of vagueness in natural language is as, or even more, important than studying ambiguity (see e.g. [9]).

It can be argued that PAPEL is inadequate for some of the tasks where a wordnet is used. For instance, without any additional information, it cannot be used as a sense inventory, nor for inferencing new relations without ambiguity-related noise. However, given the wide range of natural language processing tasks where it was used, we can say that this was quite a successful project. Since its first version, in 2008, PAPEL has been used as a gold standard for computing similarity between lexical items [10], in the adaptation of textual contents for poor literacy readers [11], in the automatic refinement of distractors for cloze questions [12], as a knowledge base for question answering [13,6] and question generation [14] systems, to validate terms describing places [15], and in the enrichment [16] and creation [7] of sentiment lexicons.

3 CARTÃO: A Larger PAPEL

Towards the creation of a broader resource, the automatic methodology for creating PAPEL was applied to two additional Portuguese dictionaries, Wiktionary. PT[1] and Dicionário Aberto [17]. Given their similar structure, based on $term_1, relation, term_2$ triples, the TBLNs obtained can be merged with PAPEL, which results in CARTÃO [18], a public TBLN for Portuguese. CARTÃO 3.5[2] contains about 147,000 terms ($\approx$93,000 nouns, $\approx$31,800 verbs, $\approx$31,000 adjectives, $\approx$3,500 adverbs), connected by about 331,000 relation instances covering a broad range of types, distributed according to table 1. This represents more 53,000 distinct terms and 140,000 relation instances than PAPEL.

---

[1] Wiktionary.PT is available from http://pt.wiktionary.org/

[2] CARTÃO is available in three files: PAPEL @ http://www.linguateca.pt/PAPEL; additional relations @ http://ontopt.dei.uc.pt

Table 1. The relations of CARTÃO

Relation	Instances	Subtypes
Synonymy	≈135,400	SINONIMO_N_DE, SINONIMO_V_DE, SINONIMO_ADJ_DE, SINONIMO_ADV_DE
Hypernymy	≈95,700	HIPERONIMO_DE
Part-of	≈9,600	PARTE_DE, PARTE_DE_ALGO_COM_PROPRIEDADE, PRO-PRIEDADE_DE_ALGO_PARTE_DE
Member-of	≈8,500	MEMBRO_DE, MEMBRO_DE_ALGO_COM_PROPRIEDADE, PROPRIEDADE_DE_ALGO_MEMBRO_DE
Contained-in	≈680	CONTIDO_EM, CONTIDO_EM_ALGO_COM_PROPRIEDADE
Material-of	≈900	MATERIAL_DE
Causation-of	≈12,800	CAUSADOR_DE, CAUSADOR_DE_ALGO_COM_PROPRIEDADE, PROPRIEDADE_DE_ALGO_QUE_CAUSA, AC-CAO_QUE_CAUSA, CAUSADOR_DA_ACCAO
Producer-of	≈2,400	PRODUTOR_DE, PRODUTOR_DE_ALGO_COM_PROPRIEDADE, PROPRIEDADE_DE_ALGO_PRODUTOR_DE
Purpose-of	≈16,600	FAZ_SE_COM, FAZ_SE_COM_ALGO_COM_PROPRIEDADE, FI-NALIDADE_DE, FINALIDADE_DE_ALGO_COM_PROPRIEDADE
Quality-of	≈2,400	TEM_QUALIDADE, DEVIDO_A_QUALIDADE
State-of	≈600	TEM_ESTADO, DEVIDO_A_ESTADO
Place-of	≈1,700	LOCAL_ORIGEM_DE
Manner-of	≈4,100	MANEIRA_POR_MEIO_DE, MANEIRA_COM_PROPRIEDADE
Manner-without	≈270	MANEIRA_SEM, MANEIRA_SEM_ACCAO
Property-of	≈37,700	DIZ_SE_SOBRE, DIZ_SE_DO_QUE;
Antonymy	≈1,500	ANTONIMO_N_DE, ANTONIMO_V_DE, ANTONIMO_ADJ_DE, ANTONIMO_ADV_DE

4   Recent Use Cases

CARTÃO was recently used in two additional tasks, briefly described here.

4.1   Answering Cloze Questions Automatically

CARTÃO and other lexical knowledge bases (LKBs), including TBLNs and wordnets, were used to automatically answer a set of 3,890 open domain cloze questions [19], generated in the scope of REAP.PT [20], an assisted language learning tutoring system for European Portuguese. Here is one of the questions:

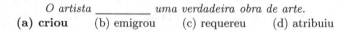

O artista _____ uma verdadeira obra de arte.
(a) **criou**   (b) emigrou   (c) requereu   (d) atribuiu

Besides the challenge of answering questions originally made to be answered by humans, this exercise can be seen as an application-oriented evaluation of Portuguese LKBs. Independently of the LKB, the best results of this task were obtained with variants of the Personalized PageRank algorithm [21], based either on a term or synset graph. The results of using this algorithm with different LKBs are presented in table 2, together with the core nodes of each LKB, and the number of covered terms. Despite being structured on synsets, TeP was transformed to a TBLN because it only covers synonymy and antonymy relations.

Though not outstanding, all the LKBs clearly outperformed the random chance baseline (25%). But CARTÃO led to the highest proportion of correct answers. Two additional conclusions were taken for this specific task: (i) the size of the LKB plays an important role; (ii) in this specific task, TBLNs led to better results, in opposition to the LKBs organised in senses/synsets (wordnets).

Table 2. Proportion of correct answers for 3,890 cloze questions, using the Personalized PageRank in different Portuguese LKBs

LKB	Nodes	Terms	Correct answers
TeP 2.0	terms	44,000	37.25%
PAPEL 3.5	terms	94,000	39.59%
CARTÃO 3.5	terms	147,000	**41.21%**
Onto.PT 0.6	synsets	169,000	37.79%
OpenWN.PT	synsets	48,000	33.11%

4.2   Automatic Generation of Poetry

In the scope of computational creativity, CARTÃO was used together with a platform for the automatic generation of poetry [22]. First, it was exploited to identify semantic relation patterns in a collection of human-written poetry, then generalised as line templates. Second, from a set of provided seed terms, a semantic domain was set by a sub-network of CARTÃO. For producing poems, CARTÃO is also used to generate new lines, based on the aforementioned templates, and to keep track of the word choices made, for later explaining them.

Additional features of the poetry generation platform include the selection of lines that suit a given metre, when possible with rhymes, and the domain expansion with relevant terms that transmit a desired sentiment. For the last part, SentiLex-PT [16], a polarity lexicon for Portuguese, is used. Figure 1 illustrates this application of CARTÃO with a generated sonnet and its contextualization, consisting of the relation(s) between the used terms and the provided seeds.

CARTÃO enables the generation of a huge variety of lines and its term-based structure is not a problem. In fact, vagueness and ambiguity are often a plus, as they increase the use of figurative language and surprise potential.

In addition to the entertainment goal, the results of this system could be used for evaluating relation instances. At least, this idea could be adapted to generate more objective sentences which, following Cruse [23], can later help assessing the relations, as an alternative to using the relation instance alone.

diurnais e diários felizmente
nem conexidade nem relação
disciplina de grande instrução
com justezas razões viu a gente

eu fiquei lésbico e relação
feita de relação e filiação
é de tino e de conhecimento
preempção é no corpo o evento

feita de gnose e sabedoria
não fica direita nem garantia
não há matéria sem disciplina

das listas que trouxeram relação
direita varuda de eleição
não fica declaração nem palavra

Contextualization:
- *diários* synonym-of *diurnais*, *relação* hypernym-of *diários*
- *conexidade* synonym-of *relação*
- *disciplina* synonym-of *instrução*
- *razões* synonym-of *justezas*, *relação* hypernym-of *razões*
- *lésbico* property-of *relação*
- *filiação* synonym-of *relação*
- *conhecimento* synonym-of *tino*
- *evento* hypernym-of *preempção*, *direita* hypernym-of *preempção*
- *sabedoria* synonym-of *gnose*
- *direita* synonym-of *garantia*
- *matéria* synonym-of *disciplina*
- *relação* hypernym-of *listas*
- *direita* synonym-of *varuda*
- *palavra* synonym-of *declaração*

Fig. 1. Automatically generated sonnet and its contextualization. Seed terms were: *conhecimento, palavra, relação*; top-5 relevant positive terms: *tratado, disciplina, direita, sabedoria, experiência*.

5   Conclusion

We have discussed the utility of TBLNs, for Portuguese. After enumerating several tasks where PAPEL has been used, we presented CARTÃO, similar but larger, and its successful utilisation for answering cloze questions and generating poetry. We believe to have shown that it is not always critical to have a wordnet, and the utilisation of less complex TBLNs should sometimes be considered. Also when there is a wordnet, TBLNs can be used for suggesting new relations, which can be added either manually, or automatically, as in the Onto.PT [4] project.

Acknowledgements. This work is currently supported by the iCIS project (CENTRO-07-ST24-FEDER-002003), co-financed by QREN, in the scope of the Mais Centro Program and European Union's FEDER.

References

1. Fellbaum, C. (ed.): WordNet: An Electronic Lexical Database (Language, Speech, and Communication). The MIT Press (1998)
2. Bond, F., Paik, K.: A survey of wordnets and their licenses. In: Procs of the 6th Global WordNet Conference, GWC 2012, pp. 64–71 (2012)

3. de Paiva, V., Rademaker, A., de Melo, G.: OpenWordNet-PT: An open Brazilian Wordnet for reasoning. In: Procs of 24th International Conference on Computational Linguistics: Demonstration Papers, COLING, pp. 353–360. The COLING 2012 Organizing Committee, Mumbai (2012)
4. Gonçalo Oliveira, H., Gomes, P.: ECO and Onto.PT: A flexible approach for creating a Portuguese wordnet automatically. Language Resources and Evaluation 48(2), 373–393 (2014)
5. Gonçalo Oliveira, H., Santos, D., Gomes, P., Seco, N.: PAPEL: A dictionary-based lexical ontology for Portuguese. In: Teixeira, A., de Lima, V.L.S., de Oliveira, L.C., Quaresma, P. (eds.) PROPOR 2008. LNCS (LNAI), vol. 5190, pp. 31–40. Springer, Heidelberg (2008)
6. Rodrigues, M., Dias, G.P.,, T.: Criação e acesso a informação semântica aplicada ao Governo Eletrónico. Linguamática 3(2), 55–68 (2011)
7. Santos, A.P., Gonçalo Oliveira, H., Ramos, C., Marques, N.C.: A bootstrapping algorithm for learning the polarity of words. In: Caseli, H., Villavicencio, A., Teixeira, A., Perdigão, F. (eds.) PROPOR 2012. LNCS (LNAI), vol. 7243, pp. 229–234. Springer, Heidelberg (2012)
8. Kilgarriff, A.: Word senses are not bona fide objects: Implications for cognitive science, formal semantics, NLP. In: Procs of 5th International Conference on the Cognitive Science of Natural Language Processing, pp. 193–200 (1996)
9. Santos, D.: The importance of vagueness in translation: Examples from English to Portuguese. Romansk Forum 5, 43–69 (1997) (revised as Santos 1998)
10. Sarmento, L.: Definition and Computation of Similarity Operations between Web-specific Lexical Items. PhD thesis. Universidade do Porto (2010)
11. Amancio, M.A., Watanabe, W.M., Candido Jr., A., de Oliveira, M., Pardo, T.A.S., Fortes, R.P.M., Alusio, S.M.: Educational FACILITA: helping users to understand textual content on the Web. In: Extended Activities Procs of 9th International Conference on Computational Processing of Portuguese Language (PROPOR 2010), Porto Alegre/RS, Brazil (April 2010)
12. Correia, R., Baptista, J., Mamede, N., Trancoso, I., Eskenazi, M.: Automatic generation of cloze question distractors. In: Second Language Studies: Acquisition, Learning, Education and Technology, Tokyo, Japan (September 2010)
13. Saias, J.M.G.: Contextualização e activação semântica na selecção de resultados em sistemas de pergunta-resposta. Master's thesis. Universidade de Évora, Évora, Portugal (2010)
14. Marques, C.J.L.: Syntactic REAP.PT. Master's thesis. Instituto Superior Técnico, Lisboa, Portugal (2011)
15. Oliveira Santos, J., Oliveira Alves, A., Câmara Pereira, F., Henriques Abreu, P.: Semantic enrichment of places for the Portuguese language. In: Procs of INFORUM 2012, Simpósio de Informática, Lisbon, Portugal, pp. 407–418 (September 2012)
16. Silva, M.J., Carvalho, P., Sarmento, L.: Building a sentiment lexicon for social judgement mining. In: Caseli, H., Villavicencio, A., Teixeira, A., Perdigão, F. (eds.) PROPOR 2012. LNCS (LNAI), vol. 7243, pp. 218–228. Springer, Heidelberg (2012)
17. Simões, A., Sanromán, Á.I., Almeida, J.J.: Dicionário-Aberto: A source of resources for the Portuguese language processing. In: Caseli, H., Villavicencio, A., Teixeira, A., Perdigão, F. (eds.) PROPOR 2012. LNCS (LNAI), vol. 7243, pp. 121–127. Springer, Heidelberg (2012)
18. Gonçalo Oliveira, H., Antón Pérez, L., Costa, H., Gomes, P.: Uma rede léxico-semântica de grandes dimensões para o português, extraída a partir de dicionários electrónicos. Linguamática 3(2), 23–38 (2011)

19. Gonçalo Oliveira, H., Coelho, I., Gomes, P.: Exploiting portuguese lexical knowledge bases for answering open domain cloze questions automatically. In: Procs of 9th Language Resources and Evaluation Conference, LREC 2014. ELRA, Reykjavik (2014)

20. Correia, R., Baptista, J., Eskenazi, M., Mamede, N.: Automatic generation of cloze question stems. In: Caseli, H., Villavicencio, A., Teixeira, A., Perdigão, F. (eds.) PROPOR 2012. LNCS (LNAI), vol. 7243, pp. 168–178. Springer, Heidelberg (2012)

21. Agirre, E., Soroa, A.: Personalizing PageRank for word sense disambiguation. In: Procs of 12th Conference of the European Chapter of the Association for Computational Linguistics, EACL 2009, pp. 33–41. ACL Press, Athens (2009)

22. Gonçalo Oliveira, H.: PoeTryMe: A versatile platform for poetry generation. In: Procs of ECAI 2012 Workshop on Computational Creativity, Concept Invention, and General Intelligence, C3GI 2012, Montpellier, France (August 2012)

23. Cruse, D.A.: Lexical Semantics. Cambridge University Press (1986)

Semantic Textual Similarity of Portuguese-Language Texts: An Approach Based on the Semantic Inferentialism Model

Vládia Pinheiro, Vasco Furtado, and Adriano Albuquerque

Programa de Pós-Graduação em Informática Aplicada – Universidade de Fortaleza,
Av. Washington Soares, 1321, Fortaleza, Ceará, Brasil
{vladiacelia,vasco,adrianoba}@unifor.br

Abstract. The Semantic Textual Similarity (STS) task aims capturing a bidirectional-graded equivalence between the pair of short texts. This work proposes a STS measure for the Portuguese Language based on Semantic Inferentialism Model (SIM) and InferenceNet.BR. We argue that the expression of inferential, causal, motivational and encyclopedic content of InferenceNet enables a more robust and efficient model for the STS task. An extrinsic evaluation in a Portuguese-language processing application - a Case-Based Reasoning system for Requirements Engineering - provided real scenario to assess how the proposed STS measure contributes to the effectiveness of NLP applications.

Keywords: Semantic Textual Similarity, Computational Semantics, Semantic Similarity Model.

1 Introduction

The Semantic Textual Similarity (STS) task [1] captures the notion that some texts are more similar than others, by measuring the degree of semantic equivalence. For example, the pair of sentences *The woman is playing the violin* and *The young lady enjoys listening to the guitar* are not equivalent, but are on the same topic. A quantifiable graded bidirectional notion of textual similarity is useful for a myriad of NLP and AI tasks, such as machine translation evaluation [2], microblog search [3], reuse of experiences [4], and knowledge acquisition [5].

In state-of-the-art STS systems, LSA Word Similarity relies on the distributional hypothesis that words occurring in the same contexts tend to have similar meanings [6]. Orthogonally, the Semantic Inferentialism Model (SIM) follows the pragmatic view of language, where semantic relations are considered according to their roles in inferences, as premises or conclusions [7]. InferenceNet's Conceptual Base of SIM [8, 9] contains world knowledge (common-sense and encyclopedic knowledge) and inferential content of concepts in the Portuguese Language. We argue that the expression of inferential, causal, motivational and encyclopedic content of

J. Baptista et al. (Eds.): PROPOR 2014, LNAI 8775, pp. 183–188, 2014.

InferenceNet enables a more robust and efficient model for the STS task. For example, knowing that guitar and violin are musical instruments helps to define that the sentences *The woman is playing the violin* and *The young lady enjoys listening to the guitar* have the same topic. On the other hand, LSA models only capture semantic relationships of a single nature – frequency of co-occurrence – making it difficult to assign a similarity score, for example, to the aforementioned pair of sentences. Furthermore, for the Portuguese language, the lack of corpora annotated with semantic information and ample domain coverage are barriers to the use of machine-learning approaches for the STS task.

Our work was developed precisely to fill these gaps, and is aimed at leveraging the STS task for Portuguese-language applications. We proposed a Semantic Textual Similarity measure that specifically uses the Word Inferential Relatedness Measure [10], proposed in SIM, plus lexical and syntactic features (grammatical class and syntactic functions). We use the proposed STS measure in one Portuguese-language processing application - a Case-Based Reasoning system for Requirements Engineering, which provided real scenario to assess how the correct computation of the similarity between two sentences or short texts contributes to the effectiveness of applications.

2   Semantic Textual Similarity Measure (SIM_t)

The generic intuition of a measure of semantic similarity between texts is that the more the concepts (represented by the words or expressions articulated in the texts) are similar to one another, the more similar such texts will be. In this paper, we propose a Semantic Textual Similarity measure (SIM_t) based on compositional semantics, which considers the structure of syntactic dependency of the sentences on the text, and on the InferenceNet' s Word Inferential Similarity Measure $\theta_{c1,c2}$.

The SIM_t measure defines the similarity between two input texts by averaging the similarities between the most similar sentences of both texts. That is, for each sentence **s** of t_1, the sentence **s'** in t_2 with greatest similarity to **s** is sought. Equation 1 shows our function for alignment of sentences, which – for a sentence **s** of t_1 (or t_2) – returns its counterpart **s'** in t_2 (or t_1) with the higher Sentence Similarity Measure **sim_s**, defined by (2).

$$s\_align(s) = \operatorname{argmax}_{s \in t_i} sim_s(s, s') \tag{1}$$

The Sentence Similarity measure **sim_s** (2) defines the similarity between two input sentences s_1 and s_2 by the weighted average of the sum of the similarities between the terms of s_1 and s_2 that meet certain lexical and syntactic constraints.

$$\boldsymbol{sim_s}(s_1, s_2) = \frac{\sum_{i=1}^{n}(\sum_{j=1}^{qi} \theta c_1,c_2) * P_i}{\sum_{i=1}^{n} q_i * P_i} \tag{2}$$

where,

- Θ_{c_1,c_2} is the value of the Word Inferential Similarity between the concepts c_1 and c_2 represented by the tokens of sentences s_1 and s_2, obtained from the set T'.
- T_1 and T_2 are the sets of significant tokens (c_1 and c_2) of sentences s_1 and s_2.
- T' is a subset of the Cartesian product T_1 x T_2, where (c_1, c_2) $\in$ T' and $c_1 \in$ T_1 and $c_2 \in T_2$.
- The approach for building T' is defined according to a criterion for the combination of tokens, which can consider lexical, syntactic and semantic characteristics of the words or expressions (tokens) of the sentences. Below are some criteria for combining tokens that can be used in T':
  1. tokens having the same grammatical class (POS tag) (e.g. NOUN x NOUN, VERB x VERB, ADVERB x ADVERB, PRONOUN x PRONOUN);
  2. tokens having the same syntactic function (e.g. SUBJECT x SUBJECT, OBJECT x OBJECT, MAIN VERB x MAIN VERB);
  3. tokens with highest value of semantic similarity;
  4. all tokens with all tokens
- n is the quantity of distinct "grammatical types" in T'. For example, using the criteria of combination by syntactic function (criterion 2), we have 3 grammatical types: SUBJECT, MAIN VERB, and OBJECT.
- q_i is the quantity of elements of T' of each "grammatical type" i.
- P_i is the weight of the "grammatical type" i;, allowing the similarity between verbs, for example, to provide a greater weight than the similarity between direct objects of the two sentences.

Flexibility in the formation of set T' allows us to adapt the measure of textual similarity proposed herein to a domain or application. However, we argue that the criterion of combining by syntactic function (criterion 2) or by POS (criterion 1) are, in order, more intuitive than the one used in [11], which combines the tokens c_1 and c_2 with greater similarity between each other (criterion 3), or the one that combines all the tokens of s_1 with all of the tokens of s_2 (criterion 4).

Finally, Formula 3 calculates **SIM$_t$** for two input texts t_1 and t_2, with p and k sentences, respectively.

$$\boldsymbol{SIM_t}(t_1, t_2) = \frac{\sum_{s \in t_1} sim_s(s,s\_align(s))}{2.p} + \frac{\sum_{s \in t_2} sim_s(s,s\_align(s))}{2.k} \qquad (3)$$

3   Extrinsic Evaluation of *SIM$_t$* Measure

In Software Engineering, textual reports are commonly used for reporting problems in software, detected in the software testing and evaluation phase. Commonly, the same defect may easily be reported multiple times, resulting in duplicate bug reports. In this context, we envision an ideal setting for evaluating the Semantic Textual Similarity *SIM$_t$* proposed in this paper.

The evaluation methodology followed the steps outlined below: (1) Formation of the base with 78 textual bug reports, in the Portuguese language, (called *cases*) of real problems occurring in the software projects of a Brazilian federal government-owned enterprise; (2) Definition of the Golden Collection – two systems analysts from the state-run enterprise analyzed each case and defined which were similar to one another; (3) Two scenarios were created, which are presented below - Scenario 1 – retrieval of cases by using a simple syntactic similarity measure proposed in [4], and Scenario 2 – retrieval of cases by using the textual similarity measure SIM_t; (4) Pre-processing of the texts of the cases; (5) Defining the test parameters of scenario 2; (6) In each scenario, five batteries of tests were carried out, in order for us to perform statistical significance testing (Student's t-test); (7) Analysis of the evaluation measures.

Tables 1 and 2 present the average results of batteries of tests in each scenario. In these tables we show the mean values of the averages and standard deviations of the measures: Recall, Precision, and F1-score. We also included an M/B column, with the average value of the mean F1-scores.

When we analyze the results, we can see that Scenario 2 (retrieval of similar cases using the SIM_t measure) brought a gain of 35.8% compared to Scenario 1 (retrieval of similar cases using syntactic similarity) in terms of average M/B. This was due primarily to the improvement in precision (decrease of false positive cases). We performed the unpaired Student's t-test with the aim of determining the degree of confidence in the gains that Scenario 2 provided, compared to Scenario 1, in all five batteries. As parameters, the averages and standard deviations of the f1-score measure were used. The result of confidence tests of the batteries for scenarios 1 and 2 was 95%.

Table 1. Test results in Scenario 1

	Average					M/B	Standard Deviation				
	B1	B2	B3	B4	B5		B1	B2	B3	B4	B5
Recall	69%	70%	70%	70%	70%		25%	20%	22%	25%	20%
Precision	11%	11%	11%	11%	11%		11%	11%	12%	11%	15%
F1-score	16%	17%	16%	17%	16%	16.4%	14%	14%	16%	16%	18%

Table 2. Test results in Scenario 2

	Average					M/B	Standard Deviation				
	B1	B2	B3	B4	B5		B1	B2	B3	B4	B5
Recall	60%	60%	60%	60%	60%		23%	23%	28%	26%	25%
Precision	50%	49%	51%	52%	51%		22%	27%	30%	26%	30%
F1-score	51%	52%	52%	53%	53%	52.2%	22%	25%	30%	26%	27%

4   Related Works

According to [11], there are three predominant approaches to computing short text similarity: (1) Vectorial Semantic techniques [12]; (2) Word Alignment techniques [13]; and (3) Machine-learning models [14]. In the latter two classes of approaches to the STS task, the semantic features are crucial and depend on the semantic similarity model used.

The *PairingWords* system [11] was the best system in the competition proposed at SemEval 2013 [1] and follows the second approach. This system was based on a term alignment score T and two heuristics for penalizing poor term alignments (e.g., antonymous words and words with a low degree of similarity). The term alignment algorithm compares the LSA Word similarity measure of all tokens of a sentence with all tokens of the second sentence, and selects the pairs of terms with the highest similarity. The LSA Word Similarity measure is boosted by eight heuristic rules based on WordNet knowledge. This system achieved an average performance of 0,6181 (Pearson correlation) for the four datasets available in Shared STS task.

The second best system in the Shared Task STS at SemEval 2013 [1] – *Galactus* system [11] - follows the aforementioned third approach. It was based on Support Vector Regression models using 52 features from different combinations of similarity metrics, parameters, ngram types and weights. The authors stated that the relatively simple *PairingWords* system has two advantages: it is faster and training is not needed thus eliminating noise induced from diverse training datasets.

5   Conclusion

In this paper, we present a Semantic Textual Similarity (STS) measure based on Semantic Inferentialism Model and InferenceNet.BR. The Word Inferential Similarity measure, proposed in SIM, considers the nature of the relationship between two concepts in order to define the similarity between them. We argue that the expression of inferential, causal, motivational and encyclopedic content of InferenceNet enables a more robust and efficient model for the STS task. Furthermore, the proposed STS measure defines the similarity between two input texts by the weighted average of the sum of the similarities between the tokens of their sentences that meet certain lexical and syntactic constraints, for example, tokens that have the same grammatical class or same syntactic function.

We use the proposed STS measure in a Portuguese-language processing application - a Case-Based Reasoning system for Requirements Engineering, which retrieves similar bug reports by comparing the similarity between the descriptive texts of the problems in software projects. The results showed that in the scenario where bug reports were retrieved by using STS measure, proposed in this work, brought a gain of 35.8% compared to the scenario where bug reports were retrieved by using a simple syntactic similarity measure. This extrinsic evaluation provided real scenario to assess how the correct computation of the similarity between two sentences or short texts contributes to the effectiveness of NLP applications. As future work, we are conducting a comparative evaluation with the LSA approaches for the STS task.

References

1. Agirre, E., Cer, D., Diab, M., Gonzalez-Agirre, A., Guo, W.: Sem 2013 shared task: Semantic Textual Similarity. In: SEM 2013: The Second Joint Conference on Lexical and Computational Semantics. Association for Computational Linguistics (2013)
2. Kauchak, D., Barzilay, R.: Paraphrasing for automatic evaluation. In: HLT-NAACL 2006, pp. 455–462 (2006)
3. Sriram, B., Fuhry, D., Demir, E., Ferhatosmanoglu, H., Demirbas, M.: Short text classification in twitter to improve information filtering. In: Proceedings of the 33rd International ACM SIGIR Conference on Research and Development in Information Retrieval, pp. 841–842. ACM (2010)
4. Albuquerque, A., Pinheiro, V., Leite, T.: Reuse of Experiences Applied to Requirements Engineering: An Approach Based on Natural Language Processing. In: Proceedings of the 24th International Conference on Software Engineering and Knowledge Engineering, SEKE 2012, São Francisco, CA (2012)
5. Pinheiro, V., Furtado, V., Pequeno, T., Franco, W.A.: Semi-Automated Method for Acquisition of Common-sense and Inferentialist Knowledge. Journal of the Brazilian Computer Society 19, 75–87 (2013), doi:10.1007/s13173-012-
6. Harris, Z.: Mathematical Structures of Language. Wiley, New York (1968)
7. Pinheiro, V., Pequeno, T., Furtado, V., Nogueira, D.: Semantic Inferentialist Analyser: Um Analisador Semântico de Sentençasem Linguagem Natural. In: Proceedings of the 7th Brazilian Symposium in Information and Human Language Technology, STIL, Brasil (2009)
8. Pinheiro, V., Pequeno, T., Furtado, V., Franco, W.: InferenceNet.Br: Expression of Inferentialist Semantic Content of the Portuguese Language. In: Pardo, T.A.S., Branco, A., Klautau, A., Vieira, R., de Lima, V.L.S. (eds.) PROPOR 2010. LNCS (LNAI), vol. 6001, pp. 90–99. Springer, Heidelberg (2010)
9. Pinheiro, V., Furtado, V., Pequeno, T., Ferreira, C.: Towards a common sense base in Portuguese for the linked open data cloud. In: Caseli, H., Villavicencio, A., Teixeira, A., Perdigão, F. (eds.) PROPOR 2012. LNCS (LNAI), vol. 7243, pp. 128–138. Springer, Heidelberg (2012)
10. Pinheiro, V., Pequeno, T., Furtado, V.: Um Analisador Semântico Inferencialis ta de Sentenças em Linguagem Natural. Linguamática 2(1), 111–130 (2010) ISSN: 1647-0818
11. Han, L., Kashyap, A.L., Finin, T., Mayfield, J., Weese, J.: UMBC_EBIQUITY-CORE: Semantic Textual Similarity Systems. In: Proceedings of the Second Joint Conference on Lexical and Computational Semantics. Association for Computational Linguistics (2013)
12. Meadow, C.T.: Text Information Retrieval Systems. Academic Press, Inc. (1992)
13. Mihalcea, R., Corley, C., Strapparava, C.: Corpus-based and knowledge-based measures of text semantic similarity. In: Proceedings of the 21st National Conference on Artificial Intelligence, pp. 775–780. AAAI Press (2006)
14. Saric, F., Glavas, G., Karan, M., Snajder, J., Basic, B.: Takelab: systems for measuring semantic text similarity. In: Proceedings of the First Joint Conference on Lexical and Computational Semantics, pp. 441–448. Association for Computational Linguistics (2012)

Temporal Aspects of Content Recommendation on a Microblog Corpus

Caio Ramos Casimiro and Ivandré Paraboni

School of Arts, Sciences and Humanities, University of São Paulo (EACH / USP),
Av. Arlindo Bettio, 1000 - São Paulo, Brazil
caiorcasimiro@gmail.com

Abstract. This paper presents a simple experiment to compare content recommendation on Twitter with and without the use of temporal information. Preliminary results suggest that the use of a particular kind of temporal information extracted from corpora (namely, the time frame within which a user browses the microblog) may lead to more accurate content recommendation.

Keywords: Content Recommendation, Microblogs, Temporal aspects.

1 Introduction

In recent years social networks have experienced a surge in user numbers. A well-known example is the microblog Twitter[1], with hundreds of millions of registered accounts[2]. On Twitter, brief posts - or tweets - are presented as a news feed in order of recency. A user may follow others, and by doing so the contents published by her followees will appear on her own news feed. This makes a convenient way of aggregating different information sources in one single place, but users may find difficult to filter out relevant information.

One possible way of dealing with potential information overload is by making use of a recommendation system (RS) [1,2,3,4]. A RS may rank a user's news feed in order to place the most relevant (and not necessarily most recent) items first according to some relevance measure. A common approach, for instance, consists of taking existing retweets as indicators of interest [3,4,5].

Different users may however have different reading habits. For instance, some readers may check their news feed repeatedly, or several times a day, whereas others may only do so once a day, or more often during the weekends [4]. Some users may find time or motivation to scan through all new (i.e., unseen) posts, whereas others may only check their latest few posts at a time. Thus, each reader has her own personal *time frame* representing how far back in time she searches for information on the microblog.

The posts published within a reader's time frame may of course represent only a fraction of her actual interests, i.e., outside the actual time frame we may still

---

[1] www.twitter.com

[2] https://blog.twitter.com/2012/twitter-turns-six

J. Baptista et al. (Eds.): PROPOR 2014, LNAI 8775, pp. 189–194, 2014.

find large amounts of potentially relevant information that she simply could not find. However, for the purpose of training and testing a RS for Twitter, it may be useful to consider that within a reader's time frame we will find all posts that she is ever likely to read and, on occasion, retweet.

In this paper we investigate the issue of how this particular temporal aspect - a reader's personal time frame - may impact content recommendation on Twitter. Using a simple RS implementation as a basis, we extract temporal information from a corpus of microblog posts to estimate the optimal time frame for each reader, and then we use this information to design a Twitter-based RS experiment to recommend posts that lie within the user's time frame only.

2   Related Work

In [4] a Twitter-based user modelling framework is presented. This framework is able to generate user profiles based on three features: profile type (based on hashtags, topics or named entities), content enrichment (i.e., the use of URL contents mentioned on tweets) and temporal restrictions (e.g., taking into account only the posts published over the weekend, or only during last month etc).

User profiles are evaluated as part of a recommendation system, and their quality is measured by the impact on the overall recommendation success. Results in [4] show that the best estimate of a user's interest in the recommendation task was obtained by using an entity-based profile with content enrichment, and by considering only posts published in the last two weeks.

The work in [5] presents a comparison between three types of user profile built from Twitter data: a profile type that models user's personal interests, a second type that models global trends on the network (e.g., using a temporal factor that assigns more weight to topics or entities that have peaks of popularity etc.), and a third type that combines these two.

According to [5], the profile that best estimates the user's interest is the profile that combines personal interests and global trends, although this was only slightly better than the personal profile. The global trends profile was shown to have the lowest performance of all.

Finally, in [1] a news recommendations system for the Huffington Post[3] is presented. The system was applied to the generation of a personalised version of the website front page. This approach makes use of a classifier trained on a news corpus to assign topics and tags to a set of posts published by each reader of the newspaper. Tags are keywords that help to describe the contents of an article, e.g., 'Obama', 'Syria', 'Football' etc. Both topics and tags are taken into account to build the reader's profile. In addition to that, a temporal factor assigns more weight to tags that are related to recent tweets.

3   Current Work

We implemented a simple Twitter-based content recommendation system to investigate the use of time frame information as seen in a corpus of microblog

---

[3] www.thehuffingtonpost.com

posts. The objective of the system is to recommend tweets published by sources 'followed by' a user, and which are deemed relevant in the sense that the user is likely to retweet this information.

As in [4], both the profile of a user's interests and the profile of the candidate tweets are simply based on hashtags (i.e., words starting with # on Twitter) as represented in the post texts. Also as pointed out in [4], we are aware that the use of hashtags is infrequent, and that hashtags do not represent an accurate estimate of user interests. However, given that the focus of our work is on the use of time frame information in the RS, and not on how to best model users interests, a simple hashtag model will suffice to illustrate this point.

The user profile is represented as a weighted vector of hashtags as described in equation 1 cf. [4]:

$$P(u) = \{(h, w(u, h)) | h \in H, u \in U\} \tag{1}$$

In the above, H represents the set of distinct hashtags contained in the corpus, U denotes the set of users of the system and $w(u, h)$ is a weight function that represents the level of interest of the user u in the hashtag h. For simplicity, we use a weight function that returns the number of times that the hashtag h was published by the user u. The values of $w(u, h)$ are then normalised to sum 1 for each user.

The profile model of the candidate tweets is also represented as a weighted vector of hashtags, but it has a distinct height function, as proposed in [4]:

$$P(t) = \{(h, w_t(t, h)) | h \in H, t \in T\} \tag{2}$$

In this equation, H represents the set of hashtags, T represents the set of tweets and $w(t, h)$ is a weight function that simply returns the number of times the hashtag h appears on tweet t. In general $w(t, h)$ will be equal to 1, however, it is possible to have duplicate hashtags in a tweet. The values of $w(t, h)$ are also normalised to sum 1 for each tweet.

Tweet recommendation proper is treated as an information retrieval problem in which the user profile represents a search query and the tweet profiles represent documents that need to be classified according to their proximity to the search query. This is done by calculating the similarity between each candidate tweet and the user profile, and then returning the candidate list sorted by similarity.

Let $p(u)$ and $p(t)$ denote the vector space representation of the user and tweet profile. The ith dimension of $p(u)$ and $p(t)$ represents $w(u, h_i)$ and $w(t, h_i)$, respectively. Thus, the similarity between a user profile and a tweet profile is calculated by the cosine similarity function described in equation 3 cf. [6]:

$$sim_{cosine}(p(u), p(t)) = \frac{p(u).p(t)}{||p(u)||.||p(t)||} \tag{3}$$

We computed the time frame of each user by measuring the elapsed time between her retweets (in the training period) and its original tweets. Next, we calculated the average time difference between these measurements. This difference is taken to be a user's time frame. Time frames range from a few hours to about a week-long period.

Time frame information is taken into account by the RS as follows. When the RS makes recommendations, it needs to extract a set of candidates, sort them according to relevance and then return the sorted recommendation list. In our approach, only candidates published within each user's time frame are considered. For instance, given a user with a 10-hours time frame, the proposed RS will extract, at the moment in which the recommendation is made, candidates published in the last 10 hours.

4   Evaluation

We collected a corpus of microblog posts from Twitter using its public API[4]. The tweets were published by 400 target users who followed two major Brazilian newspapers from January to June 2013. In addition to that, we collected all tweets published by the users followed by the target users. The set of candidate tweets therefore contains almost all tweets posted by each user during the period, the exception being the cases in which the user has retweeted information from external sources (e.g., advertisement).

This strategy produced a 90-million tweet corpus published (mainly in Portuguese) by over 40 thousand users. The last month of publications was set aside as our test data, conveying over 13 million tweets.

The proposed hashtag-based RS with personalised time frames was compared to a baseline system based on the same hashtag model, but using a fixed, 5-days recommendation time frame instead. In doing so, we would like to show that the use of personalised time frames outperforms standard recommendation.

In addition to that, both personalised and fixed time frame systems were compared to alternatives in which recommendation items are selected at random (and not based on hashtags). This comparison is intended to illustrate how the hashtag model performs in comparison with a naive baseline.

Following [1,4,5,3], we take a retweet as an indicator of a user's interest on that particular piece of information. In other words, we will assume that when a user retweets a post published by another user she is expressing interest in that post. This assumption allows us to evaluate our system by measuring the amount of retweeted material actually reproduced on the recommendation list.

The quality of the recommendation list is measured by using three metrics: Mean Reciprocal Rank (MRR), S@5 and S@10. MRR represents the average position of the most relevant item to the user. S@k denotes the probability of the relevant item being in the first k positions of the recommendation list.

During the evaluation period (i.e., the test set), each of the four systems - hashtag x random model, with either personalised or fixed time frame - was applied to generate a recommendation list for each retweet published by each target users. Each recommendation list was evaluated against its respective retweet using the metrics described above. Finally, the metrics for each model were calculated using the mean of the metrics from their recommendation lists.

---

[4] dev.twitter.com

5   Results

Results are presented in Table 1. The use of personalised candidate time frame information improves the overall quality of recommendations made by both hashtag-based and random systems alike. However, the performance of the hashtag-based system is poor, in some cases even below the random baseline. This observation is consistent with the findings in [4], and it was largely expected as only 12% of tweets in our data contained any hashtag at all (recall that the focus of the experiment was on the comparison between fixed and personalised time frames, and not on the recommendation model per se).

Table 1. Evaluation

Time frame	Hashtag model			Random baseline		
	MRR	S@5	S@10	MRR	S@5	S@10
Fixed	.0031	.0016	.0056	.0046	.0030	.0030
Personalised	.0309	.0452	.0735	.0311	.0400	.0570

6   Discussion

This paper described a preliminary experiment on the use of temporal information to improve the performance of recommender systems (RS). We extracted used-based time frames from a Twitter corpus, and we compared the use of personalised and fixed time frames in a simple RS framework. Results show that the use of personalised time frames to retrieve candidates improved the quality of recommendations generated by both models. In addition to that, the experiment showed that the hashtag-based model used as an example of RS did not generally outperform a naive baseline system.

Despite the small size of the data set and the simplicity of the models under consideration, we believe that these results suffice to illustrate the potential impact of taking temporal factors into account when generating content recommendation for microblogs or, at the very least, that different users indeed read and republish microblog posts at different rates.

As future work, we intend to investigate the use of topic-based time frames to capture the notion that different subjects (e.g., sports, politics etc.) may remain relevant (and therefore likely to be retweeted) for different lengths of time. Thus, we will attempt to improve the quality of recommendation by replacing the current user-based time frames for topic-based time frames.

Acknowledgments. The authors acknowledge support by FAPESP.

References

1. O'Banion, S., Birnbaum, L., Hammond, K.: Social media-driven news personalization. In: Proceedings of the 4th ACM RecSys Workshop on Recommender Systems and the Social Web. RSWeb 2012, pp. 45–52. ACM, New York (2012)
2. Chen, J., Nairn, R., Nelson, L., Bernstein, M., Chi, E.: Short and tweet: experiments on recommending content from information streams. In: Proceedings of the 28th International Conference on Human Factors in Computing Systems, CHI 2010, pp. 1185–1194. ACM, New York (2010)
3. Yan, R., Lapata, M., Li, X.: Tweet recommendation with graph co-ranking. In: Proceedings of the 50th Annual Meeting of the Association for Computational Linguistics: Long Papers, ACL 2012, vol. 1, pp. 516–525. Association for Computational Linguistics, Stroudsburg (2012)
4. Abel, F., Gao, Q., Houben, G.-J., Tao, K.: Semantic enrichment of twitter posts for user profile construction on the social web. In: Antoniou, G., Grobelnik, M., Simperl, E., Parsia, B., Plexousakis, D., De Leenheer, P., Pan, J. (eds.) ESWC 2011, Part II. LNCS, vol. 6644, pp. 375–389. Springer, Heidelberg (2011)
5. Gao, Q., Abel, F., Houben, G., Tao, K.: Interweaving trend and user modeling for personalized news recommendation. In: IEEE/WIC/ACM International Conference on Web Intelligence and Intelligent Agent Technology, WI-IAT 2011, vol. 1, pp. 100–103 (August 2011)
6. Manning, C., Raghavan, P., Schutze, H.: Introduction to Information Retrieval. Cambridge University Press (2008)

Development of a Lexical Resource Annotated with Semantic Roles for Portuguese

Leonardo Zilio

Universidade Federal do Rio Grande do Sul (UFRGS)
lzilio@ig.com.br

1 Introduction

Many applications in Natural Language Processing (NLP) may benefit from the use of semantic information, for example, text summarization (Yoshikawa et al. 2012), resolution of anaphora (Kong and Zhou 2012) and machine translation (Jones et al. 2012). For Portuguese, there are some important studies on semantic role labeling (Salomão 2009; Scarton 2013; Duran and Aluisio 2011, 2012), which will be presented in the following sections, but there is still much to do in this field.

This paper aims at presenting ongoing work in the development of a lexical resource for Portuguese which contains semantic role information. Semantic roles linked to verb arguments can be considered as a way to comprehend the basic semantic of a sentence without having to rely directly on lexical semantics. As an example we can use the following simple sentence:

Thomas saw the dog.

In this sentence, the syntactic arguments are the subject *Thomas* and the direct object *the dog*. With only this information, it is still not possible to comprehend whether the subject is doing something to the object, is being affected by it, or is simply experiencing it without intervention. When we describe those arguments with the roles of EXPERIENCER and THEME, respectively, then we can have a better (although not yet full) understanding of the event that is being described.

As we will see further on in this paper, semantic roles can be descriptive or numbered, and the choices made by the designer of the resource regarding the extent and grain size of the roles are of extreme importance for the final product, as well as for the development process.

The resource we present in this paper is based on VerbNet (Kipper-Schuler 2005) and PropBank (Palmer, Gildea and Kingsbury 2005), and contains sentences annotated with descriptive semantic roles.

The structure of this paper goes as follows: Section 2 will present related work, debating some of the methods used for semantic role labeling, specially the semantic role lists and the resulting resource; Section 3 will show the method used in this work, along with the materials we used; Section 4 discusses our partial results; and Section

J. Baptista et al. (Eds.): PROPOR 2014, LNAI 8775, pp. 195–200, 2014.

5 is reserved for our final remarks, including information about the availability of the resource being developed.

2   Related Work

In this section we present two semantic resources for English along with their Brazilian counterparts. We will give an overview of their methodology and type of semantic roles, and of the resulting resource.

2.1   Propositional Bank

PropBank (Palmer, Gildea and Kingsbury 2005) is a resource which presents sentences annotated with numbered semantic roles. That is, instead of using a descriptive name like AGENT or THEME, the PropBank uses semantic roles like ARG0 or ARG1. The numbered roles vary from ARG0 to ARG5 and there are also roles reserved for modifiers, which vary in name according to the type of adjunct being annotated (ARGM-TMP for temporal adjunct; ARGM-LOC for local adjuncts, etc.). These semantic roles were annotated as a semantic level for the sentences present in the Penn Tree-Bank (Marcus, Marcinkiewicz and Santorini 1993), so that the resource would have a semantic role level on top of the already existent syntactic and morphological levels. The PropBank has a total of 3,185 annotated verbs and can be consulted on-line[1].

For Portuguese, we have the PropBank.Br (Duran and Aluisio 2011, 2012) which consists of a set of annotated sentences extracted from the Bosque corpus, a part of Floresta Sintá(c)tica corpus[2]. The resource has 1,025 distinct verbs annotated with numbered semantic roles.

2.2   VerbNet

Unlike PropBank, which used corpus sentences to apply a new level of semantic information, the VerbNet (Kipper-Schuler 2005) relied on the verb classes, as organized by Levin (1993). For each class the possible meaning associated with it is provided along with an example. VerbNet uses descriptive semantic roles (such as AGENT, THEME, PACIENT, etc.) and has 272 classes, which comprise 3,769 verbs. The semantic role list consists of 36 roles, but only 30 are actually being used in the latest update (version 3.2).

VerbNet has also a Portuguese version, VerbNet.Br (Scarton 2013), which was semi-automatically imported from the English VerbNet through the existing links between VerbNet and WordNet, and between WordNet and WordNet.Br. Since the resource was semi-automatically constructed, the results may contain some noise.

---

[1] http://verbs.colorado.edu/verb-index/.
[2] http://www.linguateca.pt/Floresta/principal.html.

3   Method

For the development of our resource, we used corpora and a subcategorization frame extractor that offers a user interface for the annotation of semantic roles. We also compiled a semantic role list from the VerbNet (Kipper-Schuler 2005) and extended it with other semantic roles, as we explain below.

3.1   Corpora

As a base for the extraction of sentences, we used two corpora: one with Brazilian Cardiology papers compiled by Zilio (2009) and the other with Brazilian newspaper articles compiled by the project PorPopular[3]. The different domains of the corpora are intended for a comparison between the behavior of semantic roles in specialized and non-specialized texts. Both corpora were analyzed with the parser PALAVRAS (Bick 2000) and present dependency trees for all sentences.

3.2   Subcategorization Frames Extractor

Subcategorization Frames can be seen as simpler syntactic representations of sentences (or phrases). They show what types of phrases are present in the sentence, without regard to whether it is a subject, complement or adjunct. As such, the representation of our early example "Thomas saw the dog" would be something like NP_NP or NP_V_NP (i.e. with or without the verb position).

This type of representation shows the basic arguments of a sentence. For this reason it can be used for the purposes of semantic role labeling. In our study we used a subcategorization frame extractor developed originally by Zanette (2010) and further modified by Zilio, Zanette and Scarton (2014).

The extractor uses corpora analyzed with the parser PALAVRAS (Bick 2000) as input, and processes all the sentences, organizing them according to verb, subcategorization frame and frequency in a MySQL database.

3.3   Semantic Roles

For the annotation we selected the roles used in VerbNet and adapted them for Portuguese. The basic elements from VerbNet were preserved, but some roles for adverbial adjuncts (from PropBank) were also included, so as to make the list more robust vis-à-vis the sentences from real written texts that are present in our corpora. The list itself contains 47 semantic roles covering arguments and adjuncts. This is much more fine-grained than PropBank's list and a bit more than the VerbNet's one.

Some of the roles we add to the list, other than those for adjuncts taken from PropBank, are roles for the pronoun "se", which can have many functions in a sentence, all

---

[3]   http://www.ufrgs.br/textecc/porlexbras/porpopular/index.php.

of which are well explained in the tutorial of PropBank.Br[4]. Another role was created specifically for agents that are, in fact, places or organizations, normally used as metonymical references to the people within it, such as "New York" in the sentence: "New York does the job." All the created semantic roles either were necessary because of the language or can be joined back together with the original role from VerbNet without much effort.

3.4   Annotation Process

The annotation is made by one linguist annotator and follows these criteria:

- The most frequent verbs from the newspaper corpus are annotated first, and then follows the annotation of the same verbs in the Cardiology corpus.
- Ten sentences per subcategorization frame are annotated.
- The verbs "ser" (to be), "estar" (to be), "haver" (to exist), and "ter" (to have) were not annotated, because they present too many subcategorization frames in both corpora and would require too much work for the sake of only a few verbs.

Following this method, samples of sentences are annotated for each verb, ensuring that we annotate many verbs and many types of subcategorization frames for each of them, maximizing the reward regarding the effort.

4   Results

Until now we have annotated 85 verbs in the newspaper corpus, and 42 of them were also annotated in the Cardiology corpus (the other 43 verbs were not available or did not present ten sentences in a subcategorization frame). This represents 4,640 annotated sentences (3,470 in the newspaper corpus and 1,170 in the Cardiology corpus). From the 47 semantic roles present in our list, 45 were used so far.

The most frequent semantic roles were AGENT and THEME, but, if we look at the individual corpora, only THEME is highly frequent in both of them. The sentences in the Cardiology corpus showed a tendency to suppress the agents, highlighting the instruments, results and the objects of the domain.

A list of the most frequent semantic roles in both corpora can be seen in Table 1. In that list, one can see that semantic roles like EXPERIENCER and TOPIC, which are highly ranked in the newspaper corpus, have very few correspondents in the Cardiology corpus. We have not yet made any statistical computation of these results, since we have not yet finished the annotation process, but the results so far seem to corroborate the ones found in a study by Zilio, Ramisch and Finatto (2013).

---

[4] http://www.nilc.icmc.usp.br/portlex/images/arquivos/propbank-br/propbank.br%20tutorial.pdf.

5   Final Remarks

The partial results of the annotation were already made available for online consultation on the Jibiki platform[5] and can also be downloaded in XML format from the CAMELEON project website[6]. There is still much work to be done, though, and we will annotate many more verbs before releasing the final version of our database.

For now, the results already show some important information, like the difference in the semantic roles annotated in both corpora. If this tendency remains after more verbs are annotated, than it leads to the conclusion that semantic roles are not dependent of domain, and it seems to be possible to achieve a domain independent list. However, the frequency of each semantic role oscillates greatly from corpus to corpus, showing that their ranking may be domain specific.

Table 1. Semantic roles annotated so far in both corpora, organized by total frequency

Semantic Roles	Newspaper	Cardiology	Total
THEME	1922	841	2763
AGENT	1673	158	1831
PLACE	430	117	547
EXPERIENCER	458	46	504
RESULT	245	211	456
VERB	316	133	449
PIVOT	197	172	369
MOMENT	248	73	321
PATIENT	266	49	315
TOPIC	282	19	301
GOAL	189	96	285
INSTRUMENT	107	153	260
SITUATION	144	107	251
ATTRIBUTE	129	51	180
CAUSE	102	76	178
DESTINATION	150	7	157
BENEFICIARY	100	46	146
MANNER	71	68	139
RECIPIENT	119	3	122

Acknowledgements. We would like to thank CNPq and CAPES for funding and project CAMELEON (CAPES/Cofecub 707/11) for all the support.

---

[5] http://jibiki.univ-savoie.fr/jibiki/Home.po.
[6] http://cameleon.imag.fr/xwiki/bin/view/Main/Semantic
%20role%20labels%20corpus%20-%20Brazilian%20Portuguese.

References

1. Bick, E.: The Parsing System "Palavras" : Automatic Grammatical Analysis of Portuguese in a Constraint Grammar Framework, vol. 202. Aarhus University Press Aarhus (2000)
2. Duran, M.S., Aluísio, S.M.: Propbank-Br: A Brazilian Portuguese corpus annotated with semantic role labels. In: Proceedings of the 8th Symposium in Information and Human Language Technology, Cuiabá/MT, Brazil, October 24-26 (2011)
3. Duran, M.S., Aluísio, S.M.: Propbank-Br: A Brazilian treebank annotated with semantic role labels. In: Proceedings of the LREC 2012, Istanbul, Turquia, May 21-27 (2012)
4. Jones, B., Andreas, J., Bauer, D., Hermann, K.M., Knight, K.: Semantics-based machine translation with hyperedge replacement grammars. In: Proceedings of COLING 2012, Mumbai, India, pp. 1359–1376. The COLING 2012 Organizing Committee (2012)
5. Kipper-Schuler, K.: VerbNet: A broad-coverage, comprehensive verb lexicon. University of Pennsylvania. Tese de doutorado orientada por Martha S. Palmer (2005)
6. Kong, F., Zhou, G.: Exploring local and global semantic information for event pronoun resolution. In: Proceedings of COLING 2012, Mumbai, India, pp. 1475–1488. The COLING 2012 Organizing Committee (December 2012)
7. Levin, B.: English verb classes and alternations: A preliminary investigation, vol. 348. University of Chicago press, Chicago (1993)
8. Marcus, M.P., Marcinkiewicz, M.A., Santorini, B.: Building a large annotated corpus of english: The penn treebank. Computational linguistics 19(2), 313–330 (1993)
9. Palmer, M., Gildea, D., Kingsbury, P.: The Proposition Bank: A Corpus Annotated with Semantic Roles. Computational Linguistics Journal 31(1) (2005)
10. Salomão, M.: FrameNet Brasil: Um trabalho em progresso. Calidoscópio 7(3), 171–182 (2009)
11. Scarton, C.: VerbNet.Br: Construção semiautomática de um léxico verbal online e independente de domínio para o português do Brasil. NILC/USP. Dissertação de mestrado orientada por Sandra Maria Aluísio (2013)
12. Yoshikawa, K., Iida, R., Hirao, T., Okumura, M.: Sentence compression with semantic role constraints. In: Proceedings of the 50th Annual Meeting of the Association for Computational Linguistics, Jeju Island, Korea. Short Papers, vol. 2, pp. 349–353. Association for Computational Linguistics (2012)
13. Zanette, A.: Aquisição de Subcategorization Frames para Verbos da Língua Portuguesa. Projeto de Diplomação. UFRGS. Orientadora: Aline Villavicencio (2010)
14. Leonardo, Z., Ramischand, C., BocornyFinatto, M.J.: Desenvolvimento de um recurso léxico com papéis semânticos para o português. Linguamática 2, 23–41 (2013)
15. Leonardo, Z., Zanetteand, A., Scarton, C.: Automatic extraction of subcategorization frames from portuguese corpora. In: Aluisio, S.M., Tagnin, S.E.O. (eds.) New Languages Technologies and Linguistic Research: A Two-Way Road, pp. 78–96. Cambridge Scholars Publishing (2014)

brWaC: A WaCky Corpus
for Brazilian Portuguese*

Rodrigo Boos, Kassius Prestes, Aline Villavicencio, and Muntsa Padró

Institute of Informatics, Federal University of Rio Grande do Sul, Porto Alegre, Brazil
{rboss,kvprestes,avillavicencio,muntsa.padro}@inf.ufgrs.br

Abstract. Initiatives for constructing very large corpora have increased in recent years, especially using the Web as corpus since large corpora are crucial for many Natural Language Processing tasks. The WaCky (Web-As-Corpus Kool Yinitiative) methodology has been used to build very large corpora (over a billion words each) for languages like English, Italian and German among others. In this paper we present the ongoing work on building *brWaC*, a massive Brazilian Portuguese corpus crawled from *.br* domains. At the moment, the crawling process and the PoS tagging are finished, resulting in a tokenized and lemmatized corpus of 3 billion words. Next step is parsing the whole corpus.

Keywords: Web as Corpus, *brWaC*, WaCky, Brazilian Portuguese.

1 Introduction

Recently, big effort has been devoted to creating and processing very large corpora in order to improve performance of NLP techniques. Such big corpora are crucial for tasks like identification of multiword expressions [1,2], subcategorization frame acquisition [3] and distributional thesauri construction [4,5] among others. Most approaches to create very large corpora use the Web as a source of big amount of data. Those techniques employ crawlers to collect sets of texts which are subsequently cleaned to extract only the textual contents and filtered to remove duplicate material and noise from texts that have little human produced content. The WaCky (Web-As-Corpus Kool Yinitiative)[1] [6] in particular has been used to build very large corpora for languages like English, Italian and German. Corpus over a billion words each have been built and offered to the community for these languages. This method starts from a list of medium frequency content seed words and produces general purpose corpora which have a good level of content variation and quantity of information.

Other efforts have been made to collect comparable corpora, using focused crawling methods, starting from a domain specific page and guiding the crawler

---

* We would like to thank the support of projects CNPq (PRONEM) 003/2011, CNPq 482520/2012-4, 312184/2012-3, 551964/2011-1, PNPD 2484/2009 and Capes-Cofecub 707/11.

[1] http://wacky.sslmit.unibo.it/doku.php.

J. Baptista et al. (Eds.): PROPOR 2014, LNAI 8775, pp. 201–206, 2014.

based on the proportion of relevant words that can be found in the text of that page and of all the other pages that belong to the same host [7,8], while work to collect parallel corpora identifies sites with equivalent pages in multiple languages [9].

These initiatives represent an inexpensive way of creating a large resource, especially for languages for which freely available resources of this magnitude are still scarce. For instance, for Portuguese although a variety of different corpora is available, such as those in the Linguateca website, their sizes usually vary from 23,000 to almost 800 million tokens, table 1[2].

Table 1. Brazilian Portuguese Corpora

Corpus	Tokens	Types	Sentences
Corpus Brasileiro	792,765,372	666,509,591	29,919,794
NILC/São Carlos	42,912,644	32,461,799	1,988,621
ANCIB	1,707,731	1,257,109	83,509
OBras	1,424,014	1,133,302	37,419
ECI-EBR	922,378	723,995	44,381
ReLi	189,577	145,325	8,752
AmostRA-NILC	128,190	98,633	4,931
C-Oral-Brasil	121,092	65,303	9,329
FrasesPB	23,349	19,162	653
Total BR	840,194,347	702,414,219	32,097,389

In this work we adopt the WaCky method [6] for collecting a very large corpus for Brazilian Portuguese, the *brWaC*. We performed the crawling of more than 1,300,000 documents amounting a total of 3 billion words, resulting in the biggest Brazilian Portuguese corpus available. About 8% of the corpus has been already PoS tagged. This is an ongoing work, we plan to produce PoS tagged and dependency parsed versions of the whole corpus.

In what follows, we present a description of the WaCky initiative (§2) as well as our approach to create *brWaC* (§3). Section 4 exposes the current state of the corpus and we finish with the roadmap for future steps (§5).

2   The WaCky Approach

The WaCky approach [6] has been used to build very large corpora for English (*ukWaC*), Italian (*itWaC*) and German (*deWaC*). Lately the same methodology has been also applied to build corpora for other languages such as French (*frWaC* [10]), Croatian (*hrWaC* [11]), Slovene (*slWac* [11]) and Catalan (*caWaC* [12]). All these corpora contain more than one billion words and have been part-of-speech tagged and lemmatized with the goal of offering to the community a very large corpus to be used in several NLP tasks. For English, an additional layer

---

[2] Adapted from http://www.linguateca.pt/corpora_info.html

with full dependency parses was added using the MaltParser [13] resulting in *PukWaC* corpus.

WaCky corpora are constructed using the following steps (as defined in [6]):

1. **Seed URL collection:** to ensure content variety, bigrams generated by randomly selecting content words of medium frequency from the target language were submitted to a search engine. For English, 2,000 bigrams were used (e.g. *iraq package, soil occurs, elsewhere limit*), for German 1,653 and for Italian 1,000. For each bigram, a maximum of ten seed URLs were retrieved. From these, duplicates were removed and the remaining URLs were randomly fed to a crawler restricted to pages in the relevant language.

2. **Post-crawl cleaning:** only pages of medium size are kept and duplicates are removed. The pages remaining after the post-crawl cleaning are further cleaned to remove code and boilerplate elements and filtered according to function word density (each page must have at least 10 types, 30 tokens and 25% of function words).

3. **Near-duplicate detection and removal:** counting the n-gram overlap, using 25 5-grams, between each two documents.

4. **Annotation:** add additional information such as part-of-speech tagging and parsing to the crawled text.

3   Methods

Following WaCky method [6] the *brWaC* corpus is collected in the three main steps presented in last section: crawling, cleaning and duplicate removal. Once the plain-text corpus is built, the annotation step is performed.

3.1   Crawling

For this first step, we constructed a set of content words to use as seeds for the crawler from a list of word frequencies available for the Brazilian Portuguese corpora in Linguateca[3]. From this list, which also contains function words like prepositions, numerals and articles, we removed the stopwords[4], and applied both a high and a low frequency thresholds to obtain words with medium frequency, removing those with more than 10,000 or less than 100 occurrences. Also, we removed the ones with less than four letters, which in general are most likely to be non-content words. From this set of words, we built 1,000 random pairs to submit as queries to a search engine (Bing API). Following [6] we use bigram queries because single word queries may return uninteresting pages like company pages or definitions. Using the search engine we generate a set of URLs (10 for each query), and the crawler goes through the links in each page, storing each of the pages visited. However, these contain components beyond the human produced text that we want, such as HTML code and boilerplate.

---

[3] Linguateca Corpora Frequency List, available at
dinis2.linguateca.pt/acesso/tokens/formas.totalbr.txt.

[4] Lists of Portuguese Stopwords available at
http://www.linguateca.pt/chave/stopwords/.

3.2   Post-Crawl Cleaning

The second step included boilerplate stripping, applying a methodology based on the *boilerpipe* library [14], which uses density metrics and shallow text features. The shallow text features do not take into account what words are presented, but more general characteristics like average word and sentences length. Besides, these characteristics are analyzed in combination with the position in the text of the current analyzed text block, i.e., if it is placed near some content text blocks, then probably it is content too. The densitometric features include the link density and text density, which is the mean amount of words per line. We also experimented with additional features, and the one that produced the best results was the density of stopwords. According to [15] content texts will probably contain at least 25% of stopwords, so we removed every text block that did not satisfy this property.

3.3   Duplicate Removal

The duplicate removal step is crucial since otherwise, the amount of data collected may not reflect content variation. Before detecting duplicated texts, we exclude both small and big texts, only keeping documents between 5Kb and 200Kb. Empirically we observe that short texts generally do not contain content, and very large texts are glossaries, dictionaries and so on. We have adapted the shingling algorithm, and what we have done consisted of getting 20 5-grams sample of words from each text, and comparing them to the other texts. If there were more than 2 identical 5-grams, we assumed that the texts were duplicates. This approach evidently excludes some texts that might have content variation, since the threshold used to exclude is very low. Nevertheless, the web is a very large source of texts, so our concern is not to allow duplicates with a strict criteria, prioritizing precision, and it does not matter so much if some false negatives with good content are discarded. This duplicate filtering approach was adapted from the one proposed by [16].

3.4   Corpus Annotation

Once the corpus is created, we tokenize, lemmatize and PoS tag it, using the TreeTagger [17] trained for Portuguese[5]. TreeTagger is one of the fastest Taggers available[6], an important characteristic given the large quantities of texts that need to be processed. This is also the tagger used to tag and lemmatize *ukWaC, deWaC, frWaC* and *itWaC*.

3.5   Corpora Quality

As a post processing step to assure corpora quality we used a language detection library [18] to ensure the documents collected were written in Brazilian

---

[5] The TreeTagger supports a number of languages, including Brazilian Portuguese.
[6] According to evaluation in
http://mattwilkens.com/2008/11/08/evaluating-pos-taggers-speed/

Portuguese. From the 1,3 million documents, only 6,476 (0,49%) were detected as not in Brazilian Portuguese. This evaluation show that the corpus does not contain much noise, as were expected when automatically extracting documents from the web.

4 *brWaC* at This Moment

The process of crawling and cleaning *brWaC* is already finished. *brWaC* contains now more than 1.3 million documents with about 156 million of sentences and 3 billion of tokens. Collecting the corpus took approximately 24 hours. After that, we performed the boilerplate removal of the retrieved pages, which took around 60 hours. Additionally, duplicate page removal involved the cost of reading the texts, removing stopwords and generating n-grams, which is linear (on the number of texts), but demanding[7]. Comparing all the n-grams from different texts to one another (with a quadratic complexity) to identify the intersection between them, results in a mean cost of $20 * n^2$ ($O(n^2)$ complexity). The process of removing duplicates took in total 68 hours[8]. Currently 100% of the crawled corpus has been processed with TreeTagger.

5 Next Steps

At the moment, the crawling and cleaning steps to build *brWaC* have been already finished and the PoS tagging completed. The next steps involve parsing and distribution of the corpus[9]:

- Produce frequency list for lemmas and types.
- Add a dependency parsing layer to *brWaC* with a syntactic parser such as Palavras [19] to provide a resource similar to *PukWaC*.
- Since we are dealing with a corpus of 3 billion words, the computational cost of running previous tasks (specially the dependency parsing) is very high, thus we plan to parallelize them using Hadoop[10].

Regarding applications, a previous (smaller) version of *brWaC* has already been used for different Brazilian Portuguese applications such as identification of Multiword Expressions [20] and as a source for a simplified touchscreen typing system[11]. We plan to use the PoS tagged corpus for thesaurus construction and the parser version to build a Portuguese Distributional Memory [5], once available. So far these kind of tasks have been mainly performed for English, due to the lack of big corpus for other languages. Thus, with *brWaC* we expect to allow the community to perform such tasks for Portuguese.

---

[7] From empirical observation it takes approximately 10 million operations per text.

[8] All the processing mentioned here was performed in a machine with 3.40GHz CPU and 32Gb of RAM

[9] The Plain-text and PoS tagged versions of *brWaC* will be made available for download in the near future

[10] http://hadoop.apache.org/

[11] http://minuum.com/

References

1. Ramisch, C., Villavicencio, A., Boitet, C.: Multiword expressions in the wild? the mwetoolkit comes in handy. In: Proc. of the 23rd COLING - Demonstrations, Beijing, China. The Coling 2010 Organizing Committee (August 2010)
2. Tsvetkov, Y., Wintner, S.: Extraction of multi-word expressions from small parallel corpora. In: Coling 2010: Posters, Beijing, China, Coling 2010 (August 2010)
3. Korhonen, A., Krymolowski, Y., Briscoe, E.J.: A large subcategorization lexicon for natural language processing applications. In: Proceedings of the 5th LREC, Genova, Italy (2006)
4. Lin, D.: Automatic retrieval and clustering of similar words. In: Proceedings of the 36th ACL and 17th International COLING (1998)
5. Baroni, M., Lenci, A.: Distributional memory: A general framework for corpus-based semantics. Computational Linguistics 36(4), 673–721 (2010)
6. Baroni, M., Bernardini, S., Ferraresi, A., Zanchetta, E.: The wacky wide web: A collection of very large linguistically processed web-crawled corpora. Language Resources and Evaluation 43(3), 209–226 (2009)
7. Talvensaari, T., Pirkola, A., Järvelin, K., Juhola, M., Laurikkala, J.: Focused web crawling in the acquisition of comparable corpora. Inf. Retr. 11(5) (October 2008)
8. Granada, R., Lopes, L., Ramisch, C., Trojahn, C., Vieira, R., Villavicencio, A.: A comparable corpus based on aligned multilingual ontologies. In: Proceedings of the First Workshop on Multilingual Modeling, MM 2012, pp. 25–31. Association for Computational Linguistics, Stroudsburg (2012)
9. Barbosa, L., Sridhar, V.K.R., Yarmohammadi, M., Bangalore, S.: Harvesting parallel text in multiple languages with limited supervision. In: Kay, M., Boitet, C. (eds.) COLING, pp. 201–214. Indian Institute of Technology, Bombay (2012)
10. Ferraresi, A., Bernardini, S., Picci, G., Baroni, M.: Web corpora for bilingual lexicography: A pilot study of english/french collocation extraction and translation. In: Using Corpora in Contrastive and Translation Studies. Cambridge Scholars Publishing, Newcastle (2010)
11. Ljubešić, N., Erjavec, T.: hrwac and slwac: Compiling web corpora for croatian and slovene. In: Proceedings of 14th International Conference on Text, Speech and Dialogue, TSD (2011)
12. Ljubešić, N., Toral, A.: caWaC – a web corpus of Catalan. In: Proceedings of LREC 2014 (May 2014)
13. Nivre, J., Hall, J., Nilsson, J., Chanev, A., Eryiğit, G., Kübler, S., Marinov, S., Marsi, E.: Maltparser: A language-independent system for data-driven dependency parsing. Natural Language Engineering 13, 95–135 (2007)
14. Kohlschütter, C., Fankhauser, P., Nejdl, W.: Boilerplate detection using shallow text features. In: Proceedings of the Third ACM International Conference on Web Search and Data Mining, WSDM 2010, pp. 441–450. ACM, New York (2010)
15. Pomikálek, J.: Removing Boilerplate and Duplicate Content from Web Corpora. PhD en informatique, Masarykova univerzita, Fakulta informatiky (2011)
16. Broder, A.Z., Glassman, S.C., Manasse, M.S., Zweig, G.: Syntactic clustering of the web. Comput. Netw. ISDN Syst. 29(8-13), 1157–1166 (1997)
17. Schmid, H.: Probabilistic part-of-speech tagging using decision trees (1994)
18. Shuyo, N.: Language detection library for java (2010)
19. Bick, E.: The Parsing System Palavras. Automatic Grammatical Analysis of Portuguese in a Constraint Grammar Famework. PhD thesis, Aarhus University (2002)
20. Boos, R., Prestes, K., Villavicencio, A.: Identification of multiword expressions in the brwac. In: Proceedings of LREC 2014 (May 2014)

DeepBankPT and Companion Portuguese Treebanks in a Multilingual Collection of Treebanks Aligned with the Penn Treebank

António Branco, Catarina Carvalheiro, Francisco Costa, Sérgio Castro,
João Silva, Cláudia Martins, and Joana Ramos

Universidade de Lisboa,
Departamento de Informática, Faculdade de Ciências

Abstract. We present a new collection of treebanks for the Portuguese language, comprising five datasets that cover major types of grammatically annotated corpora: TreeBankPT, PropBankPT, DependencyBankPT, LogicalFormBankPT and DeepBankPT. This collection is the Portuguese part of a broader multilingual collection of aligned treebanks that are developed for different languages, including English, under the same methodological principles and guidelines, and whose raw text versions are translations of the Penn Treebank, a de facto standard dataset for research on language technology.

Keywords: TreeBank, PropBank, DependencyBank, LogicalFormBank, DeepBank, aligned treebanks, multilingual datasets, Portuguese.

1 Introduction

In this paper we introduce a new collection of state of the art treebanks for the Portuguese language. This collection comprises five datasets that cover major types of grammatically annotated corpora:

- TreeBankPT encodes the syntactic constituency of sentences;
- PropBankPT expands the information in the dataset above with semantic roles;
- DependencyBankPT records the information on the grammatical dependencies holding among the expressions in the sentences;
- LogicalFormBankPT associates each sentence to the representation of its semantics in a logical formalism;
- DeepBankPT associates each sentence in the dataset with its fully fledged grammatical representation, thus including also the dimensions recorded in the other four treebanks above into a single integrated representation.

The treebanks in this collection were developed under the advanced design options of dynamic treebanks (Oepen *et al.*, 2002), thus being the second collection developed under such conditions for the Portuguese language, together with the pioneer CINTIL collection (Branco *et al.*, 2010). The present collection however goes a crucial step

J. Baptista et al. (Eds.): PROPOR 2014, LNAI 8775, pp. 207–213, 2014.

further in terms of covering an important gap that existed for Portuguese: these treebanks are parallel and sentence aligned to treebanks from English and other languages, thus being suitable to support the development of a range of multilingual applications, including machine translation.

The above listed corpora for Portuguese are part of a broader multilingual collection of treebanks that are built not only over texts that are translationally equivalent among the different languages, but also under the same methodological principles and guidelines. What is more, these texts are translations from an English corpus that is a de facto standard dataset, over which most progresses on parsing have been obtained in the last decades, namely the WSJ corpus, upon which the Penn Treebank was established. That is, on the one hand, the WSJ corpus was re-annotated to develop a new dynamic English treebank, DeepBankENG (Flickinger *et al.*, 2012a); on the other hand, the WSJ corpus was translated into other languages, including Portuguese, and the corresponding collection of treebanks, ParDeepBank, has been developed, whose datasets *ipso facto* became aligned with each other and with the new English treebank.

In the next Section 2, the methodological conditions for the development of dynamic DeepBanks are introduced. The following Section 3 addresses how DeepBanks can support the establishment of a high quality collection of aligned treebanks for a given language. In Section 4, we present the ParDeepBank corpora, a multilingual collection of DeepBanks, which DeepBankPT is a component of, and in Section 5, the Portuguese DeepBankPT and companion treebanks are presented in detail. This paper closes with Section 6, with concluding remarks.

2   Advanced Treebanks

Grammatically interpreted corpora have played a major role in the progress of language technology. Such accurately annotated data sets have been of fundamental importance for the development of language processing tools and solutions with increasingly improved performance and depth of analysis. These tools range from part-of-speech taggers to semantic role labelers, and include named entity recognizers, dependency parsers or lemmatizers, among many others.

The increased sophistication and depth of analysis of these tools has required corpora with increasingly more sophisticated linguistic information. As a consequence, this has lead to increasingly more demanding conditions on the annotation process, not only in terms of the linguistic expertise required from the human annotators, but also in terms of the organization and management of the annotation process (Branco, 2009).

In general, as the information and categories being associated with linguistic items grow in complexity, the concerns about the reliability of the data set increase. Categories with more complex structure to be handled by the annotators typically imply more chances for some parts of them to be incorrectly chosen, and thus more chances for the annotated dataset to be flawed. They also bring more chances for the categories assigned by different annotators to a markable be divergent, and thus for a lower inter-annotator agreement (Artstein and Poesio, 2008).

An extreme case both of the complexity of the information to be assigned and the need and importance of supportive tools can be found in DeepBanks. These are corpora whose sentences are annotated with fully-fledged deep grammatical representations encompassing all different levels of grammatical dimensions for each sentence (from morphological analysis to meaning representation) (Cotton and Bird, 2002; Open et al., 2002, Böhmová et al., 2003, Rosén et al., 2005).

The complexity of the category to be assigned is such that, in practical terms, it is out of the range of human annotators ability to be able to compose it, even in a piecemeal fashion. In this case, the annotation process has to resort to an annotation tool, a computational grammar, which proposes a number of viable parses out of which the annotator eventually selects the one to be assigned via the selection of parse discriminants that progressively reduce the parse forest and thus the annotation space (Dipper, 2000; Oepen, 1999; Rosén et al., 2009; Rosén et al., 2012).

3   The Collection of Treebanks Extracted from a DeepBank

As it occurs many times, sophistication comes at a cost, but extra cost may bring extra benefits. That happens also in the case of DeepBanks. The construction of a DeepBank is incomparably much more demanding, in resources and organization effort, than for instance a much simpler POS annotated corpus or even a constituency treebank. Among many other things, it requires a deep processing grammar, whose development is in itself an long term endeavor of non trivial prosecution (e.g. Copestake and Flickinger, 2000; Branco and Costa, 2010). But once the development of an annotated corpus of this highly advanced type is set in motion, one is opening the way to the construction of a resource that brings a range of unique advantages.

First, since deep processing grammars are developed under a thorough grammatical framework (e.g. HPSG), in DeepBanks, sentences are annotated with information that not only is linguistically principled, but that it is also consistent across the sentences in the corpus.[1]

Second, one can extract different "vistas" from a DeepBank: for instance, an extracted data set with sentences annotated with their syntactic constituency trees (a TreeBank); or another one with sentences annotated with those trees decorated with semantic roles (a PropBank), etc. Thus, when one builds a DeepBank, it is as if in practical terms, one is getting several corpora for the cost of one (Silva et al., 2012).

Third, while they capture different grammatical dimensions of their sentences, these corpora are fully aligned among themselves as they are built on the same set of raw sentences. Thus, a DeepBank allows for a collection of monolingual corpora that are aligned among each other and that encode different grammatical dimensions.

Against this background, an important line of progression is thus to have corpora that are aligned and that represent not only different grammatical dimensions, but also consistently represent such dimensions across different languages. These certainly are

---

[1] As way of example, in the Peen Treebank, detecting inconsistencies became a topic of research in itself: see for instance Dickinson and Meurers, 2005.

assets of utmost importance to support the training and development of multilingual and machine translation solutions of increased quality.

4   A Multilingual Collection of Aligned DeepBanks

This requires the development of multilingual aligned DeepBanks. This in turn presupposes some non-trivial conditions, among which the most demanding one is perhaps that there exist deep processing grammars for the different languages at stake. To facilitate that the alignment of the multiple dimensions may carry over to the alignment across languages, these grammars are expected to be developed under some similar guidelines, principles or grammatical framework.

An initiative to develop multilingual aligned DeepBanks is under way in the scope of the DELPH-IN consortium (www.delph-in.net). This endeavor relies on grammars developed by members of the consortium for different languages. And the aligned DeepBanks are obtained by annotating with those grammars raw texts in different languages, which are aligned among each other (Flickinger *et al.*, 2012).

In order to maximize the potential of the possible research produced over these DeepBanks to be comparable to other results reported in the literature, the raw text in English is the one from Penn TreeBank (Marcus *et al.*, 1993*)*. In the other languages, the aligned texts result from the translation of that English text into those languages.

The initial languages for which there are aligned DeepBanks being developed under this arrangement are Bulgarian, English, Portuguese and Spanish. More are being prepared to join. The Portuguese version of the raw text entering this multilingual collection, that is the translation of the WSJ corpus, contains over 40,000 sentences. It is the result of the translation of the text in the Penn Treebank by a paid professional translator, a translation that was subsequently submitted to a double checking by two reviewers.

5   DeepBankPT and the *BankPT Collection of Treebanks

In its current first released version, the DeepBank for Portuguese (DeepBankPT) comprises those sentences that have been annotated so far, out of the ca. 40,000 sentences available to be treebanked. These amount to 3,406 sentences, containing 44,598 tokens. The development of the DeepBankPT dataset resorted to the deep linguistic processing grammar for Portuguese LXGram (Branco and Costa, 2010). The dynamic treebanking was supported by the annotation environment [incrs tsdb()] (Oepen, 1999).

The grammar produces the admissible grammatical representations, a so called parse forest, for each input sentence to be annotated. The annotation workbench permits the annotators to select one of the parse trees and annotate the sentence with it. The selection of the parse tree is performed by the annotators by setting up the appropriate option ("yes" vs. "no") of the set of so called binary discriminants.

These discriminants are associated to the rules of the grammar that were applied and thus supported the different trees in the parse forest.

The DeepBankPT was developed following the widely acknowledged annotation methodology that ensures the best reliability of the dataset produced, namely with a double-blind annotation followed by adjudication. Each sentence was annotated by a pair of expert annotators, graduated in linguistics, working independently of each other. The adjudication was performed by another expert researcher, with a post-graduation degree in computational linguistics. The level of inter-annotator agreement (ITA) is 0.83 in terms of the specific inter-annotator metric developed for this kind of corpora and annotation (Castro, 2011).

Besides this core data set, the collection of *BankPT treebanks includes four other data sets, namely TreeBankPT, PropBankPT, DependencyBankPT and Logical-FormBankPT. These treebanks are extracted from the DeepBankPT, following the procedures described in (Silva and Branco, 2012). They contain parts of the fully-fledged grammatical information contained in the DeepBankPT, displayed along widely acknowledged formats.

In the TreeBankPT, the sentences are associated to their syntactic constituency representations, along the lines of the Penn Treebank (Marcus *et al.*, 1993). The linguistic options adopted for this annotated corpora follow the options that were assumed for the CINTIL TreeBank and are described in detail in (Branco *et al.*, 2011a). The PropBankPT is an extension of the TreeBankPT where the constituency trees get decorated with semantic roles, along the lines of (Palmer *et al.*, 2005). The specific tag set adopted is identical to the one adopted for the CINTIL PropBank, described in (Branco *et al.*, 2012).

The DependencyBankPT, in turn, stores the sentences annotated with the representation of the grammatical dependencies among their component words. This dependency bank follows the design options adopted for the CINTIL DependencyBank, presented in (Branco *et al.*, 2011b). Finally, in the LogicalFormBankPT, the sentences are associated with their semantic representation encoded with semantic description formalism MRS (Copestake *et al.*, 2005).

Each one of the treebanks in the *BankPT collection is distributed free of charge for research purposes through the META-SHARE platform (www.meta-share.eu).

6   Concluding Remarks

In this paper, we described the new collection of advanced treebanks *BankPT for the Portuguese language, developed around the core treebank DeepBankPT. The key innovative aspect of these corpora relies on the fact that they are the first parallel corpora for Portuguese that are aligned with corpora for several other languages, including English, thus opening the way for advanced research involving Portuguese in multilingual applications, including machine translation.

References

1. Artstein, R., Poesio, M.: Inter-Coder Agreement for Computational Linguistics. Computational Linguistics 34(4) (2008)
2. Böhmová, A., Hajič, J., Hajičová, E., Hladká, B.: The Prague Dependency Treebank. In: Abeillé, A. (ed.) Treebanks. Kluwer (2003)
3. Castro, S.: Developing Reliability Metrics and Validation Tools for Datasets with Deep Linguistic Information, MA Dissertaion, Universty of Lisbon (2011)
4. Copestake, A., Flickinger, D., Pollard, C., Sag, I.A.: Minimal Recursion Semantics: An Introduction. Journal of Research on Language and Computation 3(4) (2005)
5. Copestake, A., Flickinger, D.: An open-source grammar development environment and broad-coverage English grammar using HPSG. In: LREC 2000 (2000)
6. Cotton, S., Bird, S.: An Integrated Framework for Treebanks and Multilayer Annotations. In: Proceedings of LREC 2002 (2002)
7. Branco, A., Carvalheiro, C., Pereira, S., Avelãs, M., Pinto, C., Silveira, S., Costa, F., Silva, J., Castro, S., Graça, J.: A PropBank for Portuguese: The CINTIL-PropBank. In: Proceedings of LREC 2012 (2012)
8. Branco, A., Silva, J., Costa, F., Castro, S.: CINTIL TreeBank Handbook: Design options for the representation of syntactic constituency. Department of Informatics, University of Lisbon, Technical Reports nb. di-fcul-tp-11-02 (2011)
9. António, B., Castro, S., Silva, J., Costa, F.: CINTIL DepBank Handbook: Design options for the representation of grammatical dependencies. In: Department of Informatics, University of Lisbon, Technical Reports nb. di-fcul-tr-11-03 (2011)
10. Branco, A., Costa, F., Silva, J., Silveira, S., Castro, S., Avelãs, M., Pinto, C., Graça, J.: Developing a Deep Linguistic Databank Supporting a Collection of Treebanks. In: Proceedings of LREC 2010 (2010)
11. Costa, F., Branco, A.: LXGram: A Deep Linguistic Processing Grammar for Portuguese. In: Pardo, T.A.S., Branco, A., Klautau, A., Vieira, R., de Lima, V.L.S. (eds.) PROPOR 2010. LNCS (LNAI), vol. 6001, pp. 86–89. Springer, Heidelberg (2010)
12. Branco, A.: LogicalFormBanks, the Next Generation of Semantically Annotated Corpora: key issues in construction methodology. In: Klopotek, M., Przepiorkowski, A., Wierzchón, S., Trojanowski, K. (eds.) Recent Advances in Intelligent Information Systems. Academic Publishing House EXIT, Warsaw (2009)
13. Dickinson, M., Meurers, D.: Detecting Annotation Errors in Spoken Language Corpora. In: Proceedings of the Special Session on Treebanks for Spoken Language and Discourse at the 15th Nordic Conference of Computational Linguistics (2005)
14. Dipper, S.: Grammar-based Corpus Annotation. In: Proceedings of Workshop on Linguistically Interpreted Corpora (2000)
15. Flickinger, D., Kordoni, V., Zhang, Y., Branco, A., Simov, K., Osenova, P., Carvalheiro, C., Costa, F., Castro, S.: ParDeepBank: Multiple Parallel Deep Treebanking, Proceedings. In: Proceedings of TLT 2012 (2012)
16. Flickinger, D., Kordoni, V., Zhang, Y.: DeepBank: A Dynamically Annotated Treebank of the Wall Street Journal, Proceedings. In: Proceedings of TLT 2012 (2012)
17. Marcus, M., Santorini, B., Marcinkiewicz, M.A.: Building a large annotated corpus of English: The Penn Treebank. Computational Linguistics 19(2) (1993)
18. Oepen, S., Flickinger, D., Toutanova, K., Manning, C.D., Brants, T.: The LinGO Redwoods Treebank: Motivation and Preliminary Applications. In: Proceedings of COLING 2002 (2002)

19. Oepen, S.: [incr tsdb()] — Competence and Performance Laboratory. User Manual, Technical Report, Computational Linguistics, Saarland University, Germany (1999)
20. Palmer, M., Kingsbury, P., Gildea, D.: The Proposition Bank: An Annotated Corpus of Semantic Roles. Computational Linguistics, 31 (2005)
21. Rosén, V., Meurer, P., Losnegaard, G.S., Lyse, G.I., De Smedt, K., Thunes, M., Dyvik, H.: An integrated web-based treebank annotation system. In: Proceedings of TLT 2012 (2012)
22. Rosén, V., Meurer, P., de Smedt, K.: LFG Parsebanker: A Toolkit for Building and Searching a Treebank as a Parsed Corpus. In: Van Eynde, F., Frank, A., van Noord, G., De Smedt, K. (eds.) Proceedings of TLT7 (2009)
23. Rosén, V., Meurer, P., de Smedt, K.: Constructing a Parsed Corpus with a Large LFG Grammar. In: Butt, M., King, T.H. (eds.) Proceedings of the LFG 2005 Conference. CSLI Publications (2005)
24. Silva, J., Branco, A.: Deep, consistent and also useful: Extracting vistas from deep corpora for shallower tasks. In: Proceedings of the Workshop on Advanced Treebanking, Proceedings of LREC 2012 (2012)

Gramateca: Corpus-Based Grammar of Portuguese

Diana Santos

Linguateca/University of Oslo
d.s.m.santos@ilos.uio.no

Abstract. This paper has two aims: to present Gramateca, an initiative for corpus-based grammar of Portuguese recently launched by Linguateca, discussing its rationale and aims, and to present a small pilot of an advanced corpus description for Portuguese, which we believe to be a pre-requisite to do sound grammar work, "know thy corpus".

Keywords: Portuguese, grammar, replication, corpus characterization.

1 Introduction

There are fortunately many corpus-based studies of Portuguese, from the initial studies of *Português Fundamental* in Portugal and the NURC project in Brazil, see [1,2]. We can also cite [3,4], two extensive high-quality corpus-based grammars that have recently been published. However, there is no way to directly check the data used by those studies, or, better, its interpretation.

Also, we believe that time is ripe for providing the possibility to do statistically informed larger scale grammar studies, or grammar (as a massive noun), which allow furthermore consulting the material/data used. This is why we announced Gramateca in January 2014, http://www.linguateca.pt/Gramateca/.

We believe that both NLP and linguistics have a lot to contribute to Portuguese grammar – see the recent discussion in the corpora-list on this matter – and we hope to have contributions and discussion from all interested. There are (and have been) several computational grammars of Portuguese, such as the one embedded in PALAVRAS [5] or the one used in the Tycho Brahe project [6]. Although their deployment and evaluation is naturally based on corpora, they are built to be used by computers, and their documentation is not necessarily corpus-justified.

The way we see it, there are two possible models for proceeding with a corpus-based grammar, both inspired by Biber's work: (i) Finding out, through multidimensional analysis, what are the relevant dimensions that allow one to characterize genres or text types in Portuguese, as in [7]; and (ii) Using a balanced annotated corpus of (four?) kinds of text and providing a quantitative characterization of several areas of grammar, as in [8]. However, in both cases the authors created (and annotated) a specific corpus to serve as basis for the grammar – and in the second case the corpus was not even made available for inspection. This is something we intend to improve upon, both by using available corpora, and by not restricting to one unique corpus the grammar work. We believe, the more corpora, the better, and that corpora of old language and of novel genres[9] are essential as well.

J. Baptista et al. (Eds.): PROPOR 2014, LNAI 8775, pp. 214–219, 2014.

So, we supply access to all these corpora and their annotation, not requiring that all be used. Rather, one should be able to pick exactly the genres and subsets that concern her or him. Since the particular subset will be made clear in the results and conclusions, and in the resulting annotated data (more on this below), nothing should hinder different grammarians to use different subsets. What we intend with Gramateca, by giving the community access to the annotated material, i.e., to the interpretation of the raw material, is to ensure replication, and more fine-grained experimentation from our peers.

In this initial paper, we describe the corpora and their (automatic) annotation. As a full fledged parser, PALAVRAS [10] must be acknowledged, since it is the main responsible for AC/DC annotation and even the first take of quantitative description could not be done without it. But we also intend to pursue the comparison of different automatic frameworks to analyse the same data (as already present in a few corpora).

2 Quantitative Description of the AC/DC Corpora, First Take

In order to be able to know what kind of (annotated) material we make available to the community – 1,300 million tokens, roughly 882 millions from Brazil and 244 millions from Portugal – it is important to provide a bird's eye view of all corpora, and also some quantitative features about each (all numbers are dated 14 May 2014).

It turns out that this is easier said than done, because the corpora in AC/DC have been compiled by different people in different continents and with widely differing purposes, see [11,12]. So, a common denominator often lacks in the names associated with the different parts – when the corpora have this external information at all! Also, most descriptions conflate genre, register, dissemination mode, and the oral/written dichotomy. The integrity of each corpus is not changed, anyway. The bird eye's view corresponds to the corpus todosjuntos (all toghether). What is shown in Figure 1 is therefore just a very rough approximation of the text sizes involved.

The most basic quantitative information is available from the AC/DC pages, namely the number of all tokens ("unidades"), running word tokens ("palavras"), word types ("tipos"), sentences ("frases"), and the number of PALAVRAS-attributed parts of speech: nouns (S), verbs (V), adjectives (Adj), and proper nouns (Prop),[1] illustrated for some corpora in Table 1. Using all sorts of information associated with the corpora, one could

Table 1. Basic numbers for AC/DC corpora: Museu da Pessoa, NILC/São Carlos, CHAVE and ENPCpub

Corpus	Unidades	Palavras	Tipos	Frases	S	V	Adj	Prop
MP	1,836,371	1,421,142	42,795	93,525	237,010	263,957	50,815	35,146
NSC	42,914,402	32,461,799	399,763	1,988,621	7,113,650	4,298,530	1,842,594	323,412
CHV	124,093,624	97,884,763	696,775	4,682,363	20,699,744	12,745,355	5,911,581	5,430,724
ENPC	93,160	72,374	12,874	4,371	13,273	12,774	3,853	2,542

---

[1] We also provide numbers for adverbs, personal pronouns, prepositions, conjunctions, determiners, specifiers and numerals, ommitted from the table for lack of space.

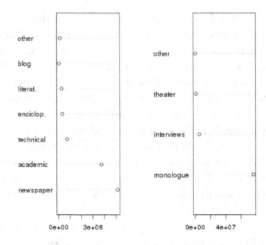

Fig. 1. Distribution of genres in the AC/DC material: written and oral

also compute e.g. number of clauses (main verbs), tense forms, passives, colours, emotions, body parts, as illustrated in Table 2. For these cases the uncertainty is considerably higher, so a measure of confidence or a confidence interval for each should be provided. As to comparing corpora with radically different sizes, blindly normalizing the different figures is dangerous, for two distinct reasons: First, one should not lightly cross-over different texts in one corpus; second, ideally one should use the smallest corpus (or corpus subdivision) as the unit, and divide the larger corpora into several of the same size, and then take the average of the resulting values.

Table 2. Other numbers for AC/DC corpora: types correspond to lemma types

Corpus	Clauses	Perfeitos	Passives	Colours	(types)	Emotions	(types)	Body	(types)
MP	229,055	46,956	6,390	651	64	12,075	437	12,075	437
NSC	3,528,855	610,856	257,849	26,845	410	231,704	1,681	88,745	336
CHV	10,429,598	1,715,006	699,868	79,659	587	711,525	2,353	249,564	372
ENPC	10,809	2,076	444	252	63	1,197	321	1158	314

3 Quantitative Description of the AC/DC Corpora, Advanced

Since corpora are not homogeneous, and often have well defined parts, corresponding to different authors, or texts, or varieties, or domains, one should be careful before amalgamating data, lest one got uninteresting and uninterpretable numbers.

So, all the statistics one might be interested in should be obtained in a form that does not illegally (or better, against linguistic sense) conflate reasons for diversity. We pursue here two examples to give concrete data relevant for making this point. For other pitfalls of quantitative characterization in Portuguese corpora, see [13].

3.1   Sentence Length

One simple issue to illustrate the dangers of amalgamation is measuring sentence length. Trying to get the average sentence length from all Portuguese (written) sentences will not give us very much. Rather, it is possible and probable that different genres, different texts, different authors, etc. will have different sentence lengths. So in Table 3 we present a set of sentence lenghts derivable from our corpora (in words, not tokens, that is, disregarding punctuation):

Table 3. Sentence length of several corpora (sections): MP stands for Museu da Pessoa, and p and b for the variety

MP	MPp	MPb	CONDIVb	CONDIVp	VercialJD	VercialEQ	ObrasMA
14.0	13.1	14.3	14.2	18.4	12.2	17.2	20.2

The distribution of sentence length in each corpus or author could also be more relevant (and characteristic) than its average, especially since averages are well-known to be prone to deformation due to outliers. In the previous table we have therefore measured average sentence length of Júlio Dinis (JD), Eça de Queirós (EQ) and Machado de Assis (MA). Back to the corpora as a whole, in figure 2 one can see the sentence lengths for different corpora in a boxplot.

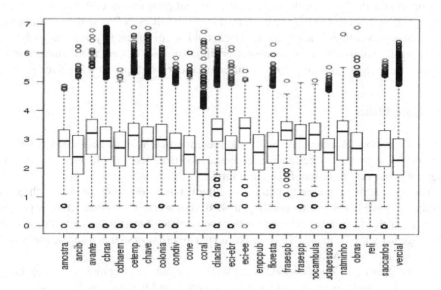

Fig. 2. Distribution of (the logarithm of) sentence lengths in several corpora

3.2 Lexical Density

Another feature that is often used as characterizing for example written vs. oral speech is lexical density, in the sense of the distribution of content vs. grammatical words in X words, or per clause or utterance. [8] find that for English the lexical density of oral speech (using also what they call "inserts") is significantly lower than for written genres.

One could even be more specific and measure the density of personal pronouns (number of personal pronouns per 100 words), or of emotion adjectives (per 1,000 words), in different genres/corpora/etc. To give a flavour of the numbers obtained in those cases, and of some surprises, see Table 4. Further data and discussion on these matters can be found in [14].

Table 4. Other indicators for several corpora (sections)

Feature	MP	MPp	MPb	CONDIVb	CONDIVp	VercialJD	VercialEQ	ObrasMA
Pers. pron.	5.9	6.1	5.0	1.7	2.2	6.7	4.2	3.6
Emotion adj.	0.9	1.1	0.8	1.4	1.8	4.7	5.3	4.0

While this is obviously a very tiny subset of all possible quantitative features, it allows us to stress both (i) the large number of statistics we can get from the corpora, and (ii) the care one has to exercise to get meaningful numbers for one's own analysis. So, while a higher density of emotion adjectives in dedicated newspaper than in interviews should be surprising at first, digging into the fashion and football subparts may explain it. Conversely, the fact that there are more personal pronouns in oral interviews from Portugal than from Brazil may be explained by more clitics in Portugal vs null objects in Brazil (a purely linguistic explanation that could be counteracted by more null subjects in Portugal), but also by a corpus characteristic, that of significant more interaction in the interviews. See a similar puzzle on different colour words per variety solved in [15].

4 Concluding

Perhaps the most important reason for the present article is to increase the Gramateca community so that it reaches a wider audience, and that more people interested in Portuguese grammar can benefit from our efforts.

There is nothing required to join the mailing list as an observer, or to use whatever results we have been able to get. If you want your work to be considered as part of Gramateca, the only thing we request is that you publish it/make it available through our site, together with the data and a detailed enough description so that it can be replicated (and hopefully improved).

At the date of writing, (independent) work on conditional connectives [16], body language [17] and emotions [18] has been launched, and the annotated data have been made available for inspection, while tools for revising annotation and for facilitating comparisons among subcorpora [19] are under development.

References

1. Santos, D.: Linguateca's infrastructure for Portuguese and how it allows the detailed study of language varieties. In: Johannessen, J. (ed.) Language Variation Infrastructure: Papers on Selected Projects, vol. 3, pp. 113–128 (2011)
2. Varejão, F. de O.A.: O português do Brasil: Revisitando a História. In: Cadernos de Letras da UFF Dossiê: Difusão da Língua Portuguesa, vol. 39, pp. 119–137 (2009)
3. Ilari, R. (ed.): Palavras de classe aberta. da Gramática do Portugués Culto Falado no Brasil, vol. 3. Editora Contexto (2014)
4. Raposo, E.B.P., do Nascimento, M.F.B., da Mota, M.A.C., Segura, L., Mendes, A., Vicente, G., Veloso, R. (eds.): Gramática do Português I e II. Gulbenkian (2013)
5. Bick, E.: Portuguese syntax: Teaching manual (2000)
6. Galves, C.: Tycho Brahe Parsed Corpus of Historical Portuguese: Syntactic Annotation System (2007-2008), http://www.tycho.iel.unicamp.br/~tycho/corpus/manual/syn-frm.html
7. Biber, D.: Variation across speech and writing. Cambridge University Press (1988)
8. Biber, D., Johansson, S., Leech, G., Conrad, S., Finegan, E.: The Longman grammar of spoken and written English. Longman (1999)
9. Crystal, D.: Internet Linguistics: A Student Guide. Routledge (2011)
10. Bick, E.: The Parsing System "Palavras": Automatic Grammatical Analysis of Portuguese in a Constraint Grammar Framework. PhD thesis, Aarhus University, Aarhus, Denmark (November 2000)
11. Santos, D., Bick, E.: Providing Internet access to Portuguese corpora: The AC/DC project. In: Gavrilidou, M., Carayannis, G., Markantonatou, S., Piperidis, S., Stainhauer, G. (eds.) Proceedings of the Second International Conference on Language Resources and Evaluation (LREC 2000), May 31-June 2, pp. 205–210 (2000)
12. Costa, L., Santos, D., Rocha, P.A.: Estudando o portugués tal como é usado: o serviço AC/DC. In: The 7th Brazilian Symposium in Information and Human Language Technology (STIL 2009), September 8-11 (2009)
13. Santos, D.: Podemos contar com as contas? In: Aluísio, S., Tagnin, S. (eds.) New Language Technologies and Linguistic Research: A Two-way Road. Cambridge Scholars Publishing (2014)
14. Santos, D.: Comparing oral (transcribed) and written corpora in portuguese, Presentation at GSCP 2014, Stockholm (2014), http://www.linguateca.pt/Diana/download/CompOralEstoc.pdf
15. Santos, D., Silva, R., Freitas, C.: Pluralidades na cor: contrastando a língua do Brasil e de Portugal. In: da Silva, A.S., Torres, A., Gonçalves, M. (eds.) Línguas Pluricêntricas: Variação Linguística e Dimensões Sociocognitivas, Braga, Aletheia, Publicações da Faculdade de Filosofia da Universidade Católica Portuguesa, pp. 555–572 (2011)
16. Marques, R.: Modalidade e condicionais em português (submitted, 2014)
17. Freitas, C., Santos, D., Sousa, R., Jansen, H., Mota, C.: Esqueleto: Body language in Portuguese (submitted, 2014)
18. Santos, D., Mota, C.: Opinions and sentiment analysis in Gramateca (submitted, 2014)
19. Simões, A.: Comparador: forma de auscultar corpos no AC/DC (2014), http://www.linguateca.pt/documentos/Comparador.pdf

Automatic Alignment of News Texts and Their Multi-document Summaries: Comparison among Methods

Verônica Agostini, Roque Enrique López Condori,
and Thiago Alexandre Salgueiro Pardo

Interinstitutional Center for Computational Linguistics (NILC),
Institute of Mathematical and Computer Sciences, University of São Paulo
{agostini,rlopez,taspardo}@icmc.usp.br

Abstract. Aligning texts and their multi-document summaries is the task of determining the correspondences among textual segments in the texts and in their corresponding summaries. The study of alignments allows a better understanding of the multi-document summarization process, which may subsidize new summarization models for producing more informative summaries. In this paper, we investigate some approaches for text-summary sentence alignment, including superficial, deep and hybrid approaches. Our results show that superficial approaches may obtain very good results.

Keywords: Sentence alignment, multi-document summarization.

1 Introduction

Multi-document summarization aims at producing a summary, *i.e.,* a condensed document, from multiple documents on the same topic. There are several methods for multi-document summarization (see, *e.g.,* [19], [24]), with most of them focusing on producing extractive summaries (by copying and pasting segments from the source texts, without rewriting them). In general, although useful results are already available, the quality of extracts are still far away from those produced by humans (which usually produce abstracts, by presenting different linguistic material in relation to the source texts) and several problems remain, as dangling anaphors, presence of redundant information, inadequate temporal ordering of events, and inappropriate treatment of contradictory information, among others.

In this scenario, the alignment of source texts/documents and their multi-document summaries is an important task. In this case, to align is to find correspondences among textual segments with different granularity levels, which means that the alignment may occur between single words, n-grams, sentences, paragraphs and even entire documents. In summarization, this task may provide information about human summarization methods, which may help understanding the nature of the phenomenon and improving the automatic process by subsidizing the creation of new summarization rules and models.

J. Baptista et al. (Eds.): PROPOR 2014, LNAI 8775, pp. 220–231, 2014.

In this paper, we report our investigation of some text-summary sentence alignment methods, evaluating both superficial and deep methods. Superficial methods are generally easier to implement and to use, while methods that use more linguistically motivated assumptions are usually more difficult to produce, especially because it is necessary to obtain language-dependent resources. In our work, we used a discourse theory (Cross-document Structure Theory (CST) [26]) in order to linguistically enhance the process. Furthermore, we combine superficial and deep information in a hybrid approach using machine learning. We run our experiments on news texts written in Brazilian Portuguese, which encompass general (day by day) language and allow us to check the robustness of the methods. Our results show that superficial approaches are enough to obtain very good results.

To the best of our knowledge, this is the first attempt of aligning texts and summaries for Brazilian Portuguese. In general, our contribution remains on developing and evaluating superficial, deep and hybrid methods.

The rest of this paper is organized as follows: in Section 2, we present the alignment task and the related work; in Section 3, we present a *corpus* that was manually annotated, used here as a gold standard for the evaluation of the alignment methods; in Section 4, we present our methods to obtain the alignments; in Section 5, we show the obtained results; and, in Section 6, we make some final remarks.

2   Basic Concepts and Related Work

The alignment originated in the machine translation area (see, *e.g.*, [11], [27]), in which the alignment occurs among textual segments (words, phrases or sentences, usually) of a document and its translated version. They are useful for both producing bilingual dictionaries and allowing the development of the current state of the art in statistical machine translation. The alignments may be 1-1, when 1 segment in a text is aligned to 1 segment in the other; 1-N (including 1-2, 1-3, etc.), when 1 segment in a text is aligned to more than 1 segment in the other; and N-N, when more than 1 segment in a text is aligned to more than 1 segment in the other. Furthermore, the alignments may be 1-0, when some information is new in a text and has no correspondence in the other. A sentence alignment example is shown in Figure 1, when two sentences, in English and Portuguese, are aligned.

Source sentence	Target sentence
I would like to humbly thank you all from the bottom of my heart.	*Eu gostaria de agradecer humildemente a todos do fundo do meu coração.*

Fig. 1. Alignment example in translation

Similarly to translation, alignment in summarization consists in finding the correspondence of segments in the source text(s) and in the summary. Figure 2 shows an example, when sentences of two texts are aligned to 1 sentence in a multi-document summary (in a 2-1 alignment). As one may see, sometimes two aligned segments (sentences, in this case) have many words in common, making it easier to find the alignment. In other cases, it is a harder task, as exemplified in Figure 3 (in a 1-1 alignment), since it is necessary to have world knowledge that "storing supplies" is a way of "preparing for a hurricane".

Sentences from the source texts	Sentence in the summary
A tocha passará por vinte países, mas o Brasil não estará no percurso olímpico. (The torch will pass through twenty countries, but Brazil will not be on the Olympic journey.)	*O Brasil não fará parte do trajeto de 20 países do revezamento da tocha.* (Brazil is not part of the path of 20 countries of the torch relay.)
O Brasil não faz parte do trajeto da tocha olímpica. (Brazil is not part of the path of the Olympic torch.)	

Fig. 2. Alignment example in summarization – a simple case

Sentence from the source text	Sentence in the summary
*Na Jamaica, muitos **estocaram alimentos, água, lanternas e velas**.* (In Jamaica, many people **stored food, water, flashlights and candles**.)	*Vários moradores e turistas nas regiões, inclusive brasileiros, foram retirados dos locais, enquanto outros estão **se preparando para a passagem do furacão**.* (Many residents and tourists in the regions, including Brazilians, were evacuated from the places, while others **are preparing themselves for the hurricane**.)

Fig. 3. Alignment example in summarization – a difficult case

There are some automatic methods proposed in the literature to find the alignments. For example, [2] used the Term-length Term-frequency (TLTF) algorithm to align sentences based on the words that these segments have in common. [21] developed an algorithm, to create an extract from a document, in which the idea is to iteratively exclude sentences from the text until the resultant extract is the most similar to the abstract. The idea of their works is to obtain extracts, which would automatically encompass the alignment information (since whole sentences are simply taken from the texts). Another example in summarization is the work of [17], in which the authors proposed a method that uses Hidden Markov Model (HMM) to perform the alignments. The HMM models heuristics based on cut-and-paste operations performed by the human summarizers, which were found by the authors. [9] and [10] use a HMM with the Expectation Maximization algorithm. The HMM is constructed using a generative history, which models how a summary is produced from a text. Other authors perform the alignments among similar texts. Their works may be replicated to the summarization area because documents and their summaries may be considered similar texts. [13] and [14] used machine learning techniques, using features as word co-occurrences, noun phrase matching, WordNet synonyms, common semantic classes for verbs, and shared proper nouns, among others. In [3], similar paragraphs are grouped and a cosine similarity measure is used to align their sentences. In multi-document summarization, we may highlight the work of [15], in which the authors used the dependency tree path using a similarity measure to perform the alignments between a single document and its summary as well as among a set of documents and their summary.

Basically, the authors used news texts in their experiments. Some perform the alignments among sentences or clauses ([2, 3, 15, 21]), others use n-grams ([17]), phrases and words ([9, 10]) or even paragraphs ([3, 13, 14]). The best results are synthesized in Table 1.

Many authors produce manual aligned corpora to be the basis for comparisons with their automatic alignment methods. In order to evaluate our methods, which will be presented in Section 4, we also created a manual aligned *corpus*, which is composed of news texts, as we describe in what follows.

Table 1. Synthesis of the main results in the literature

Work	Granularity	Precision	Recall	F-measure
[21]	clause	74.27%	80.29%	76.47%
[21]	sentence	77.45%	80.06%	78.15%
[17]	n-gram	81.50%	78.50%	79.10%
[13,14]	paragraph	49.30%	52.90%	51.00%
[3]	clause, paragraph	76.90%	55.80%	-
[9,10]	word, phrase	52.20%	71.20%	60.60%
[15]	sentence (single document)	-	-	97.70%
[15]	sentence (multi-document)	-	-	80.80%

3   The corpus

CSTNews [5] is a *corpus* consisting of 50 clusters of news texts written in Brazilian Portuguese. Each cluster contains 2 to 3 news texts on the same topic, their multi-document summaries (automatically and manually created), and many other annotations (for example, CST [26] and RST (Rhetorical Structure Theory) [20]). The clusters have, on average, 42 sentences (10 to 89 sentences), and the multi-document summaries have, on average, 7 sentences (3-14). The news texts were collected from online news agencies. We used the news texts and the human multi-document summaries (abstracts) for performing the manual alignment.

The manual alignment was conducted by 2 computational linguists, after a training phase, in daily sections of 1 to 2 hours. The task resulted in some alignment rules, which are presented in detail in [1]. The rules are important to guarantee that the alignment annotation will be consistent enough to be followed.

Overall, after the *corpus* annotation, most of the alignments showed to be of type 1-2, *i.e.*, 1 sentence in the summary aligned to 2 sentences in the texts, which was expected. This was followed by the 1-1 alignment type. The distribution of alignment types may be seen in Table 2.

Table 2. Alignment types in the *corpus*

Alignment types												
1-0	**1-1**	**1-2**	**1-3**	**1-4**	**1-5**	**1-6**	**1-7**	**1-8**	**1-9**	**1-10**	**1-11**	**1-12**
2	71	91	72	33	37	13	6	6	1	1	2	1

We obtained some extreme cases, like 1-0 types (due to summary sentences that were not annotated because these were information inferred by the human summarizers and were not in the source texts), and 1-12 types, when one sentence in the summary synthesizes 12 other sentences in the texts.

Besides that, we computed the agreement among the annotators, using the kappa agreement measure [6, 7], considering 5 clusters of the *corpus*. We obtained 0.831 of agreement, which ranges from 0 (no agreement) to 1 (total agreement). This value is very good, and reflects the reliability of the annotation. This value also indicates that, although the alignment process is subjective, the annotators know it very well.

Furthermore, we also labeled all the detected alignments, in a task that we called typification of the alignments (as shown in details in [4]). Basically, we annotated the transformations that were performed to summarize the documents. There are two major groups of types: form and content. Two aligned sentences must have 1 form type, which may be: (i) identical, when the two sentences are the same; (ii) partial, when they have many words in common, but are note identical; or (iii) different, when they have few words in common. The alignment pair may have more than one content type, which may be: (i) specification, when the summary sentence contains some more specific information related to the document sentence; (ii) generalization, when the summary sentence contains some information that is a generalization related to the document sentence; (iii) contradiction, when the sentences present some information that is contradictory; (iv) inference, when the summary sentence expresses information that was inferred from the document sentence; (v) neutral, when there is some information that does not result from a transformation; and (vi) other, when the annotators do not agree with the previous alignment. This disagreement may happen because the alignment is a subjective task. Besides that, we annotated information that is related to the onomastics (toponomastics, related to place names, and anthroponomastics, related to person names). In general, kappa measures were 0.717, 0.318 and 0.452 for form, content and joint form and content annotations, respectively, which are good considering the difficulty of the task.

One example of alignment typification may be seen in Figure 4, which shows the labels: (i) partial, because the sentences have some words in common; (ii) neutral, because some information in the pair does not have a transformation; (iii) generalization, because the names of the states were generalized to "many states" in the summary sentence; and (iv) toponomastics, because there are names of places in the document sentence.

In the CSTNews, the most frequent labels were partial and neutral for form and content labels, respectively, as Table 3 shows. Onomastics was a rare phenomenon.

Sentence in the summary	Types of the alignment	Sentence in the source text
Mais de 300 policiais federais de vários estados participaram das buscas e prisões durante a operação. (More than 300 federal police officers from many states took part in the searches and arrests during the operation.)	(i) **Partial** form overlap (ii) **Neutral** (iii) **Generalization** of content (iv) **Toponomastics**	*A PF divulgou que mais de 300 policiais federais do Amazonas, Distrito Federal, Mato Grosso, Acre e Rondônia fazem parte das investigações da "Operação Dominó".* (The FP reported that more than 300 federal police officers from Amazonas, Distrito Federal, Mato Grosso, Acre and Rondônia took part in the "Operação Dominó".)

Fig. 4. Typification example

Table 3. Occurrence of alignment types in the corpus

Category	Type	Number of occurrences	Percentage (%)
Form	Partial	871	86.06
	Different	82	8.10
	Identical	59	5.83
Content	Neutral	955	94.36
	Generalization	80	7.90
	Specification	47	4.64
	Contradiction	37	3.65
	Inference	29	2.86
	Other	6	0.59
Onomastics	Anthroponomastics	23	2.27
	Toponomastics	4	0.39

4   Methods

We followed 3 approaches to perform the automatic alignment. The first approach encompasses methods that use superficial information about the texts; the second approach is more linguistically motivated and uses a discourse theory; and the third one is a hybrid approach that combines features from the two approaches before in a machine learning solution.

The first approach consists in three superficial methods that may be used together or separately. The methods are based on: (i) word overlap, which measures the amount of common words in two sentences; (ii) relative distance (or relative position), which indicates the distance between sentences in the summary and in the text; and (iii) relative size, which measures the difference, in characters, for two sentences. Figures 5 and 6 show examples that we use to illustrate the methods.

In this work, the value of word overlap ranges from 0 to 1 and is computed with the following formula (applied for the example in Figure 5):

$$\frac{words\ in\ common\ in\ the\ two\ sentences * 2}{number\ of\ words\ in\ the\ 1st\ sentence + number\ of\ words\ in\ the\ 2nd\ sentence} = \frac{9*2}{18+13} = 0.58 \quad (1)$$

Indeed, it is easy to think that two sentences that convey the same topic will have words in common. For some cases, this method works really well.

For the relative size measure, the calculations are made as follows for the sentences in Figure 6. For this measure, 0 indicates that two sentences are equal in character size. This measure is inspired by the machine translation area, in which, if two sentences have approximate sizes, it is reasonable to assume that they convey similar meaning.

$$\frac{difference\ in\ number\ of\ characters\ in\ the\ sentences}{number\ of\ characters\ in\ the\ longest\ sentence} = \frac{98-86}{98} = 0.12 \qquad (2)$$

Sentence 1	Sentence 2
O agressor morreu, mas ainda não foi confirmado se ele foi baleado pela polícia ou se cometeu suicídio. (18 words) (The attacker died, but it is not yet confirmed whether he was shot by the police or committed suicide.)	*Ainda não se sabe se ele cometeu suicídio ou foi morto por policiais.* (13 words) (It is still unknown whether he committed suicide or was killed by police officers.)

Fig. 5. Example for word overlap

Sentence 1	Sentence 2
Na sexta-feira, choveu 12 centímetros em algumas regiões, e há previsão de mais tempestades hoje. (98 characters) (On Friday, it rained 12 cm in some areas, and there are more storms forecast today.)	*Na sexta-feira, choveu muito acima do esperado e há previsão de mais tempestades hoje.* (86 characters) (On Friday, it rained much more than expected and there are more storms forecast today.)

Fig. 6. Example for relative size

The third measure, relative position, only uses the information about the positions of the two sentences (in the summary and the text), and the size of the texts. For example, when considering a summary sentence in position "2" in a hypothetical summary with 3 sentences and a sentence in position "3" in a hypothetical text with 7 sentences, the calculations are made as follows. First, it is necessary to find a value range, to know how much a text is longer than the summary:

$$value\ range = \frac{number\ of\ sentences\ in\ the\ text}{number\ of\ sentences\ in\ the\ summary} = \frac{7}{3} = 2.33 = 2 \qquad (3)$$

Then, using the value range and the sentence positions (lower position = 2 and higher position = 3), the final calculations are made as follows.

$$\frac{lower\ position - \lceil \frac{higher\ position}{value\ range} \rceil}{value\ range\ number - 1} = \frac{2 - \lceil \frac{3}{2} \rceil}{3-1} = \frac{2-2}{2} = 0 \qquad (4)$$

The value range number is the number of ranges in a text. In this case, it was 3. For this measure, 0 indicates that the two sentences are in equivalent positions in their corresponding texts. This measure comes from the assumption that, when one or more texts are summarized, the order of the information in the texts will be respected.

This second (deep) approach uses Cross-document Structure Theory (CST) [26] to guide the alignments. CST indicates the discourse relations among passages of different texts. We assume that two sentences have an alignment if they show at least one CST relation. In this work, the CST relations among sentences in the texts and the summaries are recovered by the CSTParser [22], which has an overall performance of 68.57% for news texts. For certain cases, it is quite easy to verify that this assumption is true, as may be seen in Figure 7. In this example, two sentences have the relation "Overlap" between them, which means that the two sentences have some information in common and, at the same time, the two sentences have some information that is unique to each one. The alignment may denote exactly this kind of information, so it is reasonable to think that the existence of CST relations may indicate alignments.

Sentence in the summary	CST relation	Sentence in the source text
A outra brasileira, Joana Costa, ficou na quinta posição, com 4m20, mostrando que o nervosismo pode atrapalhar as competições em casa. (The other Brazilian, Joana Costa, was in the fifth position, with 4m20, showing that nervousness can derail competitions at home.)	Overlap	*Já a outra brasileira que participou da prova, Joana Costa, não subiu ao pódio, uma vez que não alcançou a marca da cubana.* (The other Brazilian who attended the trial, Joana Costa, didn't step to the podium, since she didn't reach the Cuban mark.)

Fig. 7. Example with CST relation

It is important to notice that this approach may possibly incorporate errors from the parser if it fails to recognize some relations.

This third approach combines the previous approaches in a machine learning solution to perform the alignments. The features are related to the superficial methods and the CST-based method for a pair of sentences (one from the texts and one from the summary), namely: word overlap, relative size, relative position, number of CST relations and type of CST relations[1]; the class values were "yes", in the case of an alignment, or "no", if an alignment does not occur. We use 4 different learning techniques in WEKA environment [12], which are J48 [25], OneR [16], SVM [8] and Naïve Bayes [18]. Our database is composed by 15689 examples related to all possible alignments in the manually annotated *corpus*. 93.55% of them (14678) are examples of pairs with "no" class, and the other 6.44% (1011) are examples of pairs with "yes" class. The database is very unbalanced, but, even with this real life scenery, the results obtained were good, as we show in the next section.

5   Results

The main results obtained by the methods for detecting the alignments are synthesized in Table 4. We show average precision, recall and f-measure computed over the

---

[1] The CST relation types are referred to the types in a typology created in [23]. They may be redundancy, complement, contradiction, source/authorship or style.

results for the clusters, showing only the results for the alignment cases. We do not show the values for the cases for which there were no alignments because all the methods performed very well in excluding "invalid" sentence pairs (*i.e.*, pairs that should not be aligned), achieving results over 90%. It is also important to say that, for the machine learning cases (the two last lines in the table), 10-fold cross-validation was performed. Therefore, it is necessary some reservation before directly comparing the results of machine learning with the other results.

Overall, one may see that our best method was the superficial method word overlap, which reached an F-measure of 66.2%. The CST method was also good, achieving an F-measure of 60.7%. Its results were lower than the ones for word overlap and machine learning methods probably because of errors from the CSTParser.

Regarding machine learning, although we have tested some other techniques too, we only show the ones that produced the best results, the J48 and OneR techniques. In particular, OneR used the word overlap feature for composing its classification rule, showing once more the discriminative power of such information.

Table 4. Main results

Method	Precision (%)	Recall (%)	F-measure (%)
Word overlap	71.8	61.3	66.2
Relative Position	12.7	68.0	21.4
Relative Size	10.1	63.3	17.5
CST method	55.0	67.8	60.7
Jing and McKeown (1999)	35.6	80.5	49.4
Machine learning (J48)	78.7	50.7	61.7
Machine learning (OneR)	86.2	47.6	61.3

Table 5. Accuracy over alignment types

Category	Type	Number of occurrences	Number of detected alignments	Percentage of detected alignments (%)
Form	Partial	871	530	60.85
	Different	82	1	1.22
	Identical	59	58	98.31
Content	Neutral	955	586	61.36
	Generalization	47	13	23.75
	Specification	80	19	27.66
	Contradiction	37	12	32.43
	Inference	29	11	37.93
	Other	6	0	0.00
Onomastics	Anthroponomastics	23	4	17.39
	Toponomastics	4	3	75.00

We also made an experiment manually balancing our database (by oversampling the minority class), and the results were very good: J48 and OneR achieved F-measures of 97.2% and 86.4%, respectively. However, we do not appreciate such solution because it introduces a large bias (since the minority class is replicated several times) and it does not correspond to the actual data that we find in the real world.

We also analyzed which features were the most important ones, using information gain for feature selection. This method ranked the features in the following order: word overlap, type of CST relations, number of CST relations, relative position and relative size, with word overlap being the best. However, running the machine learning techniques with the best features did not improve the results. It is also interesting to notice that the relative size and the relative position obtained bad results as methods, but they improved the machine learning results when used as features.

Furthermore, we reproduced the method proposed by Jing and McKeown [17], which is a very popular method. One may see that, although it achieved a good recall, its precision was low, and, overall, it was outperformed by the other methods.

Finally, Table 5 shows the percentage of alignment types that were correctly identified by our best overall method, the word overlap method. As expected, identical alignments are simple to identify. Partial and neutral alignments (which are the most common alignment types) are also correctly detected with good accuracy. "Generalization" and "specification" showed to be challenges for alignment detection, as well as the "different" cases.

6   Final Remarks

For future works, we highlight the possibility of performing the alignments among more refined textual segments, like n-grams or words. We also envision the possibility of incorporating semantic features for aiding in the alignment detection.

Acknowledgments. The authors are grateful to FAPESP and CAPES for supporting this work.

References

1. Agostini, V., Camargo, R.T., Di Felippo, A.: Manual Alignment of News Texts and their Multi-document Human Summaries. In: Aluísio, S.M., Tagnin, S.E.O. (eds.) New Language Technologies and Linguistic Research: A Two-Way Road, pp. 148–170. Cambridge Scholars Publishing (2014)
2. Banko, M., Mittal, V., Kantrowitz, M., Goldstein, J.: Generating Extraction-Based Summaries from Hand-Written Summaries by Aligning Text Spans. In: The Proceedings of the 4th Conference of the Pacific Association for Computational Linguistics, 5 p. (1999)
3. Barzilay, R., Elhadad, N.: Sentence Alignment for Monolingual Comparable Corpora. In: The Proceedings of the Empirical Methods for Natural Language, pp. 25–32 (2003)

4. Camargo, R.T., Agostini, V., Di Felippo, A., Pardo, T.A.S.: Manual Typification of Source Texts and Multi-document Summaries Alignments. Procedia – Social and Behavioral Sciences 95, 498–506 (2013)
5. Cardoso, P.C.F., Maziero, E.G., Castro Jorge, M.L.C., Seno, E.M.R., Di Felippo, A., Rino, L.H.M., Nunes, M.G.V., Pardo, T.A.S.: CSTNews - A Discourse-Annotated Corpus for Single and Multi-Document Summarization of News Texts in Brazilian Portuguese. In: The Proceedings of the 3rd RST Brazilian Meeting, October 26, pp. 88–105. Cuiabá/MT, Brazil (2011)
6. Carletta, J.: Assessing Agreement on Classification Tasks: The Kappa Statistic. Computational Linguistics 22(2), 249–254 (1996)
7. Cohen, J.: A Coefficient of Agreement for Nominal Scales. Educational and Psychological Measurement 20(1), 37–46 (1960)
8. Cortes, C., Vapnik, V.: Support-vector networks. Machine Learning 20(3), 273–297 (1995)
9. Daumé III, H., Marcu, D.: A Phrase-Based HMM Approach to Document/Abstract Alignment. In: The Empirical Methods in Natural Language Processing (EMNLP), 8 p. (2004)
10. Daumé III, H., Marcu, D.: Induction of Word and Phrase Alignments for Automatic Document Summarization. Computational Linguistics 31(4), 505–530 (2005)
11. Gale, W.A., Church, K.W.: A program for aligning sentences in bilingual corpora. Computational Linguistics 19(1), 75–102 (1993)
12. Hall, M., Frank, E., Holmes, G., Pfahringer, B., Reutemann, P., Witten, I.H.: The WEKA Data Mining Software: An Update. SIGKDD Explorations 11(1) (2009)
13. Hatzivassiloglou, V., Klavans, J.L., Eskin, E.: Detecting Text Similarity over Short Passages: Exploring Linguistic Feature Combinations via Machine Learning. In: The Proceedings of the Empirical Methods for Natural Language Processing, pp. 203–212 (1999)
14. Hatzivassiloglou, V., Klavans, J.L., Holcombe, M.L., Barzilay, R., Kan, M., McKeown, K.R.: SIMFINDER: A Flexible Clustering Tool for Summarization. In: The Proceedings of the NAACL Workshop for Summarization, pp. 41–49 (2001)
15. Hirao, T., Suzuki, J., Isozaki, H., Maeda, E.: Dependency-based Sentence Alignment for Multiple Document Summarization. In: The COLING 2004 Proceedings of the 20th International Conference on Computational Linguistics, pp. 446-452 (2004)
16. Holte, R.C.: Very simple classification rules perform well on most commonly used datasets. Machine Learning 11(1), 63–90 (1993)
17. Jing, H., McKeown, K.: The Decomposition of Human-Written Summary Sentences. In: The Proceedings of the 22nd Annual International ACMSIGIR Conference on Research and Development in Information Retrieval, pp. 129-136 (1999)
18. John, G.H., Langley, P.: Estimating continuous distributions in Bayesian classifiers. In: The Proceedings of the Eleventh Conference on Uncertainty in Artificial Intelligence, pp. 338-345 (1995)
19. Mani, I.: Automatic Summarization. Natural Language Processing, vol. 3, 285 p. John Benjamins Publishing Company, Amsterdam (2001)
20. Mann, W.C., Thompson, S.A.: Rhetorical structure theory: A theory of text organization. Tech. rep. ISI/RS-87-190, University of Southern California, 83 p. (1987)
21. Marcu, D.: The automatic construction of large-scale corpora for summarization research. In: The Proceedings of the 22nd Conference on Research and Development in Information Retrieval, pp. 137-144 (1999)

22. Maziero, E.G., Pardo, T.A.S.: Multi-Document Discourse Parsing Using Traditional and Hierarchical Machine Learning. In: The Proceedings of the 8th Brazilian Symposium in Information and Human Language Technology, Cuiabá/MT, Brazil, October 24-26, pp. 1–10 (2011)

23. Maziero, E.G., Castro Jorge, M.L.C., Pardo, T.A.S.: Identifying Multidocument Relations. In: The Proceedings of the 7th International Workshop on Natural Language Processing and Cognitive Science - NLPCS, Funchal/Madeira, Portugal, June 8-12, pp. 60–69 (2010)

24. Nenkova, A., McKeown, K.: Automatic summarization. Foundations and Trends in Information Retrieval 5(2-3), 103–233 (2011)

25. Quinlan, J.R.: C4.5: programs for machine learning, vol. 1. Morgan Kaufmann (1993)

26. Radev, D.R.: A common theory of information fusion from multiple text sources, step one: Cross-document structure. In: The Proceedings of the 1st ACL SIGDIAL Workshop on Discourse and Dialogue, pp. 74–83 (2000)

27. Yamada, K., Knight, K.: A syntax-based statistical translation model. In: The Proceedings of the 39th Annual Meeting of the Association for Computational Linguistics (ACL), Toulouse, France, pp. 523–530 (July 2001)

Using Rhetorical Structure Theory and Entity Grids to Automatically Evaluate Local Coherence in Texts

Márcio de S. Dias[1,2], Valéria D. Feltrim[1,3], and Thiago Alexandre Salgueiro Pardo[1,2]

[1] Interinstitutional Center for Computational Linguistics (NILC)
[2] University of São Paulo, São Carlos/SP, Brazil
[3] State University of Maringá, Maringá/PR, Brazil
{marciosd,taspardo}@icmc.usp.br, valeria.feltrim@din.uem.br

Abstract. This paper presents a joint model designed to measure local text coherence that uses Rhetorical Structure Theory (RST) and entity grids. The purpose is to learn patterns of entity distribution in texts by considering entity transition sequences and organizational/discourse information using RST relations in order to create a predictive model that is able to distinguish coherent from incoherent texts. In an evaluation with newspaper texts, the proposed model outperformed other methods in the area.

Keywords: Local Coherence, Rhetorical Structure Theory, Entity Grids.

1 Introduction

In text generation systems (as summarizers, question/answering systems, etc.), coherence is an essential characteristic in order to produce comprehensible texts. As such, studies and theories on coherence ([21], [12]) have supported applications that involve text generation ([29], [4], [16]).

By coherence, we mean the possibility of establishing a meaning for the text [17]. Coherence supposes that there are relationships among the elements of the text for it to make sense. It also involves aspects that are out of the text, for example, the shared knowledge between the producer (writer) and the receiver (reader/listener) of the text, inferences, intertextuality, intentionality and acceptability, among others [17].

According to Dijk and Kintsch [8], textual coherence occurs in local and global levels. Local level coherence is present by the local relationship among the parts of a text, for instance, sentences and shorter segments. On the other hand, a text presents global coherence when this text links all its elements as a whole. Psycholinguistics considers that local coherence is essential in order to achieve global coherence [25]. Thus, many researches in computational linguistics have been developed for dealing with local coherence ([1], [2], [5], [9], [10], [11], [13], [15], [18], [20]).

Examples of coherent and incoherent texts are given in Figure 1. Text A is an original text and it is considered coherent. Text B is formed by randomly permuted sentences (a change in the order of sentences) of the original text (Text A). One may see that its coherence is seriously harmed. In this text, not only the reference chain is

J. Baptista et al. (Eds.): PROPOR 2014, LNAI 8775, pp. 232–243, 2014.
© Springer International Publishing Switzerland 2014

broken (e.g., in the first sentence, "event" has no previous antecedent), but the discourse organization is also awkward, making it difficult to grasp the main idea.

In this paper, we propose a joint model for tackling local coherence in order to be able to automatically differentiate coherent from incoherent (less coherent) texts. In particular, this work is based on principles from other researches, such as that the distribution of entities in locally coherent texts presents certain regularities, which may be evidenced on the entity grids proposed by Barzilay and Lapata [2]. Another assumption is that coherent texts show certain distinct intra- and inter-discourse relation organization [20]. Combining such information in a joint model allows better dealing with the local coherence phenomenon.

Text A (original, coherent text)	Text B (permuted sentences, incoherent text)
The gymnast Jade Barbosa, who got three medals at the Pan American Games in Rio in July, won the election on the Internet and will be the Brazilian representative in the Olympic Torch Relay for Beijing 2008. The torch will travel to twenty different countries, but Brazil will not be at the Olympic route. Therefore, Jade will participate in the event in Buenos Aires, Argentina, the only city in South America to receive the symbol of the Games. The relay will be over on August 8, the first day of the Beijing Olympics.	Therefore, Jade will participate in the event in Buenos Aires, Argentina, the only city in South America to receive the symbol of the Games. The relay will be over on August 8, the first day of the Beijing Olympics. The gymnast Jade Barbosa, who got three medals at the Pan American Games in Rio in July, won the election on the Internet and will be the Brazilian representative in the Olympic Torch Relay for Beijing 2008. The torch will travel to twenty different countries, but Brazil will not be at the Olympic route.

Fig. 1. Example of coherent (Text A) and incoherent (Text B) text

For dealing with discourse, we make use of Rhetorical Structure Theory (RST) [21], whose relations are incorporated in the entity grids. A previous work [20] has already considered using discourse in such cases, but in a different way (as we present in the next section). We evaluate our proposal with newspaper texts and show that it outperforms other works in the area, showing the potential of our approach. For the evaluation, we follow the text ordering task proposed by Barzilay and Lapata [2], in which the methods must rank texts (original and sentence-permuted texts) according to their coherence.

Section 2 presents an overview of the most relevant researches related to local coherence. In Section 3, the Rhetorical Structure Theory is briefly introduced. In Section 4, the corpus used in this work is described. Section 5 details the proposed approach in this paper and other methods that were tested. Section 6 shows the experimental setup and the obtained results. Finally, Section 7 concludes this paper.

2   Related Work

In a statistical approach, Foltz et al. [10] used Latent Semantic Analysis (LSA) [19] to compute a coherence value for texts. LSA is used to produce a vector for each word or sentence so that the similarity between two words or two sentences may be measured by the cosine measure [28]. The coherence value of a text may be obtained by measuring the cosines for all pairs of adjacent sentences. Foltz et al. obtained 81%

and 87.3% of accuracy applied respectively to the set of texts related to earthquakes and accidents, in English.

Based on Centering Theory [12], Barzilay and Lapata's [2] assumption is that locally coherent texts present some regularities in entity distribution. These regularities are computed by means of an entity grid, i.e., a matrix in which the rows represent the sentences of the text and the columns the entities. For example, Figure 3 shows part of a entity grid for the text passage in Figure 2, both reproduced from [2].

1 [The Justice Department]ₛ is conducting an [anti-trust trial]ₒ against [Microsoft Corp.]ₓ with [evidence]ₓ that [the company]ₛ is increasingly attempting to crush [competitors]ₒ.
2 [Microsoft]ₒ is accused of trying to forcefully buy into [markets]ₓ where [its own products]ₛ are not competitive enough to unseat [established brands]ₒ.
...
6 [Microsoft]ₛ continues to show [increased earnings]ₒ despite [the trial]ₓ.

Fig. 2. Text with syntactic tags [2]

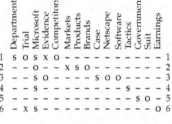

Fig. 3. Entity Grid [2]

For instance, the "Department" column in the matrix shows that the "Department" entity happens only in the first sentence in the subject (S) position. In turn, in the "Trial" column, it is shown that the "Trial" entity happens in the first sentence in the object (O) position and in the sixth sentence in some other syntactical function that is nor subject or object (indicated by X). The hyphen ('-') indicates that the entity did not happen in the corresponding sentence.

With such a matrix, it is possible to obtain the probabilities of entity transitions in texts. For example, the probability of transition [O -] (i.e., the entity happened in the object position in one sentence and did not happen in the following sentence) in the grid in Figure 3 is 0.09, computed as the ratio between its frequency of occurrence in the grid (7 occurrences) and the total number of transitions of length 2 (75 transitions). From this, a feature/characteristic vector is formed by the probabilities of all the transition types. Such vectors are used to learn the properties of coherent texts in a corpus. Figure 4 shows the feature vector representation of the grid in Figure 3.

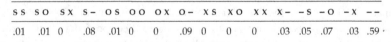

S S	S O	S X	S –	O S	O O	O X	O –	X S	X O	X X	X –	– S	– O	– X	– –
.01	.01	0	.08	.01	0	0	.09	0	0	0	.03	.05	.07	.03	.59 ·

Fig. 4. Feature-vector text representation using [2]

Barzilay and Lapata defined 3 applications to test the model prediction power: text-ordering, automatic evaluation of coherence in summaries and readability assessment. The first two are ranking problems and, according to the authors, present an efficiently learnable model that ranks the texts based on their level of local coherence.

Barzilay and Lapata's approach produced 8 models according to the use (+) or not (-) of syntactical, coreference and salience information. The syntactical information was given by a parser [23] that recognizes the grammatical function of the entities.

Coreference occurs when words refer to the same entity and, therefore, these words may be represented by a single column in the grid. For example, when the text mentions "Microsoft Corp.", "Microsoft", and "the company", such references are mapped to a single column ("Microsoft") in the grid. Salience is related to the frequency of entities in texts, allowing to build grids with the least and/or the most frequent entities in the text.

In the text-ordering task (which is the one that interests to us in this paper), for each original text considered "coherent", a set of randomly permutated versions were produced and considered "incoherent" texts. Ranking values for coherent and incoherent texts were produced by means of the predictive model trained in the SVM[light] [14] package, using a set of pairs of texts (coherent text, incoherent text). It is supposed that the ranking values of coherent texts are higher than the ones for incoherent texts. Barzilay and Lapata obtained 87.2% and 90.4% of accuracy (fraction of correct pairwise rankings in the test set) applied respectively to the set of texts related to earthquakes and accidents, in English. Such results were achieved by the model considering the three types of information (Coreference+Syntax+Salience+).

Lin et al. [20] assumed that local coherence implicitly favors certain types of discursive relation transitions. The authors used four discursive relations, based on Discourse Lexicalized Tree Adjoining Grammar (D-LTAG) [30], to develop the Discourse Role Matrix, which is composed of sentences (rows) and terms (columns), with discursive relations used over their signaling arguments. Terms were the stemmed forms of the open class words: nouns, verbs, adjectives and adverbs. For example, see the discursive grid (b) for the text (a) in Figure 5.

(S1) Japan normally depends heavily on the Highland Valley and Cananea mines as well as the Bougainville mine in Papua New Guinea. (S2) Recently, Japan has been bying copper elsewhere. (a)				

S#	Terms			
	copper	*cananea*	*depend*	...
S1	nil	Comp.Arg1	Comp.Arg1	
S2	Comp.Arg2 Comp.Arg1	nil	nil	

(b)

Fig. 5. Part of a text and its discursive grid [20]

Figure 5 shows a fragment of the matrix representation (b) of the text (a). Columns correspond to the extracted terms; rows, the contiguous sentences. A cell $C_{Ti,Sj}$ contains the set of the discourse roles of the term T_i that appears in sentence S_j. For example, the term "depend" from S1 takes part of the Comparison (Comp) relation as argument 1 (Arg1), so the cell $C_{depend,S1}$ contains the Comp.Arg1 role. A cell may be empty (nil, as in $C_{depend,S2}$) or contain multiple discursive roles (as in $C_{copper,S2}$).

Lin et al. applied their model to the same text-ordering task proposed by Barzilay and Lapata, but now the sentence-to-sentence transitions are D-LTAG relations. They obtained 89.25% and 91.64% of accuracy applied to the set of English texts related to earthquakes and accidents, respectively, improving the previous results.

Another model that used Barzilay and Lapata's approach is the one of Filippova and Strube [9], which implemented the entity model for German and conducted an entity grouping by the use of semantic relations. The Coreference+Syntax-Salience+ model developed by the authors obtained 75% of accuracy as the best result. Iida and Tokunaga [13] used the concepts of entity and coreference to evaluate the coherence of texts written in Japanese and this research obtained 76.1% of accuracy. Freitas and Feltrim [11] applied Barzilay and Lapata's entity model to evaluate coherence in newspaper texts written in Brazilian Portuguese, obtaining 74.4% of accuracy by means of the use of syntactic and salience information applied to the CSTNews corpus [6]. Besides, the authors considered the lemmas of noun phrases (NP) to minimize the lack of a coreference resolution system and used additional Type/Token information [5] to measure the lexical variety of entities in each syntactical function.

3   Rhetorical Structure Theory

The Rhetorical Structure Theory (RST) proposed by Mann and Thompson [22] considers that each text presents an underlying rhetorical structure that allows the recovery of the communicative intention of the writer.

In this model, Elementary Discourse Units (EDUs) are connected by rhetorical relations, aiming at coherently organizing discourses. The role of the nucleus (N) or satellite (S) is assigned to each EDU. The nuclei or nuclear EDUs contain the most important pieces of information in the relations and are considered more relevant than the satellites. The satellites, on the other hand, present additional information that helps the reader in the interpretation of the nuclei.

The RST relations are divided into two classes: mononuclear and multinuclear relations. The mononuclear relations are composed of pairs of EDUs that present different levels of importance: one nuclear and one satellite. On the other hand, multinuclear relations link equally important EDUs, which are classified as nuclei.

Figure 6 presents part of a text segmented in EDUs, reproduced from [27]. It is used to exemplify an RST analysis, shown in Figure 7.

---
(1) Many of Almir's – the Pernambuquinho – "courageous" attitudes were dictated by fear. (2) Few people know this, (3) but it is true.

Fig. 6. Part of a text segmented in EDUs [27]

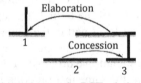

Fig. 7. Diagram that represent the RST relations for the text in Figure 6

EDU (1) illustrates the central idea of the discourse, which is the fear that influenced the way Almir (the character) acted. However, EDUs (2) and (3) indicate that few people know about the character feature and the fact that this feature is real. The relations among EDUs (1), (2) and (3) occur as they are recognized in the discourse. EDUs (2) and (3) are identified as constituents of an ELABORATION relation of the assertion in (1). In RST, this is expressed by the rhetorical relation ELABORATION. EDU (1) corresponds to the nucleus of the ELABORATION relation (the nucleus is indicated by a vertical line and is pointed by the arrow), while EDUs (2) and (3), constitute the satellite of the relation. EDU (3) is the nucleus and EDU (2) is the satellite of the CONCESSION relation.

4   The Corpus

The CSTNews corpus [6] has been created for multi-document summarization. It is composed of 140 texts distributed in 50 sets of news texts written in Brazilian Portuguese from various domains. Each set has 2 or 3 texts from different sources that address the same topic. Besides this, the corpus has, in average, 14.9 sentences per text and a total of 5,216 RST relations. Besides the original texts and their RST analyses, the corpus counts with several annotation layers. One may also find single and multi-document manually produced summaries, the identification of temporal expressions, Cross-document Structure Theory (CST) annotation [26], automatic syntactical analyses, nouns and verb senses, text-summary alignments, and, more recently, the semantic annotation of informative aspects in summaries, among other annotations. For this work, we are especially interested in the RST annotation.

For the RST annotation, Table 1 shows the obtained agreement (computed by RSTeval [24]) for the simple textual segments (for the segmentation process, therefore), complex textual segments (given by the internal nodes in the RST structure), nuclearity and relations among segments. Results are shown in F-measure values.

Table 1. Agreement results for the RST annotation in the CSTNews corpus

Criteria	F-Measure (%)
Simple textual segments	0,91
Complex textual segments	0,78
Nuclearity of each segment	0,78
RST relations among segments	0,66

According to the results, the agreement among annotators was satisfactory, and is similar to the obtained for other works ([23], [7]) for other languages. Therefore, the annotation is considered reliable and may be used for the purposes of this work.

5   Our Approach

Our approach follows Barzilay and Lapata's [2] work, but excludes the use of coreference information (as there is no widely available system for Portuguese, we used the nuclei of noun phrases as entities) and includes the use of RST relation distribution. It captures a pattern of RST discursive relations in coherent texts by using a machine learning technique, creating a predictive model that enables the evaluation of local coherence.

Our grid is formed by sentences (rows) and entities (columns), in which each cell is filled with the RST relations that the entity under focus is part of, also specifying the corresponding nuclearity. Figure 8 shows part of the grid for the text in Figure 6 along with the RST information given by the diagram in Figure 7. Relations are shown in abbreviated forms.

	attitudes	Almir	Pernambuquinho	fear	people	true
S1	elab.Nuc	elab.Nuc	elab.Nuc	elab.Nuc	-	-
S2	-	-	-	-	conces.Sat elab.Sat	conces.Nuc elab.Sat

Fig. 8. RST relation grid for the text in Figure 6

The Palavras parser [3] has been used to identify the text entities, which are all nouns and proper nouns. Therefore, our approach is prone to parsing errors. However, if such errors are systematic, useful patterns may still be learnt.

As the entity grid was created, the entity distribution was computed sentence by sentence and not by EDUs. The EDUs were not used due to the sparsity of the entity grid, making it difficult the creation of an efficient prediction model.

The sentence by sentence entity distribution was performed for all the possible RST relations in the text. RST relations transitions had length two. For example, the entity "fear" in Figure 8 is present in one relation in sentence 1 (S1) and the possible transitions are computed for all relations. In the case of the transition [**elab.Nuc, -**], there are 4 occurrences (all transitions occur from sentence 1 to 2) out of 6 length-two transitions, considering the grid in Figure 8. Thus, the transition probability [**elab.Nuc, -**] is 4/6 = 0.6666. This way, each text may be seen as a distribution defined over transition types. Now, each text may be represented as a set of transition sequences by using a standard feature vector notation, in other words, each grid j of a document d_i corresponds to a feature vector $\Phi(x_{ij}) = (P_1(x_{ij}), P_2(x_{ij}), ..., P_n(x_{ij}))$ [2], where n is the total number of possible transitions and $P_r(x_{ij})$ is the probability of transition r in grid x_{ij}. Table 2 shows part of the feature vector for the grid in Figure 8.

Table 2. The feature vector representation that uses possible length-two transitions given RST relations and nuclearity information from the grid in Figure 8

[elab.Nuc, -]	[-, conces.Nuc]	[-, elab.Sat]	[-, -]
0.6666	0.1666	0.1666	0.0

It is important to notice that our proposal is different from the one of Lin et al. [20]. We do not use only 4 D-LTAG relations, but the full relation set of RST. We also include nuclearity in the grid, because we believe that coherent texts may follow patterns of nuclei and satellites distributions, which are not considered in Lin et al. approach. Besides, Lin. et al. [20] used the stemmed open-class words and we used the nuclei of noun phrases as entities.

The feature vectors we built were used to create the coherence prediction model with the use of a machine-learning algorithm. In the next section, the evaluation of this proposed model is reported.

6   Evaluation

6.1   Experimental Setup

The text-ordering task proposed by Barzilay and Lapata [2] has been used to check and to evaluate the performance of our approach compared to other methods. In the text-ordering task, a document is given as a set of sentences and the algorithm investigates the ordering that maximizes coherence. Because of this, random permutations of the original text were generated in order to measure how often a permuted version is ranked higher than the original document. A good model should prefer an original text more often than its possible permutations.

We used 137 out of the 140 texts of the CSTNews corpus and 20 random permutations for each original text of the corpus. Three texts were not used because they did not reach the 20 different permutations defined. We decided to use 20 permutations for each text because this is also the number used by Barzilay and Lapata in their experiment. Thus, the database of this experiment is composed of 2,740 pairs of texts. The SVM^{light} [14] package has been used in the experiment with the ranking option for training and testing our coherence model. Apart from that, the 10-fold cross-validation method was used for achieving a more confident result.

The evaluation metric used was the accuracy measure, which, for a given set of pairwise rankings (an original document and one of its permutations), the accuracy is the ratio of the correct predictions made by the method over the size of the testing set for each fold. The final accuracy for each experiment is the average of the accuracies for each one of the 10 folds.

6.2   Other Methods

Besides our proposal, we implemented some other methods from literature, in order to compare our results to the current state of the art. The following methods were chosen based on their importance and used techniques to evaluate local coherence: the LSA method by Foltz et al. [10], the traditional entity grid method by Barzilay and Lapata [2] and the discourse-based method by Lin et al. [20]. All of them were adapted to Brazilian Portuguese, using the appropriate available tools and resources for this language.

The implementation of these methods carefully followed each step of the original ones. The resources used to develop the baselines were: Python 2.7[1] for all the methods, the NLTK package[2] for the stemming required by Lin et al., the Scikit-Learn

---

[1] http://www.python.org/
[2] http://www.nltk.org/

package[3] as in Foltz et al., and Palavras parser [3], mainly used in the implementation of Barzilay and Lapata and Lin et al approaches. We used the RST relations as the necessary discourse information in Lin et al.'s approach.

Barzilay and Lapata's method has been implemented without referential information, since, to the best of our knowledge, there is no available robust coreference resolution system for Brazilian Portuguese and the CSTNews corpus still does not have referential information in its annotation layers.

6.3   Results

The LSA method generates a coherence value for each original text and for its permutations. Therefore, the accuracy measure for this method was calculated by the number of times that the coherence value of the original text was greater than its permutation over the total number of text pairs (an original document and one of its permutations). Therefore, for this method, it is not necessary to perform 10-fold cross-validation.

The other methods and our approach were submitted to the text-ordering task using the CSTNews corpus. Furthermore, the implementation of Barzilay and Lapata's approach produced 4 models: (Syntactic+Salience+), (Syntactic+Salience-), (Syntactic-Salience+) and (Syntactic-Salience-), in which Salience is related to the frequency of entities, considering entities with frequency higher or equal to 2.

In the approach proposed here, two variations were created in order to check if the accuracy would improve: the RST relations were grouped according to the relation groups of Mann and Thompson [22] (Variation 1), ignoring nuclearity; the RST relations were not grouped, but the nuclearity information has been removed from the RST relations (Variation 2). Table 3 shows the accuracy of our approaches compared to the other methods, ordered by accuracy.

Table 3. Evaluation results, where diacritics ** ($p < .01$) and * ($p < .05$) indicate whether there is a significant statistical difference in accuracy compared to our approach (using t-test)

Methods	Accuracy (%)
Our Approach	**79.45**
Syntactic-Salience- from Barzilay and Lapata	78.97
Syntactic+Salience- from Barzilay and Lapata	74.10**
Discourse grids from Lin et al.	70.80*
Syntactic+Salience+ from Barzilay and Lapata	70.73**
Syntactic-Salience+ from Barzilay and Lapata	67.87**
Variation 1 of our approach	66.18**
Variation 2 of our approach	63.99**
LSA from Foltz et al.	58.40**

The t-test has been used for pointing out whether differences in accuracy are statistically significant. Comparing our approach with the other methods, one may observe that the use of all the RST relations with nuclearity information obtained better results for evaluating local coherence.

---

[3] http://scikit-learn.org/stable/

In particular, the results showed that the use of nuclearity information of RST relations significantly increased the accuracy (comparing our approach with Variation 2). For Variation 1, the grouped RST relations improved the accuracy in comparison with Variation 2. We believe that this happened because a less sparse grid (which comes from grouping the relations) results in a more efficient learning.

We believe that the use of open-class words in Lin et al.'s model may have been the cause of its significant lower accuracy compared to our approach. Regarding the open-class words, since there are more of them than the number of entities in the corpus, the generated grid was very sparse. This makes it difficult to generate a good predictive model.

The LSA model by Foltz et al. was the approach that obtained the lowest accuracy in comparison with the other analyzed models. This result may be explained by the lack of linguistic information that might have improved its accuracy.

An interesting result is given by the Syntactic-Salience- model, i.e., this approach used neither syntactic nor salience information, but it presented greater accuracy than other more complete models. This was due to few transition types obtained in the grid that only contained information of presence or not of entities in the sentences. As consequence, the grid was less sparse and helped in the generation of a good predictive model. In fact, although it produces worse results compared to our approach, the differences among these models are not statistically significant. Therefore, it is interesting to see that models of different nature may behave quite similarly.

At this point, it is important to notice that we could not directly compare our results to the ones obtained by Freitas and Feltrim [11]. Although they have used the same corpus that us, they used different text permutations (incoherent texts) and some other additional information, which makes direct comparisons unfair.

7   Final Remarks

According to the results obtained from the text-ordering task, the use of RST relations and nuclearity was the best among its variation and the other methods from literature. This approach showed to be promising and it may be used for other languages, such as English, as long as there is a corpus annotated with RST relations, a syntactic parser and specific resources that focus on the target language.

As future work, the same methodology employed in this work will be used to develop methods to evaluate local coherence for multi-document summaries with other discourse theories. For this, the focus will be on structuring the discourse to connect sentences from different documents and to establish one or more types of relations among sentences, as Cross-Document Structure Theory (CST) [26] does.

Acknowledgements. The authors are grateful to FAPESP and University of Goiás for supporting this work.

References

1. Althaus, E., Karamanis, N., Koller, A.: Computing locally coherent discourse. In: Proceedings of the 42nd Annual Meeting of the Association for Computational Linguistics, article 399, Stroudsburg, PA, USA (2004)
2. Barzilay, R., Lapata, M.: Modeling local coherence: An entity-based approach. Computational Linguistics 34, 1–34 (2008)
3. Bick, E.: The Parsing System Palavras, Automatic Grammatical Analysis of Portuguese in a Constraint Grammar Framework. Aarhus University Press (2000)
4. Bosma, W.: Query-Based Summarization using Rhetorical Structure Theory. In: Proceedings of the 15th Meetings of CLIN, LOT, Utrecht, pp. 29–44 (2004)
5. Burstein, J., Tetreault, J., Andreyev, S.: Using entity-based features to model coherence in student essays. In: Human Language Technologies: In Proceedings of the 2010 Annual Conference of the North American Chapter of the Association for Computational Linguistics, pp. 681–684 (2010)
6. Cardoso, P., Maziero, E., Jorge, M., Seno, E., di Felippo, A., Rino, L., Nunes, M., Pardo, T.: Cstnews - a discourse-annotated corpus for single and multi-document summarizationof news texts in brazilian portuguese. In: Proceedings of the 3rd RST Brazilian Meeting, pp. 88–105 (2011)
7. Cunha, I., Torres-Moreno, J.-M., Sierra, G.: On the Development of the RST Spanish Treebank. In: Proceedings of the 5th Linguistic Annotation Workshop, Portland-Oregon, pp. 1–10 (2011)
8. Dijk, T.V., Kintsch, W.: Strategics in discourse comprehension. Academic Press, New York (1983)
9. Filippova, K., Strube, M.: Extending the entity-grid coherence model to semantically related entities. In: Proceedings of the Eleventh European Workshop on Natural Language Generations, pp. 139–142 (2007)
10. Foltz, P.W., Kintsch, W., Landauer, T.K.: The Measurement of textual coherence using latent semantic analysis. Discourse Processes 25(2-3), 285–307 (1998)
11. Freitas, A.P., Feltrim, V.D.: Análise Automática de Coerência Usando o Modelo Grade de Entidades para o Português. In: Proceedings of the IX Brazilian Symposium in Information and Human Language Technology, Fortaleza, CE, Brazil, pp. 69–78 (2013)
12. Grosz, B., Aravind, K.J., Scott, W.: Centering: A framework for modeling the local coherence of discourse. Computational Linguistics 21, 203–225 (1995)
13. Iida, R., Tokunaga, T.: A metric for evaluating discourse coherence based on coreference resolution. In: Proceedings of the COLING 2012: Posters, Mumbai, India, pp. 483–494 (2012)
14. Joachims, T.: Optimizing search engines using clickthrough data. In: Proceedings of the Eighth ACM SIGKDD International Conference on Knowledge Discovery and Data Mining, New York, NY, USA, pp. 133–142 (2002)
15. Karamanis, N., Poesio, M., Mellish, C., Oberlander, J.: Evaluating centering-based metrics of coherence for text structuring using a reliably annotated corpus. In: Proceedings of the 42nd Annual Meetings of the Association for Computational Linguistics, article 391 (2004)
16. Kibble, R., Power, R.: Optimising referential coherence in text generation. Computational Linguistic 30(4), 401–416 (2004)
17. Koch, I.V., Travaglia, L.C.: A Coerência Textual, 14th edn. Contexto, São Paulo (2002)

18. Lapata, M.: Probabilistic texts structuring: Experiments with sentence ordering. In: Proceeding of the 2nd Human Language Technology Conference and Annual Meeting of the North American Chapter of the Association for Computational Linguistics, pp. 545–552 (2003)
19. Landauer, T.K., Dumais, S.T.: A solution to Plato's problem: The latent semantic analysis theory of acquisition, induction and representation to coreference resolution. In: Proceedings of the 40th Annual Meeting of the Association for Computational Linguistics, Philadelphia, PA, pp. 104–111 (1997)
20. Lin, Z., Ng, H.T., Kan, M.Y.: Automatically evaluating text coherence using discourse relations. In: Proceedings of the 49th Annual Meeting of the Association for Computational Linguistics: Human Language Technologies, Stroudsburg, PA, USA, vol. 1, pp. 997–1006 (2011)
21. Mann, W.C., Thompson, S.A.: Rhetorical Structure Theory: Toward a functional theory of text organization. Text 8(3), 243–281 (1988)
22. Mann, W.C., Thompson, S.A.: Rhetorical Structure Theory: A Theory of Text Organization. Technical Report from Information Sciences Institute (ISI), ISI/RS-87-190, pp. 1-91. University of Southern California, USA (1987)
23. Marcu, D.: The Rhetorical Parsing of Unrestricted Texts: A Surface-based Approach. Computational Linguistics 26, 396–448 (2000)
24. Maziero, E., Pardo, T.A.S.: Automatização de um método de avaliação de estruturas retóricas. In Proceedings of the RST Brazilian Meeting (2009)
25. Mckoon, G., Ratcliff, R.: Inference during reading. Psychological Review, 440-446 (1992)
26. Radev, D.: A common theory of information fusion from multiple text sources, step one: Cross-document structure. In: Proceedings of the 1st ACL SIGDIAL Workshop on Discourse and Dialogue, Hong Kong, pp. 74–83 (2000)
27. Ribeiro, G.F., Rino, L.H.M.: A Sumarização Automática com Base em Estruturas RST. Technical Reports from Interinstitutional Center for Computational Linguistics, University of São Paulo, NILC-TR-02-05. São Carlos, Brazil (2002)
28. Salton, G.: Term-Weighting Approaches in Automatic Text Retrieval. Information Processing and Management, 513–523 (1988)
29. Seno, E.R.M.: Rhesumarst: Um sumarizador automático de estruturas rst. Master Thesis. University of São Carlos. São Carlos/SP (2005)
30. Webber, B.: D-ltag: Extending lexicalized tag to discourse. Cognitive Science 28(5), 751–779 (2004)

Comparing NERP-CRF with Publicly Available Portuguese Named Entities Recognition Tools

Daniela O.F. do Amaral, Evandro Fonseca, Lucelene Lopes, and Renata Vieira

Pontifical Catholic University of Rio Grande do Sul, Computer Science,
Av. Ipiranga, 6681, Partenon, Porto Alegre, Brazil
{daniela.amaral,evandro.fonseca}@acad.pucrs.br,
{lucelene.lopes,renata.vieira}@pucrs.br

Abstract. This paper presents the evaluation of NERP-CRF, a Conditional Random Fields (CRF) based tool for Portuguese Named Entities Recognition (NER) against other publicly available NER tools. The presented evaluation is based on the comparison with three other NER tools for Portuguese. The comparison is made observing Recall and Precision measures obtained by each tool over the HAREM corpus, a golden standard for NER for Portuguese texts. The experiments were initially conducted considering ten categories and then, considering a reduced number of categories. The results show that NERP CRF outperforms the others tools when sufficiently trained for four entity categories.

Keywords: Named Entity Recognition, Conditional Random Fields, Portuguese Language.

1   Introduction

Named Entity Recognition (NER) comprises extraction and classification of named entities according to several semantic categories [1]. The entities fall under categories such as person, organization and place. This task may consider also temporal entities such as date and time. NER is an important task in many research areas, including both general and specialized domains. For instance, well known NER applications are the recognition of disease and gene names in biomedical texts [2, 3].

The number of studies on NER for the Portuguese Language [4] is quite restricted when compared to other languages such as English. HAREM is the first and only initiative for Portuguese NER [5], which had so far two editions. HAREM set out two Golden Collections: the first and the second HAREM. The corpus has annotations for NE in ten categories: Person, Place, Organization, Value, Abstraction, Time, Work, Event, Thing and Other.

This paper presents a comparative study based on the Second HAREM corpus. First, we compare our tool NERP-CRF [6] trained over the ten HAREM categories with three other tools: FreeLing [7], LTasks [8], and PALAVRAS [9]. Then we modify the training sets, considering a reduced number of categories. This paper is organized as follows: Section 2 presents the HAREM corpora; Section 3 describes the

J. Baptista et al. (Eds.): PROPOR 2014, LNAI 8775, pp. 244–249, 2014.

tools under evaluation; Section 4 presents the evaluation process and results; and Section 5 presents our conclusions.

2   Corpora

HAREM is an event for the joint assessment of NER for Portuguese, established by Linguateca [10,11]. HAREM Golden Corpus (GC), was annotated by humans and has been used as a reference for NER systems evaluation. In [5, 12] evaluations of NER systems on the basis of HAREM corpora are presented. HAREM has two editions, Table 1 shows the distribution of NE in the 10 different categories for the corresponding golden corpora.

Table 1. Number of NE in each category according to both HAREM golden corpora

Corpora	GC First HAREM 129 texts 466,355 words		GC Second HAREM 129 texts 89,241 words	
Categories				
Person	1,040	20%	2,035	28%
Place	1,258	25%	1,250	17%
Organization	946	18%	960	13%
Value	484	9%	352	5%
Abstraction	461	9%	278	4%
Time	440	9%	1,189	16%
Work	210	4%	437	6%
Event	128	2%	302	4%
Thing	79	2%	304	4%
Other	86	2%	79	2%
Total	5,132	100%	7,255	100%

3   NER Systems

We developed a system for Portuguese, called NERP-CRF. We compare it with three other tools. In general, there are few options for systems that perform NER for Portuguese. These three tools under evaluation were all that we could access and execute by ourselves. Two of them were publicly available, a third one is commonly used in Portuguese NLP groups, although it is not a freely available tool. In the following we present a brief description of the NER tools under analysis.

NERP-CRF: is a system based on the probabilistic mathematical model called Conditional Random Fields (CRF) [13]. The system was trained with First HAREM GC [14] using two input vectors. The first vector contains the POS tagging, and the Harem NE categories using BILOU notation [15]. The second is a vector of features, as described in [6]. The output is a vector with categories in BILOU notation.

Freeling: This system comprises a package of NLP tools, such as coreference, POS tagging and NER [7]. The Freeling works with texts in English, Spanish and Portuguese. It has two NER functions: the first one, simpler, is based on morphosyntactic patterns, and the second one, more elaborated, is based on machine learning algorithms. The latter form was used for comparisons in this work. This tool considers only the following NE categories: Person, Place, Organization and Other.

LTasks: LTasks is a set of web tools [8]. These tools are available but unfortunately they do not specify which techniques are used for NER. The categories are the same of HAREM with the exception of the category other.

PALAVRAS: The PALAVRAS parser is a software tool for Portuguese [9]. The output of PALAVRAS is a very rich annotation, where even syntax tree structures with all kinds of grammatical and semantic annotations are available. The system is rule-based.

4   Evaluation

The four tools described above were run over HAREM 2 by ourselves. First we present a comparison of the output for ten categories of each of the four systems (Table 2).

After that we retrained NERP-CRF, with a reduced number of categories (Tables 4 and 5). For all these experiments the output of NE (OE) was compared with the second Harem GC annotation which was used as the reference list (RL) for the calculation of t Precision ($P = | OE \cap RL | / | OE |$), Recall ($R = |OE \cap RL | / | RL |$), and F-measure (F) as the harmonic average between P and R.

Table 2 shows that there was no system that outperformed in all categories. In fact, for categories Person, Place and Organization there is a balance of precision and recall among the results.

NERP-CRF had a better performance for the Organization category with 46% of F-measure. FreeLing had the best F-measure (56%) for Place category, while LTasks and PALAVRAS got the best F-measures for Person category.

FreeLing, LTasks and PALAVRAS were not able to identify all categories. Alas, besides the Person, Place and Organization categories the performance of all systems was either quite low (below 40%) or not representative (Value category was only detectable by LTasks and NERP-CRF).

Next we considered different distributions of categories for the training of NERP-CRF, focusing only on Person, Place and Organization categories. Thus, we performed four new trainings, all using First Harem. Again we used Second Harem for testing.

Initially, we grouped all other Harem categories (Event, Work, Abstraction, Thing, Time, Value and Other) into a single one called Everything Else (EE). The results for this new situation is presented in Table 3.

Table 2. P/R/F for all categories: Person, Place, Organization, Event, Work, Abstraction, Thing, Time, Value, and Other

	Person	I RL I	= 2,035			Place	I RL I	= 1,250		
Systems	P	R	F	I OE I	IOE ∩ RL I	P	R	F	I OE I	I OE ∩ RL I
FreeLing	54%	60%	57%	2,279	1,230	52%	60%	56%	1,431	751
LTasks	62%	61%	62%	2,017	1,249	56%	53%	54%	1,170	658
PALAVRAS	60%	64%	62%	2,174	1,297	54%	55%	54%	1,264	685
NERP-CRF	56%	50%	53%	1,803	1,012	48%	53%	51%	1,382	667
	Organization	I RL I	= 960			Event	I RL I	= 302		
Systems	P	R	F	I OE I	OE ∩ RL I	P	R	F	OE I	I OE ∩ RL I
FreeLing	28%	60%	38%	2,088	575	-	-	-	-	-
LTasks	28%	60%	38%	2,043	576	12%	28%	17%	736	86
PALAVRAS	30%	51%	38%	1,630	491	53%	26%	35%	150	80
NERP-CRF	44%	48%	**46%**	1,054	460	42%	4%	7%	26	11
	Work	I RL I	= 437			Abstraction	I RL I	= 278		
Systems	P	R	F	I OE I	IOE ∩ RL I	P	R	F	I OE I	I OE ∩ RL I
FreeLing	-	-	-	-	-	-	-	-	-	-
LTasks	26%	19%	22%	321	84	19%	14%	16%	201	39
PALAVRAS	36%	30%	33%	367	132	14%	6%	8%	117	16
NERP-CRF	44%	9%	15%	93	41	14%	8%	10%	155	22
	Thing	I RL I	= 304			Time	I RL I	= 1,189		
Systems	P	R	F	I OE I	IOE ∩ RL I	P	R	F	I OE I	I OE ∩ RL I
FreeLing	-	-	-	-	-	-	-	-	-	-
LTasks	11%	5%	6%	129	14	5%	3%	4%	633	32
PALAVRAS	0%	0%	0%	22	0	-	-	-	-	-
NERP-CRF	6%	1%	1%	32	2	7%	3%	5%	624	41
	Value	I RL I	= 352			Other	I RL I	= 79		
Systems	P	R	F	I OE I	IOE ∩ RL I	P	R	F	I OE I	I OE ∩ RL I
FreeLing	-	-	-	-	-	2%	15%	3%	638	12
LTasks	46%	46%	46%	351	163	-	-	-	-	-
PALAVRAS	-	-	-	-	-	-	-	-	-	-
NERP-CRF	42%	38%	40%	321	134	100%	3%	5%	2	2

Table 3. P/R/F of NERP-CRF for categories: Person, Place, Organization and EE

NERP-CRF – Four Categories					
	P	R	F	I OE I	I OE ∩ RL I
Person	84%	60%	**70%**	1,462	1,230
Place	49%	54%	51%	1,378	671
Organization	48%	46%	47%	918	442
EE	42%	11%	18%	793	332

We then we did three new trainings and these results are presented in Table 4. These results represent three different runs with two categories. The first with Person and all other categories grouped as EE. The second with Place and EE, and the third with Organization and EE.

The results of Tables 2, 3 and 4 show an interesting evolution in both Recall and Precision for the Person category when trained over four classes. The Precision values for Organization category that went from 44% for ten categories, to 48% for four categories, until impressive 60% for two categories. A similar evolution was observed for Place category with Precision evolving 48%, 49% and 62%. The accuracy increased as expected, due to the reduced number of categories to be learned.

Table 4. P/R/F of NERP-CRF for two categories (isolating Person, Place and Organization)

	P	R	F	$|$ OE $|$	$|$ OE $\cap$ RL $|$
NERP-CRF – Two Categories – Person					
Person	70%	36%	48%	1,043	732
EE	51%	47%	49%	4,721	2,422
NERP-CRF – Two Categories – Place					
Place	62%	47%	53%	947	587
EE	66%	45%	54%	3,999	2,659
NERP-CRF – Two Categories – Organization					
Organization	60%	38%	47%	620	369
EE	61%	50%	55%	5,153	3,143

5   Final Considerations and Future Work

This paper presented an evaluation of NERP-CRF against other NER systems. The first experiment has shown balanced results for the categories Person, Place and Organization, among the systems. The second experiment show possible evolutions for NERP-CRF system by reducing the number of categories, and this improvement is more relevant in terms of Precision. The overall comparison performed led us to believe that the method used by NERP-CRF, due to the use of sets of training and testing, has a better potential for improvement than the other systems.

Additionally to the results obtained with the second experiment, NERP-CRF system also can be improved by the development of a more elaborate set of features which will be applied to the training corpus. This belief is justified by the high accuracy achieved by NERP-CRF in experiment involving Place category.

Another future work of our interest is to specialize the Place category into sub-categories considering specific domains such as Geology to perform the classification task of NE.

References

1. Jiang, J.: Information extraction from text. In: Mining Text Data, ch. 2, pp. 11–41. Springer, New York (2012)
2. Settles, B.: Biomedical named entity recognition using conditional random fields and rich feature sets. In: Proceedings of the International Joint Workshop on Natural Language Processing in Biomedicine and Its Applications, pp. 104–107 (2004)
3. Suakkaphong, N., Zhang, Z., Chen, H.: Disease Named Entity Recognition Using Semisupervised Learning and Conditional Random Fields. Journal of the American Society for Information Science and Technology, 727–737 (2011)
4. Batista, S., Silva, J., Couto, F., Behera, B.: Geographic Signatures for Semantic Retrieval. In: 6th Workshop on Geographic Information Retrieval, pp. 18–19. ACM (2010)
5. Freitas, C., Mota, C., Santos, D., Oliveira, H.G., Carvalho, P.: Second HAREM: Advancing the State of the Art of Named Entity Recognition in Portuguese. In: 7th International Conference on Language Resources and Evaluation, pp. 363–3637. LREC. European Language Resources Association. ELRA, Valletta (2010)
6. Amaral, D.O.F.: Reconhecimento de entidades nomeadas por meio de conditional random fields para a língua portuguesa. M.sc. dissertation, PUCRS, Porto Alegre, Brazil (2012)
7. Padró, L., Collado, M., Reese, S., Lloberes, M., Castellón, I.: FreeLing 2.1: Five Years of Open-Source Language Processing Tools. In: 7th International Conference on Language Resources and Evaluation, LREC, pp. 3485–3490 (2010)
8. LTasks – Language Tasks, http://ltasks.com
9. Bick, E.: Functional aspects in portuguese NER. In: Vieira, R., Quaresma, P., das Nunes, M.G.V., Mamede, N.J., Oliveira, C., Dias, M.C. (eds.) PROPOR 2006. LNCS, vol. 3960, pp. 80–89. Springer, Heidelberg (2006)
10. Santos, D., Cardoso, N.: Reconhecimento de entidades mencionadas em português: Documentação e atas do HAREM, a primeira avaliação conjunta na área. In: Santos, D., Cardoso, N. (eds.) ch. 1, pp. 1–16 (2008)
11. Santos, D.: Caminhos percorridos no mapa da portuguesificação: A linguateca em perspectiva. Linguateca 1, 25–59 (2009)
12. Carvalho, P., Oliveira, H.G., Mota, C., Santos, D., Freitas, C.: Desafios na avaliação conjunta do reconhecimento de entidades mencionadas: O Segundo HAREM. In: Mota, C., Santos, D. (eds.) Linguateca, ch. 1, pp. 11–31 (2008)
13. Lafferty, J., McCallum, A., Pereira, F.: Conditional Random Fields: Probabilistic Models for Segmenting and Labeling Sequence Data. In: 18th International Conference on Machine Learning ICML, pp. 282–289 (2001)
14. Santos, D., Cardoso, N.: Reconhecimento de entidades mencionadas em português: Documentação e atas do HAREM, a primeira avaliação conjunta na área, ch. 20, pp. 307–326 (2007)
15. Ratinov, L., Roth, D.: Design Challenges and Misconceptions in Named Entity Recognition. In: 13th Conference on Computational Natural Language Learning, CONLL, pp. 147–155 (2009)

Integrating Verbal Idioms into an NLP System

Jorge Baptista[1,3], Nuno Mamede[2,3], and Ilia Markov[1,3]

[1] Universidade do Algarve/FCHS and CECL,
Campus de Gambelas, 8005-139 Faro, Portugal
jbaptis@ualg.pt
[2] Instituto Superior Técnico, Universidade de Lisboa,
Av. Rovisco Pais, 1049-001 Lisboa, Portugal
Nuno.Mamede@ist.utl.pt
[3] INESC-ID Lisboa/L2F – Spoken Language Lab,
R. Alves Redol, 9, 1000-029 Lisboa, Portugal
{jbaptis,Nuno.Mamede,Ilia.Markov}@l2f.inesc-id.pt

Abstract. This paper describes the integration of verbal idioms into an Natural Language Processing (NLP) system, adopting a construction approach, which is based on the prior parsing stage, so that these Multi-Word Expressions (MWE) can be taken into account in subsequent tasks, such as semantic role labeling or whole-part relation extraction. The paper focuses on body-part nouns, which are often part of many verbal idioms, and uses a manually annotated corpus to evaluate its parsing strategy. Results showed a precision of 0.92, 0.83 recall, 0.87 f-measure and an accuracy 0.99.

Keywords: Verbal Idioms, Multi-Word Expressions, Body-part Nouns, Lexicon-Grammar, Natural Language Processing, Parsing, European Portuguese.

1 Introduction

Verbal idioms are idiomatic (semantically non-compositional) expressions consisting of a verb and at least one constraint argument slot, for which the overall meaning cannot be calculated from the meaning that the individual elements of the expression would present when used independently, in other contexts [10,11]: *O Pedro perdeu a cabeça* (lit: Pedro lost the=his head) 'Pedro became furious' (or 'Pedro lost his mind'). In this paper, we address the main issues raised in the process of integrating the lexicon-grammar of European Portuguese verbal idioms [2,3] into a fully-fledged natural language processing system, STRING[1] [13]. Our purpose is to highlight the detailed level of description required to identify this type of linguistic meaning units in texts, while maintaining these resources updated.

---

[1] https://string.l2f.inesc-id.pt/ [last access: 10/05/2014].

J. Baptista et al. (Eds.): PROPOR 2014, LNAI 8775, pp. 250–255, 2014.
© Springer International Publishing Switzerland 2014

2   Related Work

Work on multiword expressions (MWE) has drawn the attention of computational linguists for quite a long time [17]. Compound nouns, adverbs and other multiword lexical units pose specific problems to their automatic lexical acquisition and identification, but can be parsed using a *words-with-spaces* approach. For a recent comparison on different techniques for automatic multiword identification, see [15]. A *construction approach* [6] seems more appropriate to represent verbal idioms, provided lexical resources are available.

Extensive lists of verbal idioms, particularly the most frequent ones, have been systematically collected for Portuguese, both the European [2] and the Brazilian [18] varieties, along with their main distributional, syntactic and transformational properties, under the Lexicon-Grammar methodological and theoretical framework [10,11]. To our knowledge, so far these resources have not been integrated yet in any Portuguese NLP system, so it is difficult to ascertain the issues that may rise from the interaction of the different modules.

3   Integration of Verbal Idioms in STRING

The STRING system uses the XIP parser (Xerox Incremental Parser) [1] to segment sentences into chunks and extract dependency relations among chunks' heads. Considering that most idioms have a "normal" syntactic structure, which follows the ordinary word combinatory rules of the general grammar, STRING's strategy consists in parsing them *first* as ordinary sentences and *only then* to identify specific word combinations, whose meaning should not be calculated in a compositional way. This corresponds to the *construction approach* originally proposed by [6], and this is based on the results from the previous parsing stages, including the main syntactic dependencies such as SUBJ[ect], MOD[ifier] or direct object (CDIR), as well as auxiliary dependencies like PREPD, relating prepositions and the PP heads, or DETD, linking the determiners to the NP heads. It also involves using either surface forms or lemmas, or even restrictions in the morphological attributes of any given lemma. A large set of rules were semiautomatically built from the available lexicon-grammar of verbal idioms [2]. The idiomatic word combinations are identified by a new dependency, FIXED, which takes as arguments the verb and the frozen elements of the idiomatic expression (the number of arguments depends on the type of idiom involved). Figure 1 below illustrates the chunking tree and the relevant dependencies extracted for the sentence *O Pedro perdeu a cabeça* 'Pedro lost his mind'.

The identification of the idiom uses the previously calculated dependencies, namely the direct object (CDIR) and the main verb, and is carried out by the following rule:

```
IF (VDOMAIN(?,#2[lemma:perder]) & CDIR[post](#2,#3[surface:cabeça])) FIXED(#2,#3)
```

This rule captures any form of the lemma of the verb *perder* 'lose' (including any compound tenses) [4] and the surface form of the direct object (obligatorily

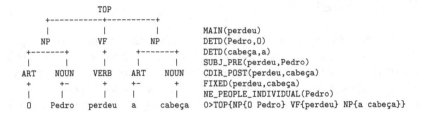

Fig. 1. Extraction of FIXED dependency for the sentence *O Pedro perdeu a cabeça* 'Pedro lost his mind'

after the verb) *cabeça* 'head'. Around 2,400 rules were semiautomatically build for 10 formal classes of verbal idioms [2]. A list of simple, manually-built examples (one for each idiom) provided with those classes was used to test the rules during the development stage.

For lack of space, we can not detail much the challenges that had to be met during the implementation of the rules, nor the solutions we provided; therefore, only the briefest overview is provided here.

Compounds. Since the idioms are being processed at a very late stage of parsing, some compound words have already been identified and the resulting parse is inadequate. For example, in: *O dinheiro subiu à cabeça do Pedro* 'The money went to Pedro's head' the compound preposition *à cabeça de* 'at the head of' has been tokenized, at an earlier stage, so the rule that would identify the verbal idiom must take this preposition into account. Notice that, for the semantic representation of the sentence, since the idiom is to be considered semantically non-compositional, an overall meaning should be attributed to the sentence, and the original meaning of the compound preposition *à cabeça de* 'at the head of' is to be discarded, in much the same way as the (potential) meaning of the verb: FIXED(subiu,à cabeça de).

Intrinsically Reflexive Constructions. In many idioms, the verb shows an intrinsically reflexive construction: *O Pedro atirou/mandou/amandou-se ao ar* (lit: Pedro threw himself to the air) 'Pedro went mad/furious'. A dependency, named CLITIC, has already been extracted between the verb and the clitic pronoun. This dependency is also extracted even if the pronoun, under certain syntactic conditions, is moved to the front of the verb (proclisis), as is *O Pedro até se atirou ao ar* (lit: Pedro even himself threw to the air). Therefore, even sentences where this fronting take place are captured by the same rule. Notice that in this intrinsically reflexive construction, the reflexive pronoun should correspond to an object NP, but for the the purpose of the semantic representation of the idiom, one can ignore it altogether: FIXED(atirar,a,ar).

Obligatory Negation. About 5% of all the Portuguese verbal idioms collected so far [7] involve an obligatory negation: *O Pedro não brinca em serviço* (lit: Pedro does not play in his job) 'Pedro is always very serious/competent in his job'.

Even if in most cases negation is carried out by negation adverb *não* 'no/not', other negation adverbs (*nunca, jamais* 'never'; *nem* 'nor') can also be used. All these cases are captured by a special feature NEG on the MOD[ifier] dependency that links these adverbs to the verb. Notice that, as the negation is considered as an intrinsic component of idiom, thus it is captured as a feature of the fixed expression: FIXED_NEG(brinca,em,serviço).

4 Evaluation

We have framed the evaluation of the idioms identification module in the task of part-whole relation extraction. We took advantage of an already existing corpus with annotated frozen expressions, in this case, with idioms involving *Nbp*. This corpus consists of a random stratified sample of 1,000 sentences, selected from a large set of around 17,000 sentences extracted from the 1st fragment (6,25 million words) of the CETEMPúblico corpus [16] using a small dictionary of around 300 *Nbp*.

The number of sentences with each *Nbp* in the sample is proportional to the number of its occurrences in the CETEMPúblico fragment. This sample had been previously annotated by 4 different native speakers, under the scope of another work on the extraction of whole-part relations involving body-part nouns and human entities [14]. The output sentences were divided into 4 subsets of 225 sentences each. Each subset was then given to a different annotator, and a common set of 100 sentences was added to each subset in order to assess inter-annotator agreement. From the 100 sentences that were annotated by all the participants in this process, we calculated the Average Pairwise Percent Agreement (0.85), the Fleiss' Kappa [8] (0.625; observed agreement 0.85/expected agreement 0.601), and the Cohen's Kappa coefficient of inter-annotator agreement [5] (0.629) using ReCal3: Reliability Calculator [2], for 3 or more annotators. According to Landis and Koch [12] these figures correspond to the lower bound of the "substantial" agreement; however, according to Fleiss [9], these results correspond to an inter-annotator agreement halfway between "fair" and "good". In view of these results, we can assume as a reasonable expectation that the remaining, independent and non-overlapping annotation of the corpus by the four annotators is sufficiently consistent, and will use it for the evaluation of the system output.

There are about 400 frozen expressions involving *Nbp* in the lexicon-grammar of European Portuguese verbal idioms, while the corpus features 40 types of these (43 instances). Table 1 shows the results from this experiment. While the observation on this sample can not be extended to the general lexicon, it may suggest the level of adequacy of the methods and resources here used.

We now describe the main errors still to be addressed by the system. Some of the false-positive cases result from the structural ambiguity between the idiom

---

[2] http://dfreelon.org/utils/recalfront/recal3/ [last access: 08.02.2014].

[3] TP: true-positives; TN: true-negatives; FP: false-positives; FN: false-negatives.

Table 1. Results

Number of sentences	TP	TN	FP	FN [3]	Precision	Recall	F-measure	Accuracy
1,000	33	957	3	7	0.92	0.83	0.87	0.99

and another construction. For example, in the next sentence, a support verb construction of the predicative noun *bofetada* 'slap' involves the same verb *dar* 'give' and the *Nbp cara* 'face': *Para mim dizer bem de Castro é o mesmo que dar uma bofetada na cara do meu pai* 'For me, to say nice things about Castro is the same as giving a slap in my father's face'. In this case, the dependency `FIXED(dar,na cara de)` has been extracted, which happens to be a frozen expression with a similar meaning.

Finally, the next sentence came from a title that is immediately followed by a quotation from the actress it mentions: *Catarina Furtado estreou Uma noite de Sonho na SIC "Fecho os olhos e oiço aplausos"* 'Catarina Furtado debuted (the show) *Uma noite de Sonho* in SIC (a tv station) "I close my eyes and I hear aplauses"'. The sentence is ambiguous with the idiom *fechar os olhos* 'close the eyes', which is an euphemism for 'to die'. Naturally, this kind of situation can not be solved without information from the context.

The false-negatives are often just idioms that were still missing in the lexicon-grammar. For example, at the end of the next sentence, the idiom *dar a cara por* (lit: to give the face for (sth.)) 'to represent (sb./sth.)' was not captured simply because it had not been listed. But, curiously, the sentence opens with another missed idiom: *Com o projeto ainda na gaveta* 'With the project still in the drawer', which is probably a reduction from the sentence form *meter na gaveta* (lit: put (sth.) in the drawer) 'to shelve (sth.); i.e., to prevent something from developing'. These complex prepositional phrases resulting from the reduction of longer sentence structures are still not described in the lexicon-grammar.

5   Conclusions

This paper set out to describe the issues raised by the integration of a large-sized database of verbal idioms of European Portuguese into a fully-fledged NLP system. The evaluation on an available corpus of sentences involving body-part nouns, a type of lexical item that is very prone to form idioms in many languages, showed promising results. Several aspects of the syntax of these idiomatic expressions have been described and taken into account for future work. Perhaps the most challenging task ahead is the complete automatization of the rule-generation process.

Acknowledgments. This work was supported by national funds through FCT – Fundação para a Ciência e a Tecnologia, under project PEst-OE/EEI/LA0021/ 2013; and Erasmus Mundus Action 2 2011-2574 Triple I - Integration, Interaction and Institutions. We would like to thank the comments of the anonymous reviewers, which helped to improve this paper.

References

1. Ait-Mokhtar, S., Chanod, J., Roux, C.: Robustness beyond shallowness: incremental dependency parsing. Natural Language Engineering 8(2/3), 121–144 (2002)
2. Baptista, J., Correia, A., Fernandes, G.: Frozen Sentences of Portuguese: Formal Descriptions for NLP. In: Workshop on Multiword Expressions: Integrating Processing. Intl. Conf. of the European Chapter of the ACL, Barcelona, Spain, pp. 72–79 (2004)
3. Baptista, J., Correia, A., Fernandes, G.: Léxico Gramática das Frases Fixas do Portugués Europeo. Cadernos de Fraseoloxía Galega 7, 41–53 (2005)
4. Baptista, J., Mamede, N., Gomes, F.: Auxiliary verbs and verbal chains in European Portuguese. In: Pardo, T.A.S., Branco, A., Klautau, A., Vieira, R., de Lima, V.L.S. (eds.) PROPOR 2010. LNCS (LNAI), vol. 6001, pp. 110–119. Springer, Heidelberg (2010)
5. Cohen, J.: A coefficient of agreement for nominal scales. Educational and Psychological Measurement 20(1), 37–46 (1960)
6. Copestake, A.: Representing idioms. Presentation at the HPSG Conference, Copenhagen (1994)
7. Fernandes, G., Baptista, J.: Frozen sentences with obligatory negation: linguistic challenges for natural language processing. In: Mellado-Blanco, C. (ed.) Colocaciones y Fraseología en Los Diccionarios, pp. 85–96. Peter Lang, Frankfurt (2008)
8. Fleiss, J.L.: Measuring nominal scale agreement among many raters. Psych. Bull. 76(5), 378–382 (1971)
9. Fleiss, J.L.: Statistical methods for rates and proportions, 2nd edn. John Wiley, New York (1981)
10. Gross, M.: Une classification des phrases "figées" du français. Revue Québécoise de Linguistique 12(2), 1–16 (1982)
11. Gross, M.: Lexicon-Grammar. In: Brown, K., Miller, J. (eds.) Concise Encyclopedia of Syntactic Theories, pp. 244–259. Pergamon, Cambridge (1996)
12. Landis, J., Koch, G.: The measurement of observer agreement for categorical data. Biometrics 33(1), 159–174 (1977)
13. Mamede, N., Baptista, J., Diniz, C., Cabarrão, V.: STRING: An Hybrid Statistical and Rule-Based Natural Language Processing Chain for Portuguese. In: Intl. Conf. on Computational Processing of Portuguese, Propor 2012, vol. Demo Session (2012), Paper available at http://www.propor2012.org/demos/DemoSTRING.pdf
14. Markov, I.: Automatic Identification of Whole-Part Relations in Portuguese. Master's thesis. U. Algarve, Faro (2014)
15. Ramisch, C., Araújo, V., Villavicencio, A.: A Broad Evaluation of Techniques for Automatic Acquisition of Multiword Expressions. In: Proceedings of the ACL 2012 Student Research Workshop, pp. 1–6. ACL (2012)
16. Rocha, P., Santos, D.: CETEMPúblico: Um corpus de grandes dimensões de linguagem jornalística portuguesa. In: Nunes, M.G. (ed.) V Encontro para o processamento computacional da língua portuguesa escrita e falada (PROPOR 2000), pp. 131–140. ICMC/USP, São Paulo (2000)
17. Sag, I.A., Baldwin, T., Bond, F., Copestake, A., Flickinger, D.: Multiword expressions: A pain in the neck for NLP. In: Gelbukh, A. (ed.) CICLing 2002. LNCS, vol. 2276, pp. 1–15. Springer, Heidelberg (2002)
18. Vale, O.: Expressões Cristalizadas do Português do Brasil: uma proposta de tipologia. Ph.D. thesis. Universidade Estadual Paulista, Araraquara, SP (2001)

Rolling out Text Categorization for Language Learning Assessment Supported by Language Technology

António Branco[1], João Rodrigues[1], Francisco Costa[1], João Silva[1], and Rui Vaz[2]

[1] Universidade de Lisboa
Departamento de Informática, Faculdade de Ciências
[2] Camões IP
Divisão de Programação, Formação e Certificação

Abstract. This paper is concerned with a tool that supports human experts in their task of classifying text excerpts suitable to be used in quizzes for learning materials and as items of exams that are aimed at assessing and certifying the language level of students taking courses of Portuguese as a second language.

We assess the performance of this tool, which is currently available as an online service and is being used by the experts of the team of Camões IP that is responsible for the elaboration of the quizzes and of exam items to be used in language certification exams.

Keywords: Language learning assessment, readability assessment, text categorization, second language skills certification, Portuguese language.

1 Introduction

As the Portuguese official language institute, Camões IP is responsible for running worldwide massive language learning courses of Portuguese as second language and for performing language certification exams. These courses and certification use quizzes and exams that resort to text excerpts belonging to one of the five language levels of skill A1, A2, B1, B2 and C1. As these excerpts can be used only once, specially in exams, the massive nature of these courses imposes a continuous task of always selecting more excerpts to be used in the upcoming exams and quizzes.

To help cope with this stringent demand, and support the instructors in their selection of these excerpts, an online tool was developed that seeks to help assigning each input excerpt to a language level. It returns the scores of a few metrics calculated with the use of natural language processing techniques. The design options and further details of this tool are presented and discussed in [1].

As this tool was made available as an online service,[1] we called LX-CEFR, it became nevertheless useful both for instructors and students. It is expected to help to improve the productivity and the consistency of the classification of candidate excerpts by the experts. And it helps also to enhance the level of interactivity of the language courses for students, as they can resort to this tool to check whether a new text they come across may be appropriate for their current language level.

---

[1] http://lxcefr.di.fc.ul.pt

J. Baptista et al. (Eds.): PROPOR 2014, LNAI 8775, pp. 256–261, 2014.

In this paper we report on a first evaluation exercise we performed over this tool, including an assessment of how inherently difficult is the task (for humans) and thus what may be the upper bound for the performance of a tool like this.

2   Language Levels and Metrics

Following CEFR[2] [2], Camões IP's language courses and language proficiency certification encompass five levels, of increasing difficulty: A1, A2, B1, B2 and C1. In order to help the user decide for the categorization of the input texts into one of these levels, the online tool displays the scores for fifteen metrics, covering for instance, number and proportion of letters, syllables, and words; number and proportion of simple, passive and subordinate clauses; number and proportion of nouns, verbs, prepositions, etc.; proportion of word types with only one occurrence; etc.

Out of these metrics, four are singled out as primary metrics and their scores offered in a radar chart (Figure 1). They are metrics that in the literature have been argued to correlate well with levels of readability [3]:

- Flesch Reading Ease index [4]
- Lexical category density in proportion of nouns
- Average word length in number of syllables per word
- Average sentence length in number of words per sentence

The Flesch metric is a wider accepted and used readability metric, which combines the third and fourth metric. Its higher scores indicate texts that are easier to read.

3   Language Processing and Projection of Scores

To obtain the scores for these metrics, the online tool resorts to a number of state of the art natural language processing tools for Portuguese, such as tokenizers, POS taggers and syntactic parsers [5][6][7]. While the scores of the eleven secondary metrics are delivered as they are obtained, the scores of the four primary metrics are further processed in order to be projected into a continuous linear scale with five major points, each corresponding to a language level.

The dataset available to support the development of this tool consisted of 125 text excerpts previously used in past exams, each classified into one of the five language levels, and all in all containing 690 sentences, with 12,231 tokens. The average scores for the four primary metrics were calculated for each one of the five language levels. For each such metric, a linear regression over those averaged values was obtained, and each language level projected into the regression line in the chart. This formed the basis for the reference scales, represented in the arms of the radar chart.

When a text is input to the online tool, its scores for these four metrics are calculated and projected into the radar chart according to the comparison with the reference values obtained with the 125 excerpt reference dataset just described above.

---

[2] Common European Framework of Reference for Languages: Learning, Teaching, Assessment.

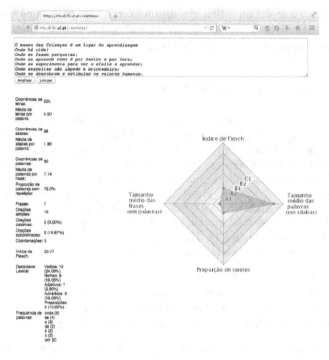

Fig. 1. Online tool partial window with an example

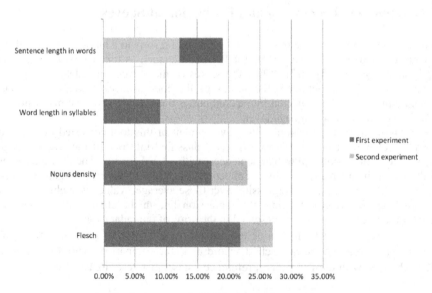

Fig. 2. Accuracy in first and second experiment

4   Evaluation of the Tool

To evaluate the tool we did a first evaluation experiment. We used this reference dataset and performed a 10-fold cross evaluation for each of the four primary metrics, obtaining accuracy values ranging from 9.00% (for word length) to 21.82% (Flesch index). The results are displayed in part of Figure 2 (in dark grey columns).

5   Assessment of the Task

In order to put these scores into perspective, it was important to assess how inherently difficult could the task be in itself, that is how well human annotators executing it could perform.

To that end, the texts in the dataset were untagged of their originally assigned level, and five language instructors were recruited, which are trained and experts in selecting and classifying texts according to the relevant five CEFR levels. These experts performed the task of classifying each one of the texts. This re-annotated dataset permitted to assess the difficulty of the task along two measures: proportion of texts upon which there is agreement among annotators in their classification; inter-annotator agreement (ITA) given by Fleiss' kappa coefficient [8].

The distribution of the classifications of the annotators per CEFR level is displayed in the chart of Figure 3.

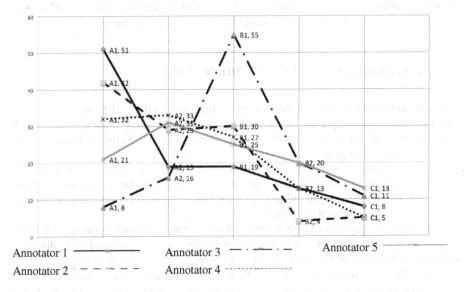

Fig. 3. Number of texts assigned to each category (A1, A2, B1, B2 and C1) by each of the five annotators (Annotator 1 to 5)

The texts that received unanimous classification were 0.90% (only one text); those receiving the same classification by a majority of at least 4 classifiers were 17.27%. There were 67.27% of the texts receiving a given classification by a majority of at least 3 annotators.

The Fleiss' kappa coefficient value obtained for ITA was 0.13, corresponding to "Slight agreement", according to [9]. This is the second worst of five levels of agreement, and very distant from the value of 0.8+ widely assumed to be the level ensuring reliability of an annotated dataset.

6   Reevaluation of the Tool

From the new reannotated dataset, we kept the 84 texts that received its classification by a majority of at least three annotators. With this subdataset, and with this new classification by several human experts, the reference scales for the four primary metrics were redone, following the same process as described above in the Section 3.

The evaluation of the ranking tool was redone, again with a 10-fold cross evaluation of the four primary metrics. The values now obtained after this fine tuning of the tool with the dataset annotated by multiple experts range from 12.16% (for sentence length) to 29.73% (word length), as displayed in part of Figure 2 (in light grey columns). There was a substantial improvement of three of the metrics with respect to the first evaluation, when the system was tuned with the dataset of texts used in previous exams, where each one was classified only by a single instructor.

7   Concluding Remarks and Future Work

The results of the assessment of the task reveal that this is a quite difficult task even for humans, and hence represent a challenge for an automated ranking tool, which nevertheless is already attaining around 1/3 of its upper bound, as this is determined by the performance of humans.

It should be noted however that the volume of the dataset we could resort to may be too small and better results may eventually be expected with a larger dataset. From the first to the second experiment, all metrics improved except one, which may be deemed to the small size of the working sample.

Also worth noting is that, with a larger enough dataset, the superposition of a linear regression upon the distribution of the CEFR levels can be challenged and enhanced, and above all, advanced machine learning methods can be eventually benefited from.

Finally, also as future work, it will be important to undertake a usability assessment in order to gain a better understanding of how the tool can possibly be better adjusted to fit the needs of the human operators.

References

[1] Branco, A., Rodrigues, J., Costa, F., Silva, J., Vaz, R.: Text Classification for Interactive Language Learning (forth)

[2] Council of Europe (2011), consulted on http://www.coe.int/t/dg4/linguistic/Cadre1_en.asp (March 31, 2014)

[3] DuBay, W.: The Principles of Readability. Impact Information, Costa Mesa (2004)

[4] Flesch, R.: How to write in plain English: A book for lawyers and consumers. Harper, New York (1979)

[5] Branco, A., Silva, J.: Evaluating Solutions for the Rapid Development of State-of-the-Art POS Taggers for Portuguese. In: Proceedings of the 4th International Conference on Language Resources and Evaluation (LREC 2004), pp. 507–510. ELRA, Paris (2004)

[6] Branco, A., Silva, J.: LX-Suite: Shallow Processing Tools for Portuguese. In: Proceedings of the 11th Conference of the European Chapter of the Association for Computational Linguistics (EACL 2006), Trento, Italy, pp. 179–182 (2006)

[7] Silva, J., Branco, A., Castro, S., Reis, R.: Out-of-the-Box Robust Parsing of Portuguese. In: Pardo, T.A.S., Branco, A., Klautau, A., Vieira, R., de Lima, V.L.S. (eds.) PROPOR 2010. LNCS (LNAI), vol. 6001, pp. 75–85. Springer, Heidelberg (2010)

[8] Fleiss, J.L.: Statistical methods for rates and proportions, 2nd edn., pp. 38–46. John Wiley, New York (1981)

[9] Landis, J.R., Koch, G.G.: The measurement of observer agreement for categorical data. Biometrics 33(1), 159–174 (1977)

Words Temporality for Improving Query Expansion

Olga Craveiro[1,2], Joaquim Macedo[3], and Henrique Madeira[2]

[1] School of Technology and Management, Polytechnic Institute of Leiria, Portugal
[2] CISUC, Department of Informatics Engineering, University of Coimbra, Portugal
{marine,henrique}@dei.uc.pt
[3] Algoritmi, Department of Informatics, University of Minho, Portugal
macedo@di.uminho.pt

Abstract. There is a lot of recent work aimed at improving the effectiveness in Information Retrieval results based on temporal information extracted from texts. Some works use all dates but others use only document creation or modification timestamps. However, no previous work explicitly focuses on the use of dates within in the document content to establish temporal relationships between words in the document. This work estimates these relationships through a temporal segmentation of the texts, exploring them to expand queries. It was achieved very promising results (13% improvement in Precision@15), especially for temporal aware queries. To the best of our knowledge, this is the first work using temporal text segmentation to improve retrieval results.

Keywords: Query Expansion, Temporal Information Retrieval, Temporal Text Segmentation.

1   Introduction

A large range of Information Retrieval (IR) tasks can benefit by incorporating temporal information [2]. The novelty or temporal incidence of information is obviously an important aspect in the assessment of its quality, importance and relevance.

In Web IR, time was proven to be valuable for crawling [8], clickstream analysis [11], and determine temporal snippets for summarize documents [1]. Even being a low percentage of total search engine queries, there is a huge number of temporal aware information needs [13]. So, there is several research work exploring temporal information to improve the effectiveness of IR results. There are time-aware language models [12] and similarity formulas with temporal components in vector space models [10]. There is work concerned with searching future events [5]. In query expansion (QE), one idea is to consider that if exists a large set of query results documents with same date, they are more likely to be relevant. The work of Amodeo et al. [4] relates the publication date of documents and their relevance for a given query to select the pseudo-relevant documents. The proposal of Whiting et al. [16] is to select better the

J. Baptista et al. (Eds.): PROPOR 2014, LNAI 8775, pp. 262–267, 2014.

terms, considering their temporal profile correlation from the pseudo-relevant documents and temporally significance.

This work establishes relationships between words based on temporal segmentation of texts. Terms co-occurring with original query terms in the same segment of pseudo-relevant documents are positively discriminated in QE process.

Section 2 describes the used temporal segmentation tool for Portuguese texts. Section 3 presents the three QE methods based on temporal segmentation. Section 4 presents the evaluation of these methods using the CHAVE collection [15]. Then, we discuss the obtained results and possible future research issues.

2   Temporal Segmentation of Texts

Since temporal segmentation of texts has a crucial importance to this work, we present a brief description of the followed approach, detailed in our prior work [9].

Prior to segmentation, temporal expressions found in the content of documents are recognized, annotated and normalized, as far as possible, linked to *chronons*. *Chronons* are dates with a given granularity (year, month, day or hour) and a mark on a timeline. Too vague temporal expressions cannot be transformed into a *chronon*.

As one can model a text as a narrative of a sequence of different topics, it makes sense to think that it describes a chronology of events that occur at different instants of time. So, an alternative to topical segmentation [7] is temporal segmentation [6], which the underlying idea is the detection of temporal discontinuities in the text. Instead of subject change, it is amended the date or time of the narrative.

Our segmentation algorithm uses temporal information extracted from the text to divide it into temporally coherent segments. These segments must be tagged with *chronons* in their sentences (including document timestamp), to obtain an association between time and document terms. The length of a segment ranges from a single sentence to a multi-paragraph text. Thus, adjacent sentences with the same *chronons* must belong to the same segment.

3   Time-Aware Query Expansion

Our proposal incorporates the word temporality in the automated QE models. Fig. 1 shows a general algorithm of QE, considering a broad view of the QE process. The two important parameters of the algorithm are: k, the number of relevant documents for the original query $q0$, and n, the maximum number of terms to in the expanded query qE. The formulae used in the computation of the scores (steps 3 and 5) are based on the formula of Rocchio [27], which need the definition of the parameter α. Roughly speaking, the algorithm can be described as follows: first, it is necessary to obtain a set of terms T_1 to be included in the expanded query qE; and then, the original query $q0$ is reformulated considering these terms and their scores.

The three proposed methods are based on the assumption that the terms that co-occur in the same temporal segment with original query terms (T_0) must assume a more important role in query reformulation. The effect of these methods is explicit *in italic* in the Fig. 1. The temporal filtering method (**TmpF**) removes all terms T_1 that not co-occur with T_0 terms in the same segment (step *TmpF of Fig. 1*), after usual stopwords removal and stemming. The temporal weighting method (**TmpW**) introduces a new formula penalizing the score of terms that not co-occur with T_0 terms, as referred in *step TmpW of Fig. 1*. Temporal Reweighting (**TmpR**) is focused on the query reformulation (see *step TmpR of Fig. 1*) without interfering with the terms selection, unlike the others. The set of terms for the reformulated query is the same. However, the weight of a term that co-occurs in the same temporal segments of the T_0 terms is increased, by using δ. α and β (used in Rocchio formula) are positive weights associated to the initial and the new query terms, respectively.

---

(1) Submit the query $q0$ with the set of terms $tq0_1, \ldots, tq0_m$ (T_0) defined by the user

(2) From the document set D composed of the top-k most relevant documents to $q0$, extract the set of terms T_1

　　　(TmpF) Remove from T_1 terms not co-occurring in same temporal segments of T_0

(3) For each term t of the terms set T_1, compute its score

　　　(TmpW) Compute for terms in T_1 a $score^*(t)=score(t) \times (1/(1+td(t)))$　$td(t) \in [0,1]$

(4) Rank the terms set T_1, in decreasing order, and pick-up the top-n

(5) Compute the weight for each term in T_0 and T_1

　　　(TmpR) Reweight using $w^*(t \mid q) = \alpha \times w(t \mid q) + (\beta + \delta) \times score\,(t)$　$\beta + \delta \leq 1$

(6) Reformulate the original query $q0$ in an expanded query qE with the terms $tqE_1, \ldots, tqE_{m+n}$ and their weights

(7) Submit the new query qE

Fig. 1. Query Expansion Algorithm.

4　Experiments and Results

The evaluation of the effectiveness of our temporal methods for improving QE was performed with the CHAVE collection created by Linguateca[1], which is the only IR test collection in Portuguese. The collection is composed of full-text from two major daily Portuguese and Brazilian newspapers, namely PUBLICO[2] and Folha de São Paulo[3], from complete editions of 1994 and 1995, a total of 210,734 documents.

In this work, we carried out 2 experiments (ExpA, ExpB) to validate the improvement achieved in time-sensitive and time-insensitive queries, respectively, when the words temporality of documents is considered. Each experiment uses a topic set. The 100 topics of CHAVE (C251-C350) were manually divided into 2

---

[1] http://www.linguateca.pt
[2] http://www.publico.pt
[3] http://www.folha.com.br

topic sets, according their time sensitivity: *TopicsSet1*[4], composed of the 40 time-sensitive topics of the collection, and *TopicsSet2*, composed of the other 60 time-insensitive topics.

Documents were temporally segmented, and then, they were indexed using Terrier 3.5[5] [14] with stopwords removal and *PortugueseSnowballStemmer* applied for stemming. The retrieval was performed using TF-IDF, considering the top 1000 documents of the results set. The results obtained were used as the baseline *noQE*.

The base model of our methods, Bose-Einstein 1 (Bo1), is the best-performing implemented by Terrier, namely improving Mean Average Precision (MAP) [3].

In all the experiments, the queries were expanded with *10* terms of the top *3* retrieved documents. The terms weight was computed by the Bo1 model's formula [3], configured with the parameter free expansion option provided by Terrier. So, we also report a run with the Bo1 model as the stronger baseline, named as *Bo1QE*.

To process a query, our methods need 2 sets of *chronons*: *QueryTermChronons*, and *CandidateTermChronons*, composed of the segments timestamps where the query terms and the candidate term occur in pseudo-relevant documents, respectively.

The number of *chronons* (j) associated to a candidate term that matches *QueryTermChronons* ranged from 1 to 10. For TmpR evaluation, β and δ varied between 0.1 and 0.9, considering that $\beta+\delta\leq1$ (see *step TmpR of Fig. 1*). α was set to 1 to give the maximum importance in the reformulated query to the initial query terms.

MAP for the top 1000 documents, Precision@10 (P@10) and Precision@15 (P@15) give the retrieval effectiveness. We also computed the number of improved (IMP) and the number of penalized (PQ) queries based on average precision of each query. *IMP; PQ (1)* and *IMP; PQ (2)* were computed taking as reference the two baselines, *noQE* and *Bo1QE*, respectively.

Table 1 shows the best results obtained by our methods in ExpA and ExpB. Due to space constraints, P@10 of ExpA is not presented, since the value of P@15 is similar.

We verified that the best results were obtained with $j\leq4$. This means that the number of *chronons* that matches the 2 sets of *chronons* should not be more than 4. This is an expected value, since the terms were obtained from only 3 documents of the result set, and most documents do not have more than 4 *chronons*.

All temporal methods in ExpA obtained better results than *noQE40* and *Bo1QE40*, with considerable gains between 2% and 13%. TmpR achieved the best MAP (0.398). TmpF obtained the best value of P@15 (0.460).

In ExpB, the time-insensitive queries were improved, but with minimal gains when compared with *Bo1QE60* (1%-2%). TmpR achieved the best MAP, and the other two methods showed similar results with a better performance in precision. They also improved more 5 queries than *Bo1QE60*. Taking *Bo1QE60* as reference, TmpF improved the average precision of 29 queries, but penalizing 27.

Table 1 shows that the values of P@10 and P@15 in ExpB are very different, unlike the values obtained in ExpA. TmpF and TmpW achieved about 2% improvement in P@10. On the contrary, these methods obtained worse values of P@15.

---

[4] *TopicsSet1*: C257, C259, C262, C265-C267, C277, C279, C280, C282, C284, C287, C290, C292, C296, C305, C308, C313, C316, C326, C327, C332-C350.
[5] http://ir.dcs.gla.ac.uk/terrier/

Table 1. Results obtained by Temporal Filtering, Temporal Weighting and Temporal Reweighting using the CHAVE collection with time-sensitive and time-insensitive queries.

ExpA	noQE40	BolQE40	TmpF	TmpW	TmpR $\beta=0.2;\delta=0.8$
MAP	0.352	0.387	0.395	0.393	**0.398**
Precision@15	0.410	0.430	**0.460**	0.453	0.443
IMP; PQ (1)		29; 9	**31; 8**	**31; 8**	31; 9
IMP; PQ (2)		22; 8	22; 8	22; 9	**28; 11**
ExpB	noQE60	BolQE60	TmpF	TmpW	TmpR $\beta=0.4;\delta=0.3$
MAP	0.302	0.344	0.344	0.344	**0.346**
Precision@10	0.502	0.515	**0.523**	**0.523**	0.512
Precision@15	0.457	**0.498**	0.493	0.496	0.495
IMP; PQ (1)		40; 20	**45; 15**	**45; 15**	39; 21
IMP; PQ (2)		29; 27	**29; 27**	29; 28	22; 38

This means that the top-10 relevant documents were promoted in the ranking of the result set. However, to obtain more 5 relevant documents, more non-relevant documents are also found. Note that, the baselines also obtained worse values of P@15.

Our proposals improved effectiveness and robustness for both time-sensitive and time-insensitive queries, although with greater evidence in time-sensitive queries. In general, our methods obtained very close results. TmpF and TmpW had a similar performance, improving the precision. TmpR obtained more relevant documents, promoting them in the ranking, and obtaining a better MAP.

5   Conclusions and Future Work

Our approach distinguishes itself from the other proposals for Temporal IR by keep the relationship between words and time given by the temporal segmentation. This work launches the foundations for further research on taking advantage of this relationship to improve the effectiveness of retrieval systems. This approach can be applied not only to QE, but also to diverse issues and components of retrieval systems. It can also be used with any other language than Portuguese, such as English; it is not dependent on language. Experimental results are very promising, mainly in time-sensitive queries, which achieved from 2% to 13% improvement in effectiveness.

For future work, we intend to combine words temporality with the words occurrence in the term-weighting formulas of the QE systems. The definition of some strategies to determine the time-sensitivity of queries is also a priority of our work.

Acknowledgements. We are especially grateful to Catarina Reis for her collaboration. This work is partially supported by Algoritmi and CISUC, financed by FCT - Fundação para a Ciência e Tecnologia, within the scope of the projects PEst-OE/EE/UI0319/2014 and Pest-OE/EE/UI0326/2014.

References

1. Alonso, O., Baeza-Yates, R., Gertz, M.: Effectiveness of temporal snippets. In: Workshop on Web Search Result Summarization and Presentation WWW 2009, Madrid, Spain (2009)
2. Alonso, O., Strötgen, J., Baeza-Yates, R., Gertz, M.: Temporal Information Retrieval: Challenges and Opportunities. In: TWAW 2011, pp. 1–8 (2011)
3. Amati, G.: Probability Models for Information Retrieval based on Divergence from Randomness. Ph.D. thesis. University of Glasgow (2003)
4. Amodeo, G., Amati, G., Gambosi, G.: On relevance, time and query expansion. In: CIKM 2011, pp. 1973–1976. ACM, New York (2011)
5. Baeza-Yates, R.: Searching the future. In: SIGIR Workshop MF/IR (2005)
6. Bramsen, P., Deshpande, P., Lee, Y.K., Barzilay, R.: Finding temporal order in discharge summaries. In: AMIA 2006: Proceedings of the American Medical Informatics Association Annual Symposium, Washington DC, USA, pp. 81–85 (2006)
7. Caillet, M., Pessiot, J.F., Amini, M.R., Gallinari, P.: Unsupervised learning with term clustering for thematic segmentation of texts. In: Fluhr, C., Grefenstette, G., Croft, W.B. (eds.) RIAO, pp. 648–657. CID (2004)
8. Cho, J., Garcia-Molina, H.: Synchronizing a database to improve freshness. SIGMOD Rec. 29(2), 117–128 (2000)
9. Craveiro, O., Macedo, J., Madeira, H.: It is the time for portuguese texts! In: Caseli, H., Villavicencio, A., Teixeira, A., Perdigão, F. (eds.) PROPOR 2012. LNCS (LNAI), vol. 7243, pp. 106–112. Springer, Heidelberg (2012)
10. Kalczynski, P.J., Chou, A.: Temporal document retrieval model for business news archives. Information Processing and Management 41(3), 635–650 (2005)
11. Kleinberg, J.: Temporal dynamics of on-line information streams. In: Garofalakis, M., Gehrke, J., Rastogi, R. (eds.) Data Stream Management: Processing High-Speed Data Streams. Springer (2006)
12. Lavrenko, V., Croft, W.B.: Relevance based language models. In: Proceedings of the 24th Annual International ACM SIGIR Conference on Research and Development in Information Retrieval, SIGIR 2001, pp. 120–127. ACM, New York (2001)
13. Nunes, S., Ribeiro, C., David, G.: Use of temporal expressions in web search. In: Macdonald, C., Ounis, I., Plachouras, V., Ruthven, I., White, R.W. (eds.) ECIR 2008. LNCS, vol. 4956, pp. 580–584. Springer, Heidelberg (2008)
14. Ounis, I., Amati, G., Plachouras, V., He, B., Macdonald, C., Lioma, C.: Terrier: A High Performance and Scalable Information Retrieval Platform. In: Proceedings of ACM SIGIR 2006 Workshop on Open Source Information Retrieval (OSIR 2006), Seattle, Washington (2006)
15. Santos, D., Rocha, P.: Chave: Topics and questions on the portuguese participation in clef. In: Peters, C., Borri, F. (eds.) Cross Language Evaluation Forum: Working Notes for the CLEF 2004 Workshop (CLEF 2004), Pisa, Italy, September 15-17, pp. 639–648. IST-CNR (2004), (revised as Santos & Rocha, 2005)
16. Whiting, S., Moshfeghi, Y., Jose, J.M.: Exploring term temporality for pseudo-relevance feedback. In: SIGIR 2011, pp. 1245–1246. ACM, New York (2011)

An Open Source Tool for Crowd-Sourcing the Manual Annotation of Texts

Brett Drury, Paula C.F. Cardoso, Jorge Valverde-Rebaza, Alan Valejo,
Fabio Pereira, and Alneu de Andrade Lopes

ICMC, University of São Paulo,
Av. Trabalhador São Carlense 400
São Carlos, SP, Brazil
C.P. 668, CEP 13560-970
http://www.icmc.usp.br/

Abstract. Manually annotated data is the basis for a large number
of tasks in natural language processing as either: evaluation or training
data. The annotation of large amounts of data by dedicated full-time annotators can be an expensive task, which may be beyond the budgets of
many research projects. An alternative is crowd-sourcing where annotations are split among many part time annotators. This paper presents a
freely available open-source platform for crowd-sourcing manual annotation tasks, and describes its application to annotating causative relations.

Keywords: crowd-sourcing, annotations, causative relations.

1 Introduction

Manually annotated data is the basis for a large number of tasks in natural language processing as either: evaluation or training data. The manual annotation
of data is a time intensive and repetitive task. It may be beyond many projects'
budget to hire full time annotators. Crowd-sourcing is an alternative annotation
policy where a large manual annotation tasks can be split among a large number
of part-time annotators. The advantage of crowd-sourcing is that each individual
spends a relatively small period of time on the annotation task.

Crowd-sourcing has become a popular approach for many natural language
processing tasks, but the popular tools such as Mechanical Turk are commercial.
To remedy the lack of freely available open sourced annotation tools we present
an freely available annotation tool which can be used to crowd-source annotation
tasks.

The remainder of the paper is organised as follows: 1. related work, where we
discuss crowd-sourcing and manual annotations, 2. platform architecture where
we discuss the underlying design principles of the application and 4 where we
discuss the application of the tool to annotate causative relations.

J. Baptista et al. (Eds.): PROPOR 2014, LNAI 8775, pp. 268–273, 2014.

2   Related Work

The related work will describe general research about manual annotation as well as crowd-sourcing.

A common strategy for manual annotation of text is to use multiple annotators, and to use inter-annotator agreement to accept common annotations. This strategy is designed to mitigate the biases of a single annotator. A systematic review of inter-annotator agreement was conducted by [3]. He conducted two tasks: POS and structure tagging on a large German news corpus. He discovered that the annotator agreement was 98.6% for the POS task and 92.4% for structure tagging. The research literature also contains results related to inter-annotator agreement for tasks such as word-sense disambiguation [9], opinion retrieval[1] and semantics[11].

The works described thus far have used dedicated annotators, an alternative is crowd-sourcing. Crowd-sourcing divides the annotation task between many annotators. The motivation for performing the annotation task may be financial[13][7] or altruistic[13]. A potential problem for crowd-sourcing is the quantity and the quality of the annotated data. Wang[13] suggests that there is a trade-off between quality and quantity because high-quality annotated data requires a comprehensive set of annotation rules. In addition crowd-sourcing may attract non-expert annotators. Hsueh et al[5] conducted a study comparing expert and non-expert annotators found that if there were sufficient numbers of non-expert annotators that they performed nearly as well as small numbers of expert annotators. This hypothesis is partially supported by [10].

A review of the research literature for crowd-sourcing natural language tasks discovered that researchers tend to use closed source commercial systems such as Mechanical Turk (www.mturk.com) and CrowdFlower (www.crowdflower.com)[12].

3   Platform Description

The traditional crowd-sourcing platforms may be insufficient for Portuguese language researchers because: 1. there may not be sufficient native Portuguese speakers with access to the Internet who are willing to work for typical crowd-sourcing pay and 2. there may not be a budget to pay people. To resolve this problem we have developed a prototype platform for crowd-sourcing annotations so that researchers can use altruistic annotators rather than ones motivated by money.

The platform is designed around an "n-tier" architecture[6]. A n-tier architecture separates: presentation, logic and storage into separate layers so that they have no inter-dependencies. The architecture of the application is displayed in Figure 1.

3.1   Presentation Layer

The display layer uses HTML to provide a rudimentary user interface. The HTML can be hard-coded into a Web Page or generated dynamically with a

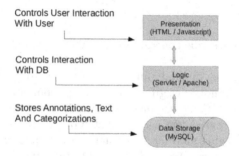

Fig. 1. N-tier architecture of application

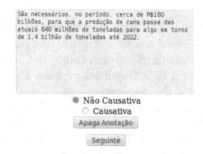

Fig. 2. Rudimentary user interface

Java Server Page (JSP). The user interface is typically: 1. a read only text field, 2. submit button and 3. category check box. Figure 2 shows an example of the user interface.

The presentation layer uses client side Javascript with the Dojo library (http://dojotoolkit.org/)) to control the annotator interaction with the Web Page. In addition the Javascript is used to communicate with the logic layer to pass and receive text information. This was achieved using an asynchronous "post" against a server side application. This technique is commonly known as AJAX.

3.2 Logic Layer

The logic layer function is to communicate with the presentation and the data layers. In this application we used Servlets which are written in Java and rely upon a Java Virtual Machine (JVM), consequently we were obliged to choose a "Servlet Container" which in this case was Tomcat.

The Servlet had two functions: 1. save information to the data layer passed to it from the presentation layer and 2. randomly select information from the database layer. The Servlet uses JDBC to communicate with the database, consequently the layer is independent of the database management system (DBMS) in the data layer.

3.3   Data Layer

The data layer responds to save and select requests from the logic layer. There is no prescribed DBMS, but we used MYSQL. The table which contains the data to be annotated must have a numeric primary key which can be either synthetic or natural. A numeric key is required because the Servlet uses a random selection strategy which relies upon a numeric range.

4   Causative Relation Case Study

The platform was used to annotate causative relations in Brazilian agricultural news. The agricultural news was scraped from news sites located on the Web, and the resulting text was segmented into sentences using the sentence splitter in the NLTK library [2]. A random selection of 2,000 sentences were reserved for annotation. These sentences were imported into a table in MySql.

There were six annotators (5 novice and 1 experienced) invited to annotate the text. The annotators had two tasks: classify the sentence into one of two categories: causative or non-causative and annotate the causative sentence. The annotation of causative sentence required the annotator to identify: 1. a cause, 2. a causative verb and 3. an effect. The annotator highlighted each part of the causative phrase in the text box using the mouse and by right clicking the highlighted text would show a context menu. The annotator would then select the requisite item (cause, effect or verb). The annotation period ran from February to March 2014. The annotation site is located at:http://goo.gl/d2UN93, where data can be downloaded. The source code is available from http://goo.gl/qTvlsx

4.1   Evaluation

The evaluation of the annotated data measured two tasks: 1. inter annotator agreement for the classification task and 2. the similarity of the annotation of causative phrases. The measures used for evaluating annotator agreement were: 1. average percentage of annotators agreeing on a single category and 2. average Cohen's kappa coefficient [4]. An average Cohen's kappa was calculated because there was a varying number of categorizations for each sentence, and consequently the chance annotator agreement varied with the number of categorizations. The results are in Table 1

Table 1. Evaluation of Causative Categorization Task

Measure	Value
Avg. Percentage	77.96% ($\pm$41.44)
Avg. Cohen's kappa coefficient	60.74 ($\pm$75.00)

The results are lower than the comparable manual annotation tasks described in the research literature, this may be due to the relative inexperience of some of the annotators as well as the subjective nature of the task.

The second evaluation was the similarity of the annotations for causative relations. The evaluation measure was an average Levenshtein distance [8] for each annotated part of the causative relation (cause, verb and effect). If there were more than two annotators which had annotated the sentence then each annotation was compared separately and an average distance calculated. We calculated the average distance for two annotator agreements : majority and full agreement. The results are in Table 2.

The values for full and majority agreement are identical which suggests that for all sentences which were annotated more than twice had full annotator agreement. This may indicate that the expert annotator agreed with the novice annotators where they annotated the same sentence. There was some confusion amongst the annotators whether to annotate resultative causative verbs as part of the effect or as the causative verb. This may account for the difference between verb and object similarity and subject similarity.

Table 2. Similarity of Causative Relation Annotation

Annotator Agreement	Subject	Object	Verb
Full	0.73 (±0.32)	0.68 (±0.29)	0.71(±0.35)
Majority	0.73 (±0.32)	0.68 (±0.29)	0.71(±0.35)

5   Conclusion

This paper describes a freely available crowd-sourcing application which is designed for distributing manual annotation tasks. The results it produced in a causative relation annotation exercise were inferior to comparable exercises described in the research literature, however this may have been due to the nature of the task rather than the distributed annotation.

We intend to increase the number of annotators to improve the quality of the annotations. We also plan to release the causative relations annotation data to the community as a benchmark to evaluate causative relation extraction strategies. In addition we will donate the tool to the community in the hope that it can be improved and assist with manual annotation tasks. This work was supported by FAPESP grants:11/20451-1,2011/22749-8 and 2013/12191-5 as well as by the CAPES funding agency.

References

1. Bermingham, A., Smeaton, A.F.: A study of inter-annotator agreement for opinion retrieval. In: Proceedings of the 32nd International ACM SIGIR Conference on Research and Development in Information Retrieval, SIGIR 2009, pp. 784–785 (2009)

2. Bird, S.: Nltk: The natural language toolkit. In: COLING, COLING-ACL 2006, pp. 69–72. Association for Computational Linguistics (2006)
3. Brants, T.: Inter-annotator agreement for a german newspaper corpus. In: Proceedings of Second International Conference on Language Resources and Evaluation, LREC 2000 (2000)
4. Cohen, J.: A Coefficient of Agreement for Nominal Scales. Educational and Psychological Measurement 20(1), 37 (1960)
5. Hsueh, P.-Y., Melville, P., Sindhwani, V.: Data quality from crowdsourcing: A study of annotation selection criteria. In: Proceedings of the NAACL HLT 2009 Workshop on Active Learning for Natural Language Processing, HLT 2009, Stroudsburg, PA, USA, pp. 27–35. Association for Computational Linguistics (2009)
6. Malkowski, S., Hedwig, M., Pu, C.: Experimental evaluation of n-tier systems: Observation and analysis of multi-bottlenecks. In: IEEE International Symposium on Workload Characterization, IISWC 2009, pp. 118–127. IEEE (2009)
7. Mason, W., Watts, D.J.: Financial incentives and the "performance of crowds". SIGKDD Explor. Newsl. 11(2), 100–108 (2010)
8. Navarro, G.: A guided tour to approximate string matching. ACM Computing Surveys 33(2001) (1999)
9. Ng, H.T., Yong, C., Foo, K.S.: A case study on Inter-Annotator agreement for word sense disambiguation. In: Proceedings of the ACL SIGLEX Workshop on Standardizing Lexical Resources (SIGLEX 1999). College Park, Maryland (1999)
10. Nowak, S., Rüger, S.: How reliable are annotations via crowdsourcing: A study about inter-annotator agreement for multi-label image annotation. In: Proceedings of the International Conference on Multimedia Information Retrieval, MIR 2010, pp. 557–566. ACM, New York (2010)
11. Passonneau, R., Habash, N.Y., Rambow, O.: Inter-annotator agreement on a multilingual semantic annotation task. In: Proceedings of the Fifth International Conference on Language Resources and Evaluation, LREC (2006)
12. Sabou, M., Bontcheva, K., Scharl, A.: Crowdsourcing research opportunities: Lessons from natural language processing. In: Proceedings of the 12th International Conference on Knowledge Management and Knowledge Technologies, I-KNOW 2012, pp. 17:1–17:8. ACM (2012)
13. Wang, A., Hoang, V.C.D., Kan, M.-Y.: Perspectives on crowdsourcing annotations for natural language processing. Language Resources and Evaluation 47(1), 9–31 (2013)

Identification of Brazilian Portuguese Causative Verbs through a Weighted Graph Classification Strategy

Brett Drury, Rafael Geraldeli Rossi, and Alneu de Andrade Lopes

ICMC, University of São Paulo,
Av. Trabalhador São Carlense 400
São Carlos, SP, Brazil
C.P. 668, CEP 13560-970
http://www.icmc.usp.br/

Abstract. Causative verbs can assist in the identification of causative relations. Portuguese has a large number of verbs that would make the manual labelling of causative verbs an manually expensive task. This paper presents a classification strategy which uses the characteristics of causative verbs co-occurring with common nouns to classify Brazilian Portuguese verbs as either: causative or non-causative. The strategy constructs a graph where verbs extracted from text are nodes. The verbs are connected if the verbs co-occur with common nouns. The classification strategy uses the unique characteristics of links between: 1. causative verbs, 2. causative verbs and non-causative verbs and 3. non-causative verbs to predict a label (causative or non-causative) for unlabelled verbs. The proposed strategy significantly outperforms a baseline and supervised learning strategies.

Keywords: Causative Verbs, Weighted Graph, Verb Classification.

1 Introduction

Causative verbs can assist in the identification of causative relations, however the manual recognition of causative verbs in Portuguese text is a non-trivial task. A weakly or semi-supervised classification approach could reduce the manual labelling effort.

This paper presents a weighted graph classification strategy which identifies causative and non-causative verbs from Brazilian Portuguese. The strategy relies upon: 1. a small hand annotated set of sentences and a 2. larger set of sentences which have no manual annotation. The strategy has five steps which are: 1. label a small number of causative and non-causative verbs, 2. extract labelled and unlabelled verbs and nouns from text, 3. construct graph from verbs and nouns, 4. remove nouns from graph to produce a graph of verbs and 5. classify unlabelled verbs in graph.

The remainder of paper will describe the following: 1. literature review, 2. description of strategy, 3. experiments and 4. conclusion.

J. Baptista et al. (Eds.): PROPOR 2014, LNAI 8775, pp. 274–279, 2014.

2   Literature Review

The literature review will discuss: 1. the nature of causative verbs and 2. graph classification strategies.

A possible definition of a causative verb is *a verb that in its transitive state express a causal relation between the subject and object or prepositional phrase of the verb* [6]. For example in the phrase, *Chuva irregular provoca diferenas nas lavouras de cana* (Irregular rain provokes differences in the cane fields) , the subject *Chuva irregular* has a causal relationship with the object *lavouras de cana*. The causal relation is provided by the the causal verb *provoca*.

Causative verbs have been categorized into three groups: simple, resultative and instrumental [4]. The groups reflect their members role in expressing a causal relationship between a subject and a object [4]. The simple causative category refers to verbs which provide a casual link between a subject and an object without reference to the cause or effects, for example *causar* (causes) and *provocar* (provokes). A resultative causative verb provides the causative link and the effect between the subject and the object, for example *reduzir* (reduces). Instrumental causative verbs provide the causal link and part of the cause, for example *envenenar* (to poison).

Causative verbs can be identified by the nouns which form part of the argument structure of the verb [7]. The hypothesis behind using verb argument structure to identify causative verbs is that they cluster around common nouns[7]. A common strategy to identify causative verbs through their association with common nouns is to construct a graph where nodes are verbs and nouns. The verbs and nouns are linked if the noun forms part of the argument structure of the verb. The verbs are classified with an association measure calculated through the characteristics of the links between the nodes [11]. This type of analysis can be referred to as *link analysis* [3]. An example of classification using link analysis is the 'LabelRank' strategy proposed by Xie and Szymanski[12]. The LabelRank algorithm propagates labels from labelled nodes to their unlabelled neighbours based upon the probability of their links.

Tang and Stevenson [11] identified causative verbs using a "Network Flow" method which weighted links using word frequencies. A distance was computed between two sub-graphs on the weighted graph which allowed for: 1. "ontological distance" between the component concepts and 2. their frequency distributions [11]. The "Network Flow" strategy superseded the "Semantic Distance Measure" which was proposed by [10]. The Semantic Distance Measure inferred word senses of nodes in WordNet based upon a weighted distance which calculated from probability distributions from a path which links word senses through the WordNet[8] hierarchy[10].

Classification of verbs is not limited to graph based methods and there have been attempts to use feature based strategies. For example, [5] proposed using semantic slots to select features to classify verbs into general categories. In addition to feature based methods clustering has been used to identify verb classes[9].

The strategy proposed by [9] introduced a small number of verbs with a verb class and clustered a larger number of verbs with no labels. The labels were propagated from the labelled examples to the unlabelled verbs.

3   Description of Strategy

The strategy is based upon the hypothesis that causative verbs co-occur with a limited number of nouns when compared with transitive non-causative verbs. We hypothesize that there are limited numbers of causes (subject) and effects (object) in any given corpus, and consequently causative verbs will co-occur frequently will these small number of nouns.

The strategy has five distinct steps: 1. label a small number of causative and non-causative verbs, 2. extract labelled and unlabelled verbs and nouns from text, 3. construct graph from verbs and nouns, 4. remove nouns from graph to produce a graph of verbs and 5. classify unlabelled verbs in graph.

The strategy used a corpus of agricultural texts which were scrapped from the Internet. The corpus contained 135080 sentences. These sentences has POS information added to it by the Aelius Tagger [2]. The proposed strategy requires a small amount of labelled data which labels a verb in a phrase as either: causative or non-causative. The labelling process presented a random sentence to an annotator who labels the verbs in the sentence as causative or non-causative. There were 285 sentences which were identified as containing causative verbs and 171 sentences which were labelled as non-causative.

The next step is a create a heterogeneous graph of nouns and verbs from both the unlabelled and labelled data. The rational of this step is for the nouns to act as a bridge between verbs. The nouns are extracted from within a fixed distance (window) of the verb. We will refer to this distance as WEZ (*window extraction size*). The co-occurring verbs and nouns are connected to form a heterogeneous graph. If a verb or noun is a stop word it is removed from the graph. Stop words are removed to minimize a characteristic of Portuguese where a non-causative verb can co-occur with the infinitive form of another verb which may be either: causative or non-causative.

The third step is to transform the heterogeneous graph of nouns and verbs to a graph of verbs. This is achieved by removing the nouns, and joining the verbs which are connected to the noun. This step is fundamental to the proposed method as it allows us to evaluate the strength of the connection in the graph between pairs of verbs. The strength of the connection between a pair of verbs (NodeA, NodeB) is estimated by Equation 1. The higher the value produced by Equation 1 the stronger the affinity between the two verbs. This value is central to our proposed method because we estimate that the affinity between causative verbs will be stronger than: 1. the affinity between non-causative verbs and 2. between non-causative and causative verbs. The links between pairs of verbs are replaced by one bi-directional link with the value calculated by Equation 1, we will refer to this value as the relative link value.

$$RL = \frac{1}{2} \left(\frac{CLa}{TLa} + \frac{CLb}{TLb} \right) \tag{1}$$

CLa = Number of out-links to Node A from Node B, TLa= Total number of out-links for Node A, CLb = Number of out-links to Node B from Node A, TLb= Total number of out-links for Node B.

The last step in our proposed method is to classify the unlabelled verbs in the verb graph. Unlabelled verbs may have links to: 1. unlabelled verbs, 2. labelled causative verbs or 3.non-causative verbs. The classification strategy for an unlabelled verb relies upon the difference between its relative link values for connections to labelled causative verbs and its relative link values for connections to unlabelled causative verbs.

The rational for this classification strategy is based on a hypothesis for link characteristics between: 1. causative verb to causative verb (CC), 2. causative verb to non-causative verb (CN) and 3. non-causative verb to non-causative verb (NN). Our hypothesis is that CC relationship will have a significantly higher relative-link values than either the: CN or NN relationships.

We do not prescribe a statistical measure for estimating the difference between the distributions of relative-link values, but standard statistical tests can applied such as student t-test.

4   Experiments

The evaluation experiments were in two stages: 1. determine the best configuration of variables for the proposed strategy and 2. evaluate the best configuration of the proposed against competing classification methods.

The configuration experiments evaluated the effect of the following variables: 1. comparison of statistical measure to compute relative-link values sum, student t-test, and mean), 2. WEZ values, 3. maximum number of verb-noun pairs used in graph construction and 4. minimum co-occurrence frequency. The experiments were 10 X 80:20 holdout evaluation which used: labelled data for evaluation and in-conjunction with labelled data to construct the graph. The hold-out evaluation used a mean accuracy to judge the effectiveness of each configuration.

The configuration experiments demonstrated that: 1. the mean was the most effective statistical measure for comparing the distributions of relative link values, 2. WEZ of 1 was the most effective configuration, 3. the number of noun-pairs did not effect the accuracy figure and 4. minimum co-occurrence frequency of 1 was the most effective.

4.1   Comparison of Proposed Strategy with Competing Strategies

This experiment compared the best configuration of the proposed strategy with competing methods. There were three competing methods: supervised learning with Naive Bayes and Entropy Maximization classifier, and a baseline which randomly labels verbs as either causative or non-causative. The supervised learning strategies treated the process as a classification task where a verb would be classified as either causative or non-causative.

The experiment used a 80:20 hold-out evaluation similar to the previous experiment, but the evaluation was run a 1000 times. The supervised methods had access to the 80% of labelled data which was randomly selected for an individual iteration of the hold-out evaluation. The features for the supervised learning were: 1. name of verb, 2. its POS tag, and 3. co-occurring nouns within a fixed distance of the verb and their POS tags.

The results for the proposed strategy and its competitors are shown in Table 1 and show a clear advantage for the proposed strategy over the both the baseline and supervised strategy.

Table 1. Comparison of proposed measure with competing strategies by accuracy

Strategy	NC.	C.
Proposed	1.00 ($\pm$0.00)	0.97 ($\pm$0.07)
Supervised (NB)	0.67 ($\pm$0.15)	0.70 ($\pm$0.18)
Supervised (ME)	0.80 ($\pm$0.07)	0.72 ($\pm$0.14)
Baseline	0.50 ($\pm$0.20)	0.50 ($\pm$0.15)

5   Conclusion

The results demonstrate that the proposed strategy produces superior results than: 1. a baseline which randomly labels verbs as causative or non-causative and 2. a supervised learning strategy. The best configuration of the proposed strategy produces far superior results than the average of any single configuration. This result suggests that a combination of factors significantly improves the performance of the strategy rather than any single factor.

It is possible that the proposed strategy outperforms the supervised strategies because it relies the unique argument structure of causative verbs and consequently does not require large amounts of training data. The lack of labelled data may explain the relatively poor performance of the supervised strategy (NB classifier).

The priority of any future work is to increase the quantity of evaluation data. We do not have the resources to annotate large amounts of data, therefore we have created a web-site to crowd source the annotation of causative relations. We will release the annotated data to the community.

The proposed strategy is a viable strategy where there is small amounts of labelled data, and therefore may be used in future label propagation strategies.

The work presented in this paper was supported by FAPESP grant no: 11/20451-1 and 2011/12823-6.

References

1. Bird, S.: Nltk: The natural language toolkit. In: Proceedings of the ACL Workshop on Effective Tools and Methodologies for Teaching Natural Language Processing and Computational Linguistics. Association for Computational Linguistics, Philadelphia (2002)
2. de Alencar, L.F.: Aelius: Uma ferramenta para anotação automática de corpora usando o nltk. In: ELC 2010, The 9th Brazilian Corpus Linguistics Meeting (2010)
3. Getoor, L., Diehl, C.P.: Link mining: A survey. SIGKDD Explor. Newsl. 7(2), 3–12 (2005)
4. Girju, R.: Automatic detection of causal relations for question answering. In: Proceedings of the ACL 2003 Workshop on Multilingual Summarization and Question Answering, vol. 12, pp. 76–83. Association for Computational Linguistics (2003)
5. Joanis, E., Stevenson, S., James, D.: A general feature space for automatic verb classification. Natural Language Engineering 14(3), 337–367 (2008)
6. Khoo, C.S.G., Chan, S., Niu, Y.: Extracting causal knowledge from a medical database using graphical patterns. In: Proceedings of the 38th Annual Meeting on Association for Computational Linguistics, ACL 2000, pp. 336–343. ssociation for Computational Linguistics (2000)
7. Merlo, P., Stevenson, S.: Automatic verb classification based on statistical distributions of argument structure. Comput. Linguist. 27(3), 373–408 (2001)
8. Miller, G.A.: Wordnet: A lexical database for english. Communications of the ACM 38, 39–41 (1995)
9. Stevenson, S., Joanis, E.: Semi-supervised verb class discovery using noisy features. In: Proceedings of the Seventh Conference on Natural Language Learning (2003)
10. Tsang, V., Stevenson, S.: Calculating semantic distance between word sense probability distributions. In: Proceedings of the Eighth Conference on Computational Natural Language Learning (CoNLL 2004), pp. 81–88 (2004)
11. Tsang, V., Stevenson, S.: A graph-theoretic framework for semantic distance. Comput. Linguist. 36(1), 31–69 (2010)
12. Xie, J., Szymanski, B.K.: Labelrank: A stabilized label propagation algorithm for community detection in networks. CoRR, abs/1303.0868 (2013)

Lausanne: A Framework for Collaborative Online NLP Experiments

Douglas Iacovelli, Michelle Reis Galindo, and Ivandré Paraboni

School of Arts, Sciences and Humanities, University of São Paulo (EACH / USP)
Av. Arlindo Bettio, 1000 - São Paulo, Brazil
{douglas.iacovelli,michelle.galindo,ivandre}@usp.br

Abstract. This paper introduces Lausanne - a tool for collaborative online NLP experiments. Lausanne has been successfully applied to the implementation of a practical experiment to collect human-produced natural language descriptions, and it is freely available for research purposes.

Keywords: Data collection, Experimental NLP.

1 Introduction

Text corpora are central to many Natural Language Processing (NLP) studies, including both language interpretation and generation. In some cases, however, knowing the surface strings that represent the outcome of a communication process may not be sufficient, and it may be necessary to obtain their non-linguistic context as well. For instance, in order to understand how humans give directions (e.g., 'turn left') it may be necessary to consider not only a text corpus, but also the visual context perceived by the speaker.

In NLP and related fields, a common way of acquiring information beyond text is by making use of controlled experiments involving human subjects. In experiments of this kind, stimulus materials (e.g., text, pictures etc.) are presented in a controlled context, and subject's reactions (e.g., written or spoken answers, mouse clicks, response times, etc.) are recorded.

A particularly useful kind of NLP experiment takes the form of a collaborative task, i.e., involving two subjects - a speaker and a hearer - in a dialogue situation. In this case, an initial stimulus (e.g., text or image) is presented to the speaker to elicit a response (e.g., a textual description). This information is sent to the hearer and will, in turn, elicit a linguistic or non-linguistic response that may be used as feedback to the speaker.

Implementing a collaborative experiment setting may be a labour-intensive task, with little opportunity for reuse in subsequent experiments. As a first step to mitigate some of these difficulties, this papers introduces *Lausanne*, a reusable framework for the development of collaborative online NLP experiments, and its application in a practical research project.

J. Baptista et al. (Eds.): PROPOR 2014, LNAI 8775, pp. 280–285, 2014.

2   Background

We are not aware of any reusable framework for the design of NLP experiments in the literature. In what follows we discuss related work reporting the kinds of experiment that we may implement using the *Lausanne* framework. These may be divided into three categories: hearer-oriented, speaker-oriented and collaborative settings.

Speaker-oriented experiment settings elicit linguistic responses based on a given context (e.g., text or images) presented as stimuli. Experiments of this kind are arguably simple to implement (i.e., basically consisting of software for displaying the stimuli and storing the input data) and are particularly popular in studies on language production. Examples of corpora collected in this way incude *TUNA* [1], *GRE3D3* and *GRE3D7* [2,3], *Drawer* [4], *Stars* [5], and others.

Hearer-oriented experiment settings elicit non-linguistic responses (e.g., mouse clicks, actions, response times etc.) based on the context provided. Depending on the kind of information to be collected, implementation may however become costly. For instance, collecting navigation steps in a virtual world will require custom software. Experiments of this kind are common in language production and interpretation alike. Examples of non-linguistic data sets collected in this way include experiments on reference resolution [6] and interpretation [7], the use of vague language [8], navigation on web pages [9], and instruction-giving in virtual worlds [10].

Both speaker- and hearer-oriented experiment settings constitute valuable data collection techniques for NLP research, but human-computer interaction may not always represent a sufficiently natural model of human-to-human communication. A more robust alternative for these cases would be the use of a collaborative experiment setting involving both a human speaker and a human hearer as in Figure 1.

hearer   **speaker**

Fig. 1. A collaborative experiment setting

Collaborative experiment settings are useful for studies on language production, interpretation, or both. Implementation may however be significantly more complex than the previous cases, requiring communication facilities between participant pairs, and distinct interfaces for speaker and hearer. Examples of corpora collected in this way include *iMap* [11], and *GIVE-2* [12].

3   Current Work

3.1   Overview

Lausanne has been developed as a framework for the design of NLP experiments for data collection in the study of language production and interpretation. Although our primary focus is on the collaborative setting, simpler speaker- and hearer-oriented experiments may be easily deployed by setting the appropriate configuration.

Communication between speaker and hearer has been implemented through online message exchange provided by Meteor[1] and stored on MongoDB database[2]. Experiment participants may therefore be geographically distant from each other (e.g., in different institutions). The user interface layer was built using *HTML5*, *CSS3* and Javascript, allowing researchers to design experiments to run on ordinary desktop computers and popular mobile and TV sets.

The design of an experiment using *Lausanne* consists of (a) providing the desired stimuli (e.g., a collection of images to be presented to the participants, including a number of practice instances to be presented before the actual experiment), (b) providing the expected answers to be provided by speaker and hearer, and (c) setting a number of parameters intended to model how the experiment should proceed. For instance, we may choose the order in which the stimuli are presented, or the action to be performed when the hearer selects the wrong answer, among many other possibilities.

In the current version of the system, much of the configuration work still requires a certain level of hard coding. Most changes are however simple or even intuitive, and are generally indicated in the system code and documentation.

Running an experiment involves recruiting a participant pair at an agreed time, and letting them log onto the system by means of a unique password for each trial. Speaker and hearer follow the instructions provided on screen and, upon completion of the trial, may be requested to switch roles and start all over again, if desired.

The output of an experiment is a text table conveying all details of the experiment and the data collected at every step performed by each participant. In order to make the discussion more concrete, in the next section we will illustrate a number of features of *Lausanne* through an example of practical experiment that has been developed using the framework.

3.2   Lausanne in Practice: The Stars2 Experiment

Lausanne has been originally developed for the design of an online collaborative experiment in language production called *Stars2*. The experiment was successfully implemented and, subsequently, *Lausanne* has been expanded and generalised in a number of ways to facilitate reuse.

---

[1] http://www.meteor.com/

[2] http://www.mongodb.org/

Stars2 was an online experiment designed to test a number of hypotheses concerning how human speakers describe geometric objects in simple visual scenes. The experimental conditions and underlying hypotheses in *Stars2* are beyond the scope of this paper and will be described elsewhere. Data of this kind is intended to inform referring expression generation algorithms as in [13]. In what follows we focus on how the experiment was configured and carried out.

Figure 2 shows examples of speaker and hearer screens. Speaker are instructed to describe an object pointed by an arrow, which is not shown on the hearer screen. Instead, the hearer screen shows object labels, and hearers are required to select the letter that corresponds to the text message received from the speaker.

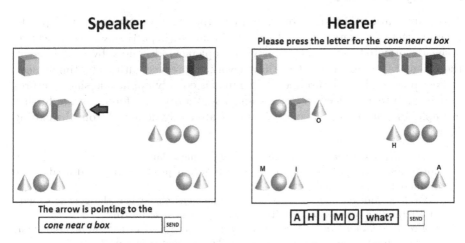

Fig. 2. Speaker and hearer screens used in the Stars2 experiment

The goal of the *Stars2* experiment was to collect descriptions provided by the speaker under conditions of interest. The role of the hearer was limited to validating the descriptions provided, so as to minimize ambiguity and other undesired effects that are common in speaker-oriented experiments.

In *Stars2* we chose to present the stimuli in random order. In addition to that, *Lausanne* allows the implementation of several kinds of variation to disguise the purpose of the experiment or to prevent undesired learning effects. For instance, each image may be presented in reverse, using multiple sets of object and different colour schemes. Variation, if allowed, may be randomized and evenly distributed to guarantee a completely balanced experiment setting. For instance, the system may be set to randomly display exactly half stimuli images in normal orientation, and the other half in reverse orientation.

In the present setting the speaker provides a description of the object on screen, which the hearer may or may not be able to identify. If necessary, the hearer may press a 'what?' button to request the speaker to clarify. Should the hearer choose the wrong answer, the corresponding condition will be reintroduced at a subsequent stage (possibly in a different layout). All these policies were specifically tailored for the *Stars2* experiment and may be reconfigured.

The system makes a number of checks to ensure that descriptions provided by the speaker are minimally consistent with the experiment rules. For instance, the system issues a warning for overly brief descriptions as these were unlikely to occur in the *Stars2* domain. Rules of this kind are not configured by parameters, and will require a certain amount of Javascript programming to be implemented.

The experiment ran for a 4-months period, during which 76 participants produced 1216 definite descriptions in Portuguese. Among these, only 12 descriptions were discarded due to ambiguity. This result, in our view, makes a strong case for the use of a collaborative setting as a means to collect high-quality data.

3.3   Beyond Stars2

There are many other kinds of experiment that may be envisaged and put to practice using *Lausanne*. For a start, recall that single participant (i.e., speaker- or hearer-oriented) experiments may be implemented simply by switching off the collaborative function of the framework. Moreover, although stimuli need to be represented as picture files, text input is equally applicable since pictures may display text (indeed in any colour or font size). The following are further examples of routine NLP experiments that are easily deployed from the exiting framework:

- *Evaluating a system that outputs text*: e.g., using a language generator [14,15], a document summarizer or any other NLP tool that produces text and needs to be evaluated by human subjects.
- *Collecting data on how humans produce language*: e.g., when the goal is to gain understanding on how to describe objects in a visual scene [16].
- *Collecting data on how humans interpret language*: e.g., if we present an ambiguous instruction on screen, will the speaker make the right decision?
- *Text or image annotation*: e.g., the speaker may be required to translate sentences from one language to another and, if in collaborative mode, the translations may be validated by the hearer.

4   Conclusions

This paper has introduced *Lausanne* - a simple tool for collaborative online NLP experiments. *Lausanne* has been successfully applied to the implementation of a practical experiment to collect human-produced natural language descriptions, and it is now freely available for download and reuse for research purposes.

Although in the present release a certain amount of hard coding is still required, we expect *Lausanne* to be sufficiently general for a number of standard NLP tasks on data collection for the study of language production and/or interpretation. As future work, we intend to make further adjustments to improve reusability and documentation, and run a follow-up experiment that takes advantage of the existing framework.

Acknowledgments. The authors acknowledge support by FAPESP and the University of São Paulo.

References

1. Gatt, A., van der Sluis, I., van Deemter, K.: Evaluating algorithms for the generation of referring expressions using a balanced corpus. In: 11th European Workshop on Natural Language Generation, ENLG 2007 (2007)
2. Dale, R., Viethen, J.: Referring expression generation through attribute-based heuristics. In: Proceedings of the 12th European Workshop on Natural Language Generation, ENLG 2009, pp. 58–65. Association for Computational Linguistics, Stroudsburg (2009)
3. Viethen, J., Dale, R.: GRE3D7: A corpus of distinguishing descriptions for objects in visual scenes. In: Proceedings of the UCNLG+Eval: Language Generation and Evaluation Workshop, Edinburgh, Scotland, pp. 12–22 (July 2011)
4. Viethen, J., Dale, R.: Algorithms for generating referring expressions: Do they do what people do? In: Proceedings of the 4th International Conference on Natural Language Generation (INLG), Sydney, Australia, pp. 63–70 (2006)
5. Teixeira, C.V.M., Paraboni, I., da Silva, A.S.R., Yamasaki, A.K.: Generating relational descriptions involving mutual disambiguation. In: Gelbukh, A. (ed.) CICLing 2014, Part I. LNCS, vol. 8403, pp. 492–502. Springer, Heidelberg (2014)
6. Paraboni, I.: Uma arquitetura para a resolução de referências pronominais possessivas no processamento de textos em língua portuguesa. Master's thesis. PUCRS, Porto Alegre (1997)
7. Paraboni, I.: Generating references in hierarchical domains: The case of Document Deixis. PhD thesis. University of Brighton (2003)
8. van Deemter, K.: Finetuning NLG through experiments with human subjects: The case of vague descriptions. In: Belz, A., Evans, R., Piwek, P. (eds.) INLG 2004. LNCS (LNAI), vol. 3123, pp. 31–40. Springer, Heidelberg (2004)
9. Paraboni, I., Masthoff, J., van Deemter, K.: Overspecified reference in hierarchical domains: Measuring the benefits for readers. In: Proceedings of the Fourth International Natural Language Generation Conference, pp. 55–62 (2006)
10. Paraboni, I., van Deemter, K.: Reference and the facilitation of search in spatial domains. Language and Cognitive Processes (online) (2013)
11. Guhe, M.: Generating referring expressions with a cognitive model. In: Proceedings of PRE-CogSci 2009: Production of Referring Expressions: Bridging the Gap Between Computational and Empirical Approaches to Reference (2009)
12. Gargett, A., Garoufi, K., Koller, A., Striegnitz, K.: The GIVE-2 corpus of giving instructions in virtual environments. In: Proceedings of LREC 2010 (2010)
13. Ferreira, T.C., Paraboni, I.: Classification-based referring expression generation. In: Gelbukh, A. (ed.) CICLing 2014, Part I. LNCS, vol. 8403, pp. 481–491. Springer, Heidelberg (2014)
14. Pereira, D.B., Paraboni, I.: Statistical surface realisation of portuguese referring expressions. In: Nordström, B., Ranta, A. (eds.) GoTAL 2008. LNCS (LNAI), vol. 5221, pp. 383–392. Springer, Heidelberg (2008)
15. de Novais, E.M., Paraboni, I.: Portuguese text generation using factored language models. Journal of the Brazilian Computer Society, 1–12 (2012)
16. de Lucena, D.J., Pereira, D.B., Paraboni, I.: From semantic properties to surface text: The generation of domain object descriptions. Inteligencia Artificial, Revista Iberoamericana de Inteligencia Artificial 14(45), 48–58 (2010)

Alignment-Based Sentence Position Policy in a News Corpus for Multi-document Summarization

Fernando Antônio Asevedo Nóbrega[1], Verônica Agostini[1], Renata T. Camargo[2],
Ariani Di Felippo[2], Thiago Alexandre Salgueiro Pardo[1]

Núcleo Interinstitucional de Linguística Computacional (NILC)
[1] Instituto de Ciências Matemáticas e de Computação, Universidade de São Paulo
[2] Departamento de Letras, Universidade Federal de São Carlos
{fasevedo,agostini,taspardo}@icmc.usp.br,
renatatironi@hotmail.com, arianidf@gmail.com

Abstract. This paper presents an empirical investigation of sentence position relevance in a corpus of news texts for generating abstractive multi-document summaries. Differently from previous work, we propose to use text-summary alignment information to compute sentence relevance.

1 Introduction

Multi-Document Summarization (MDS) is the task of automatically producing a unique summary from a group of source texts (documents) on the same topic [11][14]. It is a relatively new area (dating back to 1995 [13]) and brings old and well-known scientific challenges from the first studies in summarization in the 50s as well as introduces new and exciting challenges, e.g., to deal with redundant, complementary and contradictory information, to normalize different writing styles and referring expression, to balance different perspectives and sides of the same events and facts, to properly deal with evolving events and their narration in different moments, and to arrange information pieces from different texts to produce coherent and cohesive summaries, among others.

MDS, as many other Natural Language Processing tasks, may benefit from specialized corpora, as the ones built for the tasks of the Text Analysis Conferences. Such corpora usually contain large groups of source texts and human summaries, subsidizing researches on the nature and the phenomena that happen in summaries as well as allowing the development/training and comparative evaluation of state-of-the-art summarization systems.

In this paper, we report an empirical study of sentence position relevance for summarization, using a corpus of news texts and their abstractive multi-document summaries – the CSTNews corpus [1][5] – to learn summarization preferences, building on some previous work [10][9]. Giving one more step from where these works stopped, we use one of the corpus available annotations – the text-summary alignment information – to determine, in a more precise way, a robust sentence

J. Baptista et al. (Eds.): PROPOR 2014, LNAI 8775, pp. 286–291, 2014.

position policy to the selection of sentences that may compose the summaries. For now, we are only worried on characterizing such policy, in a theoretical perspective. To the best of our knowledge, this is the first attempt carried out for a corpus in Brazilian Portuguese.

This paper is organized as follows. Section 2 introduces the CSTNews corpus and its annotation layers. Section 3 presents our study on sentence position and the achieved results.

2   The CSTNews Corpus

The CSTNews corpus [1][5] is a reference corpus for MDS composed of 50 clusters of news texts in Brazilian Portuguese (BP). Each cluster contains two or three source texts on the same topic, which were manually selected from on-line mainstream Brazilian news agencies as *Folha de São Paulo, Estadão, O Globo, Gazeta do Povo,* and *Jornal do Brasil.* Besides the original texts, each cluster conveys a manual (abstractive) single-document summary (with 30% compression rate) for each document in the cluster, a manual (abstractive) multi-document summary for the cluster and its corresponding manual extractive summary, and an automatic multi-document summary, produced by a state-of-the-art system for Portuguese [7].

The corpus also has annotated versions of the source texts and multi-document summaries in different linguistic levels, and according to different linguistic theories and models. Specifically, the source texts are manually annotated in different ways for discourse organization, following both the Rhetorical Structure Theory [12] and Cross-document Structure Theory [16]. They also have other manual annotations: their temporal expressions annotated and resolved, their most frequent nouns indexed to their corresponding senses in Princeton Wordnet, and subtopic segmentations and the keywords for each subtopic. Automatic annotations are also available, as the syntactical analyses for each sentence, produced by the PALAVRAS parser [3].

More recently, the corpus had the source text sentences aligned to the sentences of the manual multi-document summaries that shared some information with the formers. Therefore, since each summary comes from 1 or more texts, each sentence in the summaries might be aligned to more than 1 sentence in the texts. Most of the sentences in the summaries were aligned up to 5 sentences from the texts. In general, 42% of the sentences of the texts were aligned to some sentence in the summaries. As an example of alignment, in Table 1 we show a sentence in a summary that was aligned to 2 sentences from different source texts (translated from Portuguese):

Table 1. Example of alignment

Sentence from the summary	Sentences from source texts
Brazil will not be part of the torch relay, which includes 20 countries.	The torch will pass by twenty countries, but Brazil is not in the Olympic way.
	Brazil is not part of the path of the Olympic torch.

The alignment was performed by two computational linguistics, showing a 0.831 inter-annotator agreement kappa value [6], indicating that the annotation is reliable.

3   Sentence Position Policy

In [10] it was defined what was called "sentence position policy" for summary composition. The authors attributed a score for each sentence position in the texts and ranked the sentence relevance in terms of this score. Therefore, a good summary should be composed of the sentences from the best ranked positions. To compute the score of each sentence position, the authors counted the number of topic keywords in all the sentences in each specific position in a group of texts and averaged such number by the number of sentences in that position. They evaluated the resulting sentence position policy for single-document summarization and achieved good results. [9] produced more refined results by counting and averaging, for each sentence position, the number of Summary Content Units (SCUs) in the sentences. The SCUs were those available according to the pyramid method [15]. The authors evaluated the resulting policy for multi-document summarization and produced state of the art results. The above works demonstrated that position policies are worthy to pursue for corpus characterization and summarization.

In this paper, we build on the previous work by refining even more the calculation of the sentence position policy. We use the text-summary alignment information in the CSTNews corpus to better compute the sentence position rank for multi-document summarization. For each sentence position, we count and average the number of alignments they have with the corresponding manual multi-document summary.

Alignments may be more informative than topic keywords or SCUs for the envisioned task. While topic keywords are at the lexical level and SCUs are more conceptual, the alignments may represent any of this information. We consider two versions of the alignment-based policy: one counting the total number of alignments among the sentences and other counting only once the alignments for a pair of sentences, does not mattering how many alignments they have.

For comparison purposes, we also created an alternative, and simpler, position policy. As our corpus does not present topic keywords or pyramid SCUs, we used the own words (excluding stopwords and punctuation marks) in the manual multi-document summaries to score each sentence position. In this case, each sentence position is scored as the average of different words from the summary that it contains.

Figures 1 and 2 show the graphics for sentence position policy by counting summary words, using the absolute sentence positions and their normalized versions, respectively. The normalized graphic allows to make text sizes uniform, ranging from 0 to 1, and, therefore, resulting in fairer analyses. In each graphic, the blue line represents the average values for each sentence position, while the green and red lines incorporate the standard deviation above and below the blue line, respectively.

One may see from the graphics that the first sentences are more relevant than the others. There is also a variation among the positions 20 and 30, approximately, accompanied by a variation of the standard deviation too, showing that it is not possible to fully trust in such specific sentence positions to compose good summaries. It is also possible to realize that the normalized sentence positions show the same behavior of the non-normalized version.

Figures 3 and 4 show the graphics for the sentence position policy with the two versions of alignment information – considering the alignments only once for each sentence position and all the alignments for each sentence, respectively. In relation to the graphics for the summary words, similar behavior may be observed for the alignments. However, one may notice that the curves fall softer, indicating that the first sentences of the texts have more relevance than the others, but are closer to one another than the word-based computation revealed. We do not show the normalized versions of the graphics because they also show similar behavior.

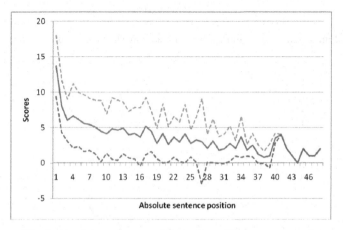

Fig. 1. Absolute sentence position policy with summary words

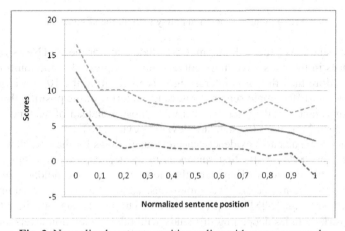

Fig. 2. Normalized sentence position policy with summary words

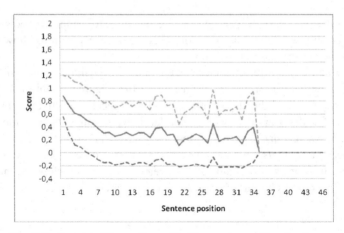

Fig. 3. Sentence position policy with alignments considered only once

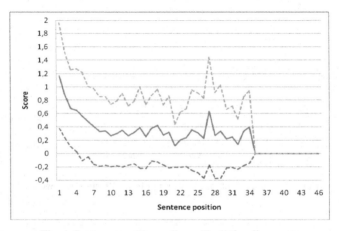

Fig. 4. Sentence position policy with all the alignments

From this study, as expected, one may conclude that, for the CSTNews corpus, the first sentences in the texts have more content that is expressed in the multi-document summaries. More than this, it is interesting to see that, from the 10th position on, it is more difficult to differentiate sentence relevance in terms of the position. In fact, after the most important initial sentences, there is almost a plateau of sentence relevance marked by disturbing areas (as standard deviation evidences).

The achieved results also evidence our previous findings for the CSTNews. In [4], it was verified that 89% of the first sentences in the source texts were aligned to the sentences in the summary. They are, therefore, very good candidate sentences to compose multi-document extractive summaries, as several works on summarization have showed (see, e.g., [2][8]) and [9] demonstrated to produce state-of-the-art results.

Future work includes deepening this study with other annotation layers available in the corpus (as the discourse annotations) and applying these strategies to produce automatic summarization systems, which would probably be strong baselines in the area.

Acknowledgments. The authors are grateful to FAPESP, CAPES and CNPq for supporting this work.

References

[1] Aleixo, P., Pardo, T.A.S.: CSTNews: Um Córpus de Textos Jornalísticos Anotados segundo a Teoria Discursiva Multidocumento CST (Cross-document Structure Theory). ICMC-USP Technical Report N. 326, p.12 (2008)

[2] Baxendale, P.B.: Machine-made index for technical literature – an experiment. IBM Journal, 354–361 (1958)

[3] Bick, E.: The Parsing System Palavras - Automatic Grammatical Analysis of Portuguese in a Constraint Grammar Famework. PhD Thesis. Aarhus University Press (2000)

[4] Camargo, R.T.: Investigação de Estratégias de Sumarização Humana Multidocumento. MSc Dissertation. Departamento de Letras, Universidade Federal de São Carlos, p.133 (2013)

[5] Cardoso, P.C.F., Maziero, E.G., Jorge, M.L.C., Seno, E.M.R., Di Felippo, A., Rino, L.H.M., Nunes, M.G.V., Pardo, T.A.S.: CSTNews - A Discourse-Annotated Corpus for Single and Multi-Document Summarization of News Texts in Brazilian Portuguese. In: Proceedings of the 3rd RST Brazilian Meeting, pp. 88–105 (2011)

[6] Carletta, J.: Assessing Agreement on Classification Tasks: The Kappa Statistic. Computational Linguistics 22(2), 249–254 (1996)

[7] Castro Jorge, M.L.R., Pardo, T.A.S.: Experiments with CST-based Multidocument Summarization. In: Proceedings of the ACL Workshop TextGraphs-5: Graph-based Methods for Natural Language Processing, pp. 74–82 (2010)

[8] Edmundson, H.P.: New methods in automatic extracting. Journal of the ACM 16(2), 264–285 (1969)

[9] Katragadda, R., Pingali, P., Varma, V.: Sentence Position revisited: A robust light-weight Update Summarization 'baseline' Algorithm. In: Proceedings of the Third International Cross Lingual Information Access Workshop, pp. 46–52 (2009)

[10] Lin, C.Y., Hovy, E.: Identifying Topics by Position. In: Proceedings of the Fifth Conference on Applied Natural Language Processing, pp. 283–290 (1997)

[11] Mani, I.: Automatic Summarization. John Benjamins Publishing Co., Amsterdam (2001)

[12] Mann, W.C., Thompson, S.A.: Rhetorical Structure Theory: A Framework for the Analysis of Texts. ISI Reprint Series ISI/RS-87-190. Information Sciences Institute (1987)

[13] McKeown, K., Radev, D.R.: Generating summaries of multiple news articles. In: Proceedings of the 18th Annual International ACM-SIGIR Conference on Research and Development in Information Retrieval, pp. 74–82 (1995)

[14] Nenkova, A., McKeown, K.: Automatic Summarization. Foundations and Trends in Information Retrieval Series. Now Publishers Inc. (2011)

[15] Nenkova, A., Passonneau, R., McKeown, K.: The pyramid method: Incorporating human content selection variation in summarization evaluation. ACM Transactions on Speech and Language Processing 4(2), 1–23 (2007)

[16] Radev, D.R.: A common theory of information fusion from multiple text sources, step one: Cross-document structure. In: Proceedings of 1st ACL SIGDIAL Workshop on Discourse and Dialogue, pp. 74–83 (2000)

Identification of Related Brazilian Portuguese Verb Groups Using Overlapping Community Detection

Alan Valejo, Brett Drury, Jorge Valverde-Rebaza, and Alneu de Andrade Lopes

ICMC, University of São Paulo,
Av. Trabalhador São Carlense 400
São Carlos, SP, Brazil
C.P. 668, CEP 13560-970
http://www.icmc.usp.br/
{alan,brett.drury,jvalverr,alneu}@icmc.usp.br

Abstract The grouping of related verbs is a mature problem in linguistics and natural language processing. There have been a number of resources which have grouped together English verbs, for example VerbNet. In comparison Portuguese has fewer resources, some of which have been based upon English verb studies. The manual grouping of Portuguese verbs would be a manually intensive task, consequently this paper presents a strategy for grouping Portuguese verbs. The strategy connects verbs through common arguments and uses overlapping community detection algorithm to identify related verbs.

Keywords: Text Clustering, Verb Groups, Community Detection, Overlapping Communities, Complex Networks.

1 Introduction

The clustering of related verbs is a mature task in the study of language. There have been systematic attempts to classify verbs in English [9], but there have been a small number of equivalent studies of Portuguese verbs. A manual approach would be a time intensive approach which may be unfeasible due to the smaller number of researchers working in Portuguese compared to English.

Graphs can represent the interconnections between words in a sentence and therefore groups of related words can have strong interconnections. Strongly connected vertices in a graph are known as communities. Communities of vertices may share similar characteristics and/or play similar roles [6]. Community detection has been used to: predict quality of relationships in social networks [19], identify similar proteins [3] and the identification of word senses [8].

This paper applies the Link Clustering algorithm [1] to detect related groups of verbs in a graph. The results produced by the clustering approach are compared with the synsets from PTE[1]. Our initial study suggests that the use of the

---

[1] http://www.nilc.icmc.usp.br/tep2/

J. Baptista et al. (Eds.): PROPOR 2014, LNAI 8775, pp. 292–297, 2014.

community detection algorithm is a promissor way for identification of related Brazilian Portuguese verb groups.

The remainder of the paper will cover the following: 2. related work, 3. description of method, 4. experimental results and 5. conclusion and future work.

2   Related Work

The seminal work in verb grouping was produced by Beth Levin [9]. She grouped related English verbs into 57 classes. These groups are now known as "Levin Classes". The lexical resource, VerbNet [14], extended the verb classification proposed by Levin. There have been a number of attempts in Portuguese to group verbs. VerbNet-Br [13] was semi-automatic grouping of Portuguese verbs which was based upon the translation of English verb diathesis alternations into Portuguese and existing lexical resources in Portuguese and English. Onto.PT [11] in comparison was constructed automatically. Onto.PT has a general verb synset which contains a number of verb synsets. WordNet-Br [5] is similar to WordNet because it groups nouns, adverbs, etc into synsets. WordNet-Br relied upon a manual approach which used information in pre-existing language resources.

There have been number of attempts to classify related verbs into categories, for example, [7] used semantic slots to extract features for a learner which classified verbs. Stevenson [16] used argument structure of verbs to classify them into groups. A semi-supervised clustering approach has been proposed [15]. The approach use a small set of verbs labelled with a class name. These verbs were clustered with verbs with no labels. The distance measure was based upon argument structure. The labels from the known verbs were propagated to unlabelled verbs within the same group.

There have been attempts to use graph based methods to classify verbs. Sun[17] used a hierarchical graph based clustering method to group verbs. Another graph-based method was proposed by [18] who computed an "ontological distance" between concepts (including verbs) based upon their argument frequency distributions. Graphs which represent "natural" or "real" systems have concentrated groups of edges. These groups of edges can be known as communities [6]. The research in this area concentrates on splitting a set of nodes into disjoint subsets so that the number of edges that connect vertices of partitions is minimized. There are number of algorithms which have designed to do this for example: betweenness [10], and fastgreedy [4], label propagation [12]. A survey of community detection and graph partitioning is given by [6].

3   Method Description

The method has two distinct steps: 1. graph construction and 2. community detection. The method creates undirected graphs. We will first briefly describe the concepts of an undirected graph which will be used in the method description.

An undirected graph is represented by $G = (V, E, W)$ where: 1. V represents the set of vertices, 2. E represents the set of edges, and 3. W the set of relationship weights. An edge of two vertices v and u in an undirected graph can be represented as: $e = \{(v, u) \mid u \in V, v \in V\}$. The weight of the same is represented as: $w(v, u)$. The quantity of edges and vertices is represented as $n = |V|$ and $m = |E|$ respectively, whereas "$|.|$" is the cardanility of the set. The graphs produced in this work are considered undirected, therefore edges are not order dependent and can be represented by a non ordered pair $(v, u) = (u, v)$ $\forall\, v, u \in V$.

The similarity between a pair of vertices (v, u) is defined by score $S_{v,u}$. The basic structural definition of a vertex $v \in V$ in the neighbourhood $\Gamma(v) = \{u \mid (v, u) \in E \lor (u, v) \in E\}$ which displays the set of neighbours of v. Show $\Lambda_{v,u} = \Gamma(u) \cap \Gamma(v)$ the set of neighboring base pair v e u. We will refer to the similarity between a pair of vertices by Equation 1.

$$S_{v,u} = \sum_{cn \in \Lambda_{v,u}} \frac{w(cn, v) + w(cn, u)}{2} \tag{1}$$

3.1   Graph Construction

The graph construction step has two sub-steps: 1. construct a hybrid graph of verbs and their arguments and 2. remove verb arguments to construct a graph of verbs.

The construction of the hybrid graph has a pre-processing step which removes the stopwords and lemmatizes the remaining text. The next step extracts verbs and their alternations from the corpus. Verbs are identified by their POS tag[2]. Their alternations are extracted with a "window". The window is a fixed distance from the target verb from which words are extracted. For example, if the "window" size is 1 then words which are directly to the right and left of the target verb are extracted. The verbs and their alternations are added as nodes to the graph. The verbs are connected to their alternations.

A simple hybrid graph constructed from the sentence "Segundo levantamento do Comite Binacional de Brasiguaios existem hoje 350 mil sem-terra no Paraguai 50% seriam brasileiros." which was extracted from the Mac-Morpho Corpus is demonstrated in Figure 1.

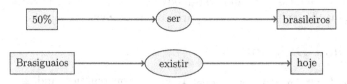

Fig. 1. Simple hybrid graph of verbs and their alternations. The ellipses represent verbs and boxes represent the alternations. The arrows indicate the position of the alternations in relation to the verb.

---

[2] POS tag identify the word class, such as verb or substantive.

The next step removes the verb alternations and connects verbs with common alternations. An example of this process is provided from the sentense "Pastre afirma que a grande safra incentiva os agricultores a investirem na produção" which was extracted from the Mac-Morpho Corpus is demonstrated in Figure 2. The verbs "investir" and "incentivar" have common alternations, and therefore are linked, whilst "afirmar" does not have any common alternations, and therefore is not linked to any of the other verbs.

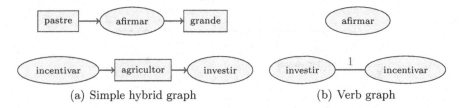

(a) Simple hybrid graph   (b) Verb graph

Fig. 2. Constructing a graph of verbs from the hybrid graph. The verbs "investir" and "incentivar" have a common neighbor, thus they have a edge with weight 1.

The final step in the method is to use community detection algorithms [6] to identify highly connected groups of verbs and interconnected groups. There are no prescribed community algorithm for this step.

4   Experiments

A manual analysis of existing verb group resources demonstrate that a verb on average is a member of two distinct verb groups. In addition small verb groups can be sub-groups of larger verb groups, which can be an indicator of high granularity (Table 1). These observations influenced the choice of community algorithm which had to able to detect communities which overlapped and were hierarchical, consequently we choose the Hierarchical Link Clustering (HCL) algorithm [1].

The experiments in this paper used the Mac-Morpho Corpus which is incorporated in the NLTK library [2]. The results of the experiments were evaluated using the PTE 2.0, which is a thesaurus for Brazilian Portuguese, composed of synsets from Wordnet-Br. An example of a selection of synsets can be found in Table 1.

The experiments used a extraction window size of 1, and each experiment used intervals of 5000 words. The experimental results were evaluated with two evaluation metrics: density and f-measure. The density measures the ratio between the number of links and the number of nodes belonging to the same groups [1]. The f-measure calculates the harmonic mean of recall and precision of the extracted communities compared to the synsets from PTE.

Figure 3 illustrates the relationship between the number of words taking into account in the experiment and two quality measures which were extracted from our experiment. The experiments show that increasing number of words in the graph construct process: 1. the graph is more dense, it implicate more consistent relationships between related verbs, and 2. the f-measure obtained shows that the extracted verb groups have a higher similarity to the synsets from PTE.

Table 1. Details on the verb synsets of the PTE 2.0

Number of verbs	10910
Number of synsets	4145
Average verbs per synset	2
Number of verbs in the larger synset	53
Number of synsets with a verb	167
Number of synsets with two verbs	939
Average number of synsets membership per verb (overlaps)	2

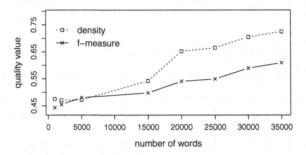

Fig. 3. Relationship between the number of considered words and two quality measures, density and f-measure

5   Conclusion

This paper presented the use of a community detection algorithm to detect related Brazilian Portuguese verbs. This paper demonstrates that this is a viable approach. In the experiments the "window" size was set to 1. Alternatively, we are currently studying the influence of the "window" size on the solution quality. We observed that HCL have computational cost considerably high (quadratic time complexity). Therefore, it was not possible to evaluate it on a larger number of words. Based on this, we plan to evaluate other community detection algorithms and verify their impact on the quality of the communities and runtime.

This work was partially supported by CNPq grant: 151836/2013-2, FAPESP grants: 2011/22749-8, 11/20451-1 and 2013/12191-5 as well as by the CAPES funding agency.

References

1. Ahn, Y., Bagrow, J.P., Lehmann, S.: Link communities reveal multiscale complexity in networks. Nature 466(7307), 761–764 (2010)
2. Bird, S.: Nltk: The natural language toolkit. In: Proceedings of the ACL Workshop on Effective Tools and Methodologies for Teaching Natural Language Processing and Computational Linguistics. Association for Computational Linguistics, Philadelphia (2002)
3. Chen, J., Yuan, B.: Detecting functional modules in the yeast protein–protein interaction network. Bioinformatics 22(18), 2283–2290 (2006)

4. Clauset, A., Newman, M.E.J., Moore, C.: Finding community structure in very large networks. Physical Review (December 2004)
5. Dias-da Silva, B.C.: Brazilian portuguese wordnet: A computational linguistic exercise of encoding bilingual relational lexicons. International Journal of Computational Linguistics and Applications 1(1-2), 137–150 (2010)
6. Fortunato, S.: Community detection in graphs. Physics Reports 486(3-5), 75–174 (2010)
7. Joanis, E., Stevenson, S., James, D.: A general feature space for automatic verb classification. Natural Language Engineering 14(3), 337–367 (2008)
8. Jurgens, D.: Word sense induction by community detection. In: Proceedings of TextGraphs-6: Graph-based Methods for Natural Language Processing, TextGraphs-6, Stroudsburg, PA, USA, pp. 24–28. Association for Computational Linguistics (2011)
9. Levin, B.: English verb classes and alternations: A preliminary investigation. The University of Chicago Press (1993)
10. Newman, M.E.J., Girvan, M.: Finding and evaluating community structure in networks. Physical Review (2004)
11. Gonçalo Oliveira, H., Gomes, P.: Onto.pt: Automatic construction of a lexical ontology for portuguese. In: STAIRS. Frontiers in Artificial Intelligence and Applications, pp. 199–211. IOS Press (2010)
12. Raghavan, U.N., Albert, R., Kumara, S.: Near linear time algorithm to detect community structures in large-scale networks. Physical Review (September 2007)
13. Scarton, C.E.: Construção semiautomática de um léxico computacional de verbos para o português do brasil. In: The 8th Brazilian Symposium in Information and Human Language Technology, 2011 (2011)
14. Schuler, K.K.: Verbnet: A Broad-coverage, Comprehensive Verb Lexicon. PhD thesis. University of Pennsylvania, Philadelphia, PA, USA, AAI3179808 (2005)
15. Stevenson, S., Joanis, E.: Semi-supervised verb class discovery using noisy features. In: Proceedings of the Seventh Conference on Natural Language Learning (2003)
16. Stevenson, S., Merlo, P.: Automatic verb classification using distributions of grammatical features. In: Proceedings of the Ninth Conference on European Chapter of the Association for Computational Linguistics, EACL 1999, pp. 45–52. Association for Computational Linguistics (1999)
17. Sun, L., Korhonen, A.: Hierarchical verb clustering using graph factorization. In: Proceedings of the Conference on Empirical Methods in Natural Language Processing, EMNLP 2011, pp. 1023–1033 (2011)
18. Tsang, V., Stevenson, S.: A graph-theoretic framework for semantic distance. Comput. Linguist. 36(1), 31–69 (2010)
19. Valverde-Rebaza, J., Lopes, A.A.: Exploiting behaviors of communities of twitter users for link prediction. Social Network Analysis and Mining 3(4), 1063–1074 (2013)

Author Index

Printed in the United States
By Bookmasters